Quantum Optics

D.F. Walls   Gerard J. Milburn

Quantum Optics

D.F. Walls, F.R.S.
(1942 – 1999)

Gerard J. Milburn
Physics Department
University of Queensland
St Lucia, Brisbane QLD 4072, Australia
milburn@physics.uq.edu.au

ISBN: 978-3-642-06676-4   e-ISBN: 978-3-540-28574-8

Cover design: eStudio Calamar S.L.

Printed on acid-free paper

9 8 7 6 5 4 3 2 1

springer.com

Preface to Second Edition

The field of quantum optics today is very different form the field that Dan Walls and I surveyed in 1994 for the first Edition of this book. Some of the new fields that have emerged over the years were hinted at in the earlier edition: quantum information has at least some roots in the study of Bell's Inequalities, while the fields of ion trapping and quantum condensed gases have their roots in the old chapter on light forces. However such is the growth of activity in each of these areas that I have found it necessary to write four new chapters for this edition. In order to keep the book to a reasonable size this has meant cutting some of the material from the first edition. The old chapter on Intracavity Atomic Systems is largely gone with parts distributed in the new chapter on Cavity QED and elsewhere. Likewise the old chapter on Resonance Fluorescence has been redistributed across Chaps. 10 and 11 in this edition. No doubt more cutting could have been made but I have tried to keep some continuity with the previous edition. In any case an emphasis on experimental realisations has been retained in the new material. Preparing this edition was not as much fun as the first. With Dan Walls untimely death in 1999, I have been denied the consolations of a shared task and soldiered on alone (although I must admit to hearing his voice from time to time as I cut and pasted). I can only hope that I have not lost his vision for the book in my unchallenged role of sole author.

Brisbane, Australia,   G.J. Milburn
October 2007.

Contents

Chapter 1
Introduction

The first indication of the quantum nature of light came in 1900 when Planck discovered he could account for the spectral distribution of thermal light by postulating that the energy of a simple harmonic oscillator was quantized. Further evidence was added by Einstein who showed in 1905 that the photoelectric effect could be explained by the hypothesis that the energy of a light beam was distributed in discrete packets later known as photons.

Einstein also contributed to the understanding of the absorption and emission of light from atoms with his development of a phenomenological theory in 1917. This theory was later shown to be a natural consequence of the quantum theory of electromagnetic radiation.

Despite this early connection with the quantum theory, physical optics developed more or less independently of quantum theory. The vast majority of physical-optics experiments can be adequately explained using classical theory of electromagnetism based on Maxwell's equations. An early attempt to find quantum effects in an optical interference experiment by G.I. Taylor in 1909 gave a negative result. Taylor's experiment was an attempt to repeat Young's famous two slit experiment with one photon incident on the slits. The classical explanation based in the interference of electric field amplitudes and the quantum explanation based on the interference of probability amplitudes both correctly explain the phenomenon in this experiment. Interference experiments of Young's type do not distinguish between the predictions of the classical theory and the quantum theory. It is only in higher order interference experiments, involving the interference of intensities, that differences between the predictions of classical and quantum theory appear. In such an experiment the probability amplitudes to detect a photon from two different fields interfere on a detector. Whereas classical theory treats the interference of intensities, in quantum theory the interference is still at the level of probability amplitudes. This is one of the most important differences between the classical and the quantum theory.

The first experiment in intensity interferometry was the famous experiment of R. Hanbury Brown and R.Q. Twiss. This experiment studied the correlation in the photocurrent fluctuations fro two detectors. Later experiments were based on photon counting, and the correlation between photon number was studied.

The Hanbury–Brown and Twiss experiment observed an enhancement in the two-time correlation function of short time delays for a thermal light source, known as photon bunching. This was a consequence of the large intensity fluctuations in the thermal source. Such photon bunching phenomenon may be adequately explained using a classical theory with a fluctuating electric field amplitude. For a perfectly amplitude stabilized light field, such as an ideal laser operating well above threshold, there is no photon bunching. A photon counting experiment where the number of photons arriving in an interval of time T are counted, shows that there is still randomness in the arrival time of the photons. The photon number distribution for an ideal laser is Poissonian. For thermal light a super-Poissonian photocount distribution results.

While the these results may be derived form a classical and quantum theory, the quantum theory makes additional unique predictions. This was first elucidated by R.J. Glauber in his quantum formulation of optical coherence theory in 1963. Glauber was jointly awarded the 2005 Nobel Prize in physics for this work. One such prediction is photon anti bunching, in which the initial slope of the two-time photon correlation function is positive. This corresponds to an enhancement, on average, of the temporal separation between photo counts at a detector, or photon anti-bunching. The photo-count statistics may also be sub-Poissonian. A classical theory of fluctuating field amplitudes would require negative probability in order to give anti-bunching. In the quantum picture it is easy to visualize photon arrivals more regular than Poissonian.

It was not until 1975 that H.J. Carmichel and D.F. Walls predicted that light generated in resonance fluorescence fro a two-level atom would exhibit photon anti-bunching that a physically accessible system exhibiting non-classical behaviour was identified. Photon anti-bunching in this system was observed the following year by H.J. Kimble, M. Dagenais and L. Mandel. This was the first non classical effect observed in optics and ushered in a new era of quantum optics.

The experiments of Kimble et al. used an atomic beam and hence the photon anti-bunching was convoluted with the atomic number fluctuations in the beam. With the development of ion trap technology it is now possible to trap a single ion for many minute and observe fluorescence. H. Walther and co workers in Munich have studied resonance fluorescence from a single ion in a trap and observed both photon bunching and anti-bunching.

In the 1960s improvements in photon counting techniques proceeded in tandem with the development of new laser light sources. Light from incoherent (thermal) and coherent (laser) sources could now be distinguished by their photon counting statistics. The groups of F.T. Arecchi in Milan, L. Mandel in Rochester and R. Pike in Malvern measured the photo count statistics of the laser. These experiments showed that the photo-count statistics went from super-Poissonian below threshold to Poissonian far above threshold. Concurrently the quantum theory of the laser was being developed by H. Haken in Stuttgart, M.O. Scully and W. Lamb in Yale and M. Lax and W.H. Louisell in New Jersey. In these theories both the atomic variables and the electromagnetic field were quantized. The results of these calculations were that the laser functioned as an essentially classical device. In fact H. Risken showed that it could be modeled as a van der Pol Oscillator.

In the late 80s the role of noise in the laser pumping process was shown to obscure the quantum aspects of the laser. If the noise in the pump can be suppressed the laser may exhibit sub-Poissonian statistics. In other words the intensity fluctuations may be reduced below the shot noise level of normal lasers. Y. Yamamoto first in Tokyo and then Stanford has pioneered experimental developments of semiconductor lasers with suppressed pump noise. More recently, Yamamoto and others have pioneered the development of the single photon source. This is a source of transform-limited pulsed light with one and only one photon per pulse: the ultimate limit of an anti-bunched source. The average field amplitude of such a source is zero while the intensity is definite. Such sources are highly non classical and have applications in quantum communication and computation.

It took another nine years after the first observation of photon anti-bunching for another prediction of the quantum theory of light to be observed – squeezing of quantum fluctuations. The electric field of a nearly monochromatic plane wave may be decomposed into two quadrature component amplitudes of an oscillatory sine term and a cosine term. In a coherent state, the closest quantum counter-part to a classical field, the fluctuations in the two quadrature amplitudes are equal and saturate the lower bound in the Heisenberg uncertainty relation. The quantum fluctuations in a coherent state are equal to the zero point fluctuations of the vacuum and are randomly distributed in phase. In a squeezed state the fluctuations are phase dependent. One quadrature phase amplitude may have reduced fluctuations compared to the vacuum while, in consequence, the other quadrature phase amplitude will have increased fluctuations, with the product of the uncertainties still saturating the lower bound in the Heisenberg uncertainty relation.

The first observation of squeezed light was made by R.E. Slusher in 1985 at AT&T Bell Laboratories in four wave mixing. Shortly after squeezing was demonstrated using optical parametric oscillators, by H.J. Kimble and four wave mixing in optical fibres by M.D. Levenson. Since then, greater and greater degrees of quantum noise suppression have been demonstrated, currently more than 7 dB, driven by new applications in quantum communication protocols such as teleportation and continuous variable quantum key distribution.

In the nonlinear process of parametric down conversion, a high frequency photon is absorbed and two photons are simultaneously produced with lower frequencies. The two photons produced are correlated in frequency, momentum and possibly polarisation. This results in very strong intensity correlations in the down converted beams that results in strongly suppressed intensity difference fluctuations as demonstrated by E. Giacobino in Paris and P. Kumar in Evanston.

Early uses of such correlated twin beams included accurate absorption measurements in which the sample was placed in one arm with the other beam providing a reference. when the twin beams are detected and the photo currents are subtracted, the presence of very weak absorption can be seen because of the small quantum noise in the difference current. More recently the strong intensity correlations

have been used to provide an accurate calibration of photon detector efficiency by
A. Migdall at NIST and also in so called quantum imaging in which an object paced
in one path changes the spatial pattern of intensity correlations between the two
twin beams.

The high degree of correlation between the down converted photons enables
some of the most stringent demonstrations of the violation of the Bell inequalities
in quantum physics. In 1999 P. Kwiat obtained a violation by more than 240 standard deviations using polarisation correlated photons produced by type II parametric
down conversion. The quadrature phase amplitudes in the twin beams generated in
down conversion carry quantum correlations of the Einstein-Podolsky-Rosen type.
This enabled the continuous variable version of quantum teleportation, proposed by
L. Vaidmann, to be demonstrated by H.J. Kimble in 1998. More recently P.K. Lam,
using the same quadrature phase correlations, demonstrated a continuous variable
quantum key distributions.

These last examples lie at the intersection of quantum optics with the new field
of quantum information. Quantum entanglement enables new communication and
computational tasks to be performed that are either difficult or impossible in a classical world. Quantum optics provides an ideal test bed for experimental investigations
in quantum information, and such investigations now form a large part of the experimental agenda in the field.

Quantum optics first entered the business of quantum information processing
with the proposal of Cirac and Zoller in 1995 to use ion trap technology. Following pioneering work by Dehmelt and others using ion traps for high resolution spectroscopy, by the early 1990s it was possible to trap and cool a single ion
to almost the ground state of its vibrational motion. Cirac and Zoller proposed a
scheme, using multiple trapped ions, by which quantum information stored in the
internal electronic state of each ion could be processed using an external laser to
correlate the internal states of different ions using collective vibrational degrees of
freedom. Ion traps currently provide the most promising approach to quantum information processing with more than eight qubits having been entangled in the labs
of D. Wineland at NIST in Colorado and R. Blatt in Innsbruck.

Quantum computation requires the ability to strongly entangle independent degrees of freedom that are used to encode information, known as qubits. It was initially thought however that the very weak optical nonlinearities typically found in
quantum optics would not be powerful enough to implement such entangling operations. This changed in 2001 when E. Knill, R. Laflamme and G.J. Milburn, followed
shortly thereafter by T. Pittman and J. Franson, proposed a way to perform conditional entangling operations using information encoded on single photons, and photon counting measurements. Early experimental demonstrations of simple quantum
gates soon followed.

At about the same time another measurement based protocol for quantum computing was devised by R. Raussendorf and H. Breigel. Nielsen showed how this approach could be combined with the single photon methods introduced by Knill et al.,
to dramatically simplify the implementation of conditional gates. The power of this
approach was recently demonstrated by A. Zeilinger's group in Vienna. Scaling up

this approach to more and more qubits is a major activity of experimental quantum optics.

These schemes provide a powerful incentive to develop a totally new kind of light source: the single photon pulsed source. This is a pulsed light source that produces one and only one photon per pulse. Such sources are in development in many laboratories around the world. A variety of approaches are being pursued. Sources based on excitons in semiconductor quantum dots are being developed by A. Imamoglu in Zurich, A. Shields in Toshiba Cambridge, and Y. Yamamoto and J. Vukovic in Stanford. NV centres in diamond nanocrystal are under development by S. Prawer in Melbourne. An interesting approach based on down conversion in optical fibers is being studied by A. Migdall in NIST. Sources based on single atoms in optical cavities have been demonstrated by H. Walther in Munich and P. Grangier in Paris. Once routinely available, single photon sources will enable a new generation of experiments in single photon quantum optics.

Beginning in the early 1980s a number of pioneers including G. Ashkin, C. Cohen Tannoudji and S. Chu began to study the forces exerted on atoms by light. This work led to the ability to cool and trap ensembles of atoms, or even single atoms, and culminated in the experimental demonstration by E. Cornell and C. Weimann of a Bose Einstein condensate using a dilute gas of rubidium atoms at NIST in 1995, followed soon thereafter by W. Ketterle at Harvard. Discoveries in this field continue to enlighten our understanding of many body quantum physics, quantum information and non linear quantum field theory. We hardly touch on this subject in this book, which is already well covered in a number of recent excellent texts, choosing instead to highlight some aspects of the emerging field of quantum atom optics.

Chapter 2
Quantisation of the Electromagnetic Field

Abstract The study of the quantum features of light requires the quantisation of the electromagnetic field. In this chapter we quantise the field and introduce three possible sets of basis states, namely, the Fock or number states, the coherent states and the squeezed states. The properties of these states are discussed. The phase operator and the associated phase states are also introduced.

2.1 Field Quantisation

The major emphasis of this text is concerned with the uniquely quantum-mechanical properties of the electromagnetic field, which are not present in a classical treatment. As such we shall begin immediately by quantizing the electromagnetic field. We shall make use of an expansion of the vector potential for the electromagnetic field in terms of cavity modes. The problem then reduces to the quantization of the harmonic oscillator corresponding to each individual cavity mode.

We shall also introduce states of the electromagnetic field appropriate to the description of optical fields. The first set of states we introduce are the number states corresponding to having a definite number of photons in the field. It turns out that it is extremely difficult to create experimentally a number state of the field, though fields containing a very small number of photons have been generated. A more typical optical field will involve a superposition of number states. One such field is the coherent state of the field which has the minimum uncertainty in amplitude and phase allowed by the uncertainty principle, and hence is the closest possible quantum mechanical state to a classical field. It also possesses a high degree of optical coherence as will be discussed in Chap. 3, hence the name coherent state. The coherent state plays a fundamental role in quantum optics and has a practical significance in that a highly stabilized laser operating well above threshold generates a coherent state.

A rather more exotic set of states of the electromagnetic field are the squeezed states. These are also minimum-uncertainty states but unlike the coherent states the quantum noise is not uniformly distributed in phase. Squeezed states may have less noise in one quadrature than the vacuum. As a consequence the noise in the other quadrature is increased. We introduce the basic properties of squeezed states in this chapter. In Chap. 8 we describe ways to generate squeezed states and their applications.

While states of definite photon number are readily defined as eigenstates of the number operator a corresponding description of states of definite phase is more difficult. This is due to the problems involved in constructing a Hermitian phase operator to describe a bounded physical quantity like phase. How this problem may be resolved together with the properties of phase states is discussed in the final section of this chapter.

A convenient starting point for the quantisation of the electromagnetic field is the classical field equations. The free electromagnetic field obeys the source free Maxwell equations.

$$\nabla \cdot \boldsymbol{B} = 0 \,, \tag{2.1a}$$

$$\nabla \times \boldsymbol{E} = -\frac{\partial \boldsymbol{B}}{\partial t} \,, \tag{2.1b}$$

$$\nabla \cdot \boldsymbol{D} = 0 \,, \tag{2.1c}$$

$$\nabla \times \boldsymbol{H} = \frac{\partial \boldsymbol{D}}{\partial t} \,, \tag{2.1d}$$

where $\boldsymbol{B} = \mu_0 \boldsymbol{H}$, $\boldsymbol{D} = \varepsilon_0 \boldsymbol{E}$, μ_0 and ε_0 being the magnetic permeability and electric permittivity of free space, and $\mu_0 \varepsilon_0 = c^{-2}$. Maxwell's equations are gauge invariant when no sources are present. A convenient choice of gauge for problems in quantum optics is the Coulomb gauge. In the Coulomb gauge both $\boldsymbol{B}$ and $\boldsymbol{E}$ may be determined from a vector potential $\boldsymbol{A}(\boldsymbol{r}, t)$ as follows

$$\boldsymbol{B} = \nabla \times \boldsymbol{A} \,, \tag{2.2a}$$

$$\boldsymbol{E} = -\frac{\partial \boldsymbol{A}}{\partial t} \,, \tag{2.2b}$$

with the Coulomb gauge condition

$$\nabla \cdot \boldsymbol{A} = 0 \,. \tag{2.3}$$

Substituting (2.2a) into (2.1d) we find that $\boldsymbol{A}(\boldsymbol{r}, t)$ satisfies the wave equation

$$\nabla^2 \boldsymbol{A}(\boldsymbol{r},t) = \frac{1}{c^2} \frac{\partial^2 \boldsymbol{A}(\boldsymbol{r},t)}{\partial t^2} \,. \tag{2.4}$$

We separate the vector potential into two complex terms

$$\boldsymbol{A}(\boldsymbol{r},t) = \boldsymbol{A}^{(+)}(\boldsymbol{r},t) + \boldsymbol{A}^{(-)}(\boldsymbol{r},t) \,, \tag{2.5}$$

where $\boldsymbol{A}^{(+)}(\boldsymbol{r}, t)$ contains all amplitudes which vary as $\mathrm{e}^{-i\omega t}$ for $\omega > 0$ and $\boldsymbol{A}^{(-)}(\boldsymbol{r}, t)$ contains all amplitudes which vary as $\mathrm{e}^{i\omega t}$ and $\boldsymbol{A}^{(-)} = (\boldsymbol{A}^{(+)})^*$.

It is more convenient to deal with a discrete set of variables rather than the whole continuum. We shall therefore describe the field restricted to a certain volume of space and expand the vector potential in terms of a discrete set of orthogonal mode functions:

$$A^{(+)}(r,t) = \sum_k c_k u_k(r) e^{-i\omega_k t} , \qquad (2.6)$$

where the Fourier coefficients c_k are constant for a free field. The set of vector mode functions $u_k(r)$ which correspond to the frequency ω_k will satisfy the wave equation

$$\left(\nabla^2 + \frac{\omega_k^2}{c^2}\right) u_k(r) = 0 \qquad (2.7)$$

provided the volume contains no refracting material. The mode functions are also required to satisfy the transversality condition,

$$\nabla \cdot u_k(r) = 0 . \qquad (2.8)$$

The mode functions form a complete orthonormal set

$$\int_V u_k^*(r) u_{k'}(r) dr = \delta_{kk'} . \qquad (2.9)$$

The mode functions depend on the boundary conditions of the physical volume under consideration, e.g., periodic boundary conditions corresponding to travelling-wave modes or conditions appropriate to reflecting walls which lead to standing waves. For example, the plane wave mode functions appropriate to a cubical volume of side L may be written as

$$u_k(r) = L^{-3/2} \hat{e}^{(\lambda)} \exp(i k \cdot r) \qquad (2.10)$$

where $\hat{e}^{(\lambda)}$ is the unit polarization vector. The mode index k describes several discrete variables, the polarisation index ($\lambda = 1, 2$) and the three Cartesian components of the propagation vector k. Each component of the wave vector k takes the values

$$k_x = \frac{2\pi n_x}{L}, \quad k_y = \frac{2\pi n_y}{L}, \quad k_z = \frac{2\pi n_z}{L}, \qquad n_x, n_y, n_z = 0, \pm 1, \pm 2, \ldots \qquad (2.11)$$

The polarization vector $\hat{e}^{(\lambda)}$ is required to be perpendicular to k by the transversality condition (2.8).

The vector potential may now be written in the form

$$A(r,t) = \sum_k \left(\frac{\hbar}{2\omega_k \varepsilon_0}\right)^{1/2} \left[a_k u_k(r) e^{-i\omega_k t} + a_k^\dagger u_k^*(r) e^{i\omega_k t}\right] . \qquad (2.12)$$

The corresponding form for the electric field is

$$E(r,t) = i\sum_k \left(\frac{\hbar\omega_k}{2\varepsilon_0}\right)^{1/2} \left[a_k u_k(r)\,\mathrm{e}^{-i\omega_k t} - a_k^\dagger u_k^*(r)\mathrm{e}^{i\omega_k t}\right] . \qquad (2.13)$$

The normalization factors have been chosen such that the amplitudes a_k and $a_k^\dagger$ are dimensionless.

In classical electromagnetic theory these Fourier amplitudes are complex numbers. Quantisation of the electromagnetic field is accomplished by choosing a_k and $a_k^\dagger$ to be mutually adjoint operators. Since photons are bosons the appropriate commutation relations to choose for the operators a_k and $a_k^\dagger$ are the boson commutation relations

$$[a_k, a_{k'}] = \left[a_k^\dagger, a_{k'}^\dagger\right] = 0, \quad \left[a_k, a_{k'}^\dagger\right] = \delta_{kk'} . \qquad (2.14)$$

The dynamical behaviour of the electric-field amplitudes may then be described by an ensemble of independent harmonic oscillators obeying the above commutation relations. The quantum states of each mode may now be discussed independently of one another. The state in each mode may be described by a state vector $|\Psi\rangle_k$ of the Hilbert space appropriate to that mode. The states of the entire field are then defined in the tensor product space of the Hilbert spaces for all of the modes.

The Hamiltonian for the electromagnetic field is given by

$$H = \frac{1}{2}\int \left(\varepsilon_0 E^2 + \mu_0 H^2\right)\mathrm{d}r . \qquad (2.15)$$

Substituting (2.13) for E and the equivalent expression for H and making use of the conditions (2.8) and (2.9), the Hamiltonian may be reduced to the form

$$H = \sum_k \hbar\omega_k \left(a_k^\dagger a_k + \frac{1}{2}\right) . \qquad (2.16)$$

This represents the sum of the number of photons in each mode multiplied by the energy of a photon in that mode, plus $\frac{1}{2}\hbar\omega_k$ representing the energy of the vacuum fluctuations in each mode. We shall now consider three possible representations of the electromagnetic field.

2.2 Fock or Number States

The Hamiltonian (2.15) has the eigenvalues $\hbar\omega_k(n_k + \frac{1}{2})$ where n_k is an integer ($n_k = 0, 1, 2, \ldots, \infty$). The eigenstates are written as $|n_k\rangle$ and are known as number or Fock states. They are eigenstates of the number operator $N_k = a_k^\dagger a_k$

$$a_k^\dagger a_k |n_k\rangle = n_k |n_k\rangle . \qquad (2.17)$$

The ground state of the oscillator (or vacuum state of the field mode) is defined by

$$a_k|0\rangle = 0 . \tag{2.18}$$

From (2.16 and 2.18) we see that the energy of the ground state is given by

$$\langle 0|H|0\rangle = \frac{1}{2}\sum_k \hbar\omega_k . \tag{2.19}$$

Since there is no upper bound to the frequencies in the sum over electromagnetic field modes, the energy of the ground state is infinite, a conceptual difficulty of quantized radiation field theory. However, since practical experiments measure a change in the total energy of the electromagnetic field the infinite zero-point energy does not lead to any divergence in practice. Further discussions on this point may be found in [1]. a_k and $a_k^\dagger$ are raising and lowering operators for the harmonic oscillator ladder of eigenstates. In terms of photons they represent the annihilation and creation of a photon with the wave vector k and a polarisation $\hat{e}_k$. Hence the terminology, annihilation and creation operators. Application of the creation and annihilation operators to the number states yield

$$a_k|n_k\rangle = n_k^{1/2}|n_k - 1\rangle, \quad a_k^\dagger|n_k\rangle = (n_k + 1)^{1/2}|n_k + 1\rangle . \tag{2.20}$$

The state vectors for the higher excited states may be obtained from the vacuum by successive application of the creation operator

$$|n_k\rangle = \frac{\left(a_k^\dagger\right)^{n_k}}{(n_k!)^{1/2}}|0\rangle, \quad n_k = 0, 1, 2 \dots . \tag{2.21}$$

The number states are orthogonal

$$\langle n_k|m_k\rangle = \delta_{mn} , \tag{2.22}$$

and complete

$$\sum_{n_k=0}^{\infty} |n_k\rangle\langle n_k| = 1 . \tag{2.23}$$

Since the norm of these eigenvectors is finite, they form a complete set of basis vectors for a Hilbert space.

While the number states form a useful representation for high-energy photons, e.g. γ rays where the number of photons is very small, they are not the most suitable representation for optical fields where the total number of photons is large. Experimental difficulties have prevented the generation of photon number states with more than a small number of photons (but see 16.4.2). Most optical fields are either a superposition of number states (pure state) or a mixture of number states (mixed state). Despite this the number states of the electromagnetic field have been used as a basis for several problems in quantum optics including some laser theories.

2.3 Coherent States

A more appropriate basis for many optical fields are the coherent states [2]. The coherent states have an indefinite number of photons which allows them to have a more precisely defined phase than a number state where the phase is completely random. The product of the uncertainty in amplitude and phase for a coherent state is the minimum allowed by the uncertainty principle. In this sense they are the closest quantum mechanical states to a classical description of the field. We shall outline the basic properties of the coherent states below. These states are most easily generated using the unitary displacement operator

$$D(\alpha) = \exp\left(\alpha a^\dagger - \alpha^* a\right) , \tag{2.24}$$

where α is an arbitrary complex number.

Using the operator theorem [2]

$$e^{A+B} = e^A e^B e^{-[A,B]/2} , \tag{2.25}$$

which holds when

$$[A,[A,B]] = [B,[A,B]] = 0,$$

we can write $D(\alpha)$ as

$$D(\alpha) = e^{-|\alpha|^2/2} e^{\alpha a^\dagger} e^{-\alpha^* a} . \tag{2.26}$$

The displacement operator $D(\alpha)$ has the following properties

$$D^\dagger(\alpha) = D^{-1}(\alpha) = D(-\alpha), \qquad D^\dagger(\alpha) a D(\alpha) = a + \alpha,$$
$$D^\dagger(\alpha) a^\dagger D(\alpha) = a^\dagger + \alpha^* . \tag{2.27}$$

The coherent state $|\alpha\rangle$ is generated by operating with $D(\alpha)$ on the vacuum state

$$|\alpha\rangle = D(\alpha)|0\rangle . \tag{2.28}$$

The coherent states are eigenstates of the annihilation operator a. This may be proved as follows:

$$D^\dagger(\alpha) a |\alpha\rangle = D^\dagger(\alpha) a D(\alpha)|0\rangle = (a + \alpha)|0\rangle = \alpha|0\rangle . \tag{2.29}$$

Multiplying both sides by $D(\alpha)$ we arrive at the eigenvalue equation

$$a|\alpha\rangle = \alpha|\alpha\rangle . \tag{2.30}$$

Since a is a non-Hermitian operator its eigenvalues α are complex.

Another useful property which follows using (2.25) is

$$D(\alpha + \beta) = D(\alpha) D(\beta) \exp\left(-i \operatorname{Im}\{\alpha\beta^*\}\right) . \tag{2.31}$$

The coherent states contain an indefinite number of photons. This may be made apparent by considering an expansion of the coherent states in the number states basis.

Taking the scalar product of both sides of (2.30) with $\langle n|$ we find the recursion relation

$$(n+1)^{1/2} \langle n+1|\alpha\rangle = \alpha \langle n|\alpha\rangle . \tag{2.32}$$

It follows that

$$\langle n|\alpha\rangle = \frac{\alpha^n}{(n!)^{1/2}} \langle 0|\alpha\rangle . \tag{2.33}$$

We may expand $|\alpha\rangle$ in terms of the number states $|n\rangle$ with expansion coefficients $\langle n|\alpha\rangle$ as follows

$$|\alpha\rangle = \sum |n\rangle\langle n|\alpha\rangle = \langle 0|\alpha\rangle \sum_n \frac{\alpha^n}{(n!)^{1/2}} |n\rangle . \tag{2.34}$$

The squared length of the vector $|\alpha\rangle$ is thus

$$|\langle\alpha|\alpha\rangle|^2 = |\langle 0|\alpha\rangle|^2 \sum_n \frac{|\alpha|^{2n}}{n!} = |\langle 0|\alpha\rangle|^2 e^{|\alpha|^2} . \tag{2.35}$$

It is easily seen that

$$\langle 0|\alpha\rangle = \langle 0|D(\alpha)|0\rangle$$
$$= e^{-|\alpha|^2/2} . \tag{2.36}$$

Thus $|\langle\alpha|\alpha\rangle|^2 = 1$ and the coherent states are normalized.

The coherent state may then be expanded in terms of the number states as

$$|\alpha\rangle = e^{-|\alpha|^2/2} \sum \frac{\alpha^n}{(n!)^{1/2}} |n\rangle . \tag{2.37}$$

We note that the probability distribution of photons in a coherent state is a Poisson distribution

$$P(n) = |\langle n|\alpha\rangle|^2 = \frac{|\alpha|^{2n} e^{-|\alpha|^2}}{n!} , \tag{2.38}$$

where $|\alpha|^2$ is the mean number of photons ($\bar{n} = \langle\alpha|a^\dagger a|\alpha\rangle = |\alpha|^2$).

The scalar product of two coherent states is

$$\langle\beta|\alpha\rangle = \langle 0|D^\dagger(\beta) D(\alpha)|0\rangle . \tag{2.39}$$

Using (2.26) this becomes

$$\langle\beta|\alpha\rangle = \exp\left[-\frac{1}{2}\left(|\alpha|^2 + |\beta|^2\right) + \alpha\beta^*\right] . \tag{2.40}$$

The absolute magnitude of the scalar product is

$$|\langle \beta | \alpha \rangle|^2 = e^{-|\alpha - \beta|^2} . \tag{2.41}$$

Thus the coherent states are not orthogonal although two states $|\alpha\rangle$ and $|\beta\rangle$ become approximately orthogonal in the limit $|\alpha - \beta| \gg 1$. The coherent states form a two-dimensional continuum of states and are, in fact, overcomplete. The completeness relation

$$\frac{1}{\pi} \int |\alpha\rangle\langle\alpha| d^2\alpha = 1 , \tag{2.42}$$

may be proved as follows.

We use the expansion (2.37) to give

$$\int |\alpha\rangle\langle\alpha| \frac{d^2\alpha}{\pi} = \sum_{n=0}^{\infty} \sum_{m=0}^{\infty} \frac{|n\rangle\langle m|}{\pi\sqrt{n!m!}} \int e^{-|\alpha|^2} \alpha^{*m} \alpha^n d^2\alpha . \tag{2.43}$$

Changing to polar coordinates this becomes

$$\int |\alpha\rangle\langle\alpha| \frac{d^2\alpha}{\pi} = \sum_{n,m=0}^{\infty} \frac{|n\rangle\langle m|}{\pi\sqrt{n!m!}} \int_0^{\infty} r dr e^{-r^2} r^{n+m} \int_0^{2\pi} d\theta e^{i(n-m)\theta} . \tag{2.44}$$

Using

$$\int_0^{2\pi} d\theta e^{i(n-m)\theta} = 2\pi \delta_{nm} , \tag{2.45}$$

we have

$$\int |\alpha\rangle\langle\alpha| \frac{d^2\alpha}{\pi} = \sum_{n=0}^{\infty} \frac{|n\rangle\langle n|}{n!} \int_0^{\infty} d\varepsilon \, e^{-\varepsilon} \varepsilon^n , \tag{2.46}$$

where we let $\varepsilon = r^2$. The integral equals $n!$. Hence we have

$$\int |\alpha\rangle\langle\alpha| \frac{d^2\alpha}{\pi} = \sum_{n=0}^{\infty} |n\rangle\langle n = 1 , \tag{2.47}$$

following from the completeness relation for the number states.

An alternative proof of the completeness of the coherent states may be given as follows. Using the relation [3]

$$e^{\zeta B} A e^{-\zeta B} = A + \zeta [B,A] + \frac{\zeta^2}{2!} [B,[B,A]] + \cdots , \tag{2.48}$$

it is easy to see that all the operators A such that

$$D^\dagger(\alpha) A D(\alpha) = A \tag{2.49}$$

are proportional to the identity.

We consider

$$A = \int d^2\alpha |\alpha\rangle\langle\alpha|$$

then

$$D^\dagger(\beta) \int d^2\alpha |\alpha\rangle\langle\alpha| D(\beta) = \int d^2\alpha |\alpha-\beta\rangle\langle\alpha-\beta| = \int d^2\alpha |\alpha\rangle\langle\alpha| \,. \qquad (2.50)$$

Then using the above result we conclude that

$$\int d^2\alpha |\alpha\rangle\langle\alpha| \propto I \,. \qquad (2.51)$$

The constant of proportionality is easily seen to be π.

The coherent states have a physical significance in that the field generated by a highly stabilized laser operating well above threshold is a coherent state. They form a useful basis for expanding the optical field in problems in laser physics and nonlinear optics. The coherence properties of light fields and the significance of the coherent states will be discussed in Chap. 3.

2.4 Squeezed States

A general class of minimum-uncertainty states are known as *squeezed states*. In general, a squeezed state may have less noise in one quadrature than a coherent state. To satisfy the requirements of a minimum-uncertainty state the noise in the other quadrature is greater than that of a coherent state. The coherent states are a particular member of this more general class of minimum uncertainty states with equal noise in both quadratures. We shall begin our discussion by defining a family of minimum-uncertainty states. Let us calculate the variances for the position and momentum operators for the harmonic oscillator

$$q = \sqrt{\frac{\hbar}{2\omega}}(a+a^\dagger), \qquad p = i\sqrt{\frac{\hbar\omega}{2}}(a-a^\dagger) \,. \qquad (2.52)$$

The variances are defined by

$$V(A) = (\Delta A)^2 = \langle A^2\rangle - \langle A\rangle^2 \,. \qquad (2.53)$$

In a coherent state we obtain

$$(\Delta q)^2_{coh} = \frac{\hbar}{2\omega}, \quad (\Delta p)^2_{coh} = \frac{\hbar\omega}{2} \,. \qquad (2.54)$$

Thus the product of the uncertainties is a minimum

$$(\Delta p \, \Delta q)_{\text{coh}} = \frac{\hbar}{2} \, . \tag{2.55}$$

Thus, there exists a sense in which the description of the state of an oscillator by a coherent state represents as close an approach to classical localisation as possible. We shall consider the properties of a single-mode field. We may write the annihilation operator a as a linear combination of two Hermitian operators

$$a = \frac{X_1 + iX_2}{2} \, . \tag{2.56}$$

X_1 and X_2, the real and imaginary parts·of the complex amplitude, give dimensionless amplitudes for the modes' two quadrature phases. They obey the following commutation relation

$$[X_1, X_2] = 2i \tag{2.57}$$

The corresponding uncertainty principle is

$$\Delta X_1 \, \Delta X_2 \geq 1 \, . \tag{2.58}$$

This relation with the equals sign defines a family of minimum-uncertainty states. The coherent states are a particular minimum-uncertainty state with

$$\Delta X_1 = \Delta X_2 = 1 \, . \tag{2.59}$$

The coherent state $|\alpha\rangle$ has the mean complex amplitude α and it is a minimum-uncertainty state for X_1 and X_2, with equal uncertainties in the two quadrature phases. A coherent state may be represented by an "error circle" in a complex amplitude plane whose axes are X_1 and X_2 (Fig. 2.1a). The center of the error circle lies at $\frac{1}{2}\langle X_1 + iX_2\rangle = \alpha$ and the radius $\Delta X_1 = \Delta X_2 = 1$ accounts for the uncertainties in X_1 and X_2.

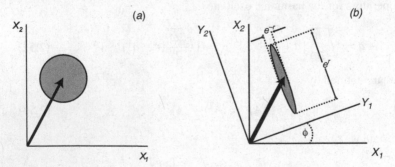

Fig. 2.1 Phase space representation showing contours of constant uncertainty for (**a**) coherent state and (**b**) squeezed state $|\alpha, \varepsilon\rangle$

There is obviously a whole family of minimum-uncertainty states defined by $\Delta X_1 \Delta X_2 = 1$. If we plot ΔX_1 against ΔX_2 the minimum-uncertainty states lie on a hyperbola (Fig. 2.2). Only points lying to the right of this hyperbola correspond to physical states. The coherent state with $\Delta X_1 = \Delta X_2$ is a special case of a more general class of states which may have reduced uncertainty in one quadrature at the expense of increased uncertainty in the other ($\Delta X_1 < 1 < \Delta X_2$). These states correspond to the shaded region in Fig. 2.2. Such states we shall call *squeezed states* [4]. They may be generated by using the unitary squeeze operator [5]

$$S(\varepsilon) = \exp\left(1/2\varepsilon^* a^2 - 1/2\varepsilon a^{\dagger 2}\right) . \tag{2.60}$$

where $\varepsilon = re^{2i\phi}$.

Note the squeeze operator obeys the relations

$$S^{\dagger}(\varepsilon) = S^{-1}(\varepsilon) = S(-\varepsilon) , \tag{2.61}$$

and has the following useful transformation properties

$$
\begin{aligned}
S^{\dagger}(\varepsilon) a S(\varepsilon) &= a\cosh r - a^{\dagger} e^{-2i\phi}\sinh r, \\
S^{\dagger}(\varepsilon) a^{\dagger} S(\varepsilon) &= a^{\dagger}\cosh r - a e^{-2i\phi}\sinh r , \\
S^{\dagger}(\varepsilon)(Y_1 + iY_2)S(\varepsilon) &= Y_1 e^{-r} + iY_2 e^{r},
\end{aligned} \tag{2.62}
$$

where

$$Y_1 + iY_2 = (X_1 + iX_2)e^{-i\phi} \tag{2.63}$$

is a rotated complex amplitude. The squeeze operator attenuates one component of the (rotated) complex amplitude, and it amplifies the other component. The degree of attenuation and amplification is determined by $r = |\varepsilon|$, which will be called the *squeeze factor*. The squeezed state $|\alpha, \varepsilon\rangle$ is obtained by first squeezing the vacuum and then displacing it

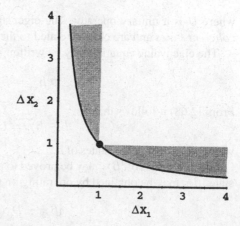

Fig. 2.2 Plot of ΔX_1 versus ΔX_2 for the minimum-uncertainty states. The *dot* marks a coherent state while the *shaded region* corresponds to the squeezed states

$$|\alpha, \varepsilon\rangle = D(\alpha) S(\varepsilon) |0\rangle . \tag{2.64}$$

A squeezed state has the following expectation values and variances

$$\langle X_1 + iX_2 \rangle = \langle Y_1 + iY_2 \rangle e^{i\phi} = 2\alpha,$$

$$\Delta Y_1 = e^{-r}, \qquad \Delta Y_2 = e^{r},$$

$$\langle N \rangle = |\alpha^2| + \sinh^2 r,$$

$$(\Delta N)^2 = |\alpha \cosh r - \alpha^* e^{2i\phi} \sinh r|^2 + 2\cosh^2 r \sinh^2 r . \tag{2.65}$$

Thus the squeezed state has unequal uncertainties for Y_1 and Y_2 as seen in the error ellipse shown in Fig. 2.1b. The principal axes of the ellipse lie along the Y_1 and Y_2 axes, and the principal radii are ΔY_1 and ΔY_2. A more rigorous definition of these error ellipses as contours of the Wigner function is given in Chap. 3.

2.5 Two-Photon Coherent States

We may define squeezed states in an alternative but equivalent way [6]. As this definition is sometimes used in the literature we include it for completeness.

Consider the operator

$$b = \mu a + v a^\dagger \tag{2.66}$$

where

$$|\mu|^2 - |v|^2 = 1 .$$

Then b obeys the commutation relation

$$[b, b^\dagger] = 1 . \tag{2.67}$$

We may write (2.66) as

$$b = U a U^\dagger \tag{2.68}$$

where U is a unitary operator. The eigenstates of b have been called *two-photon coherent states* and are closely related to the squeezed states.

The eigenvalue equation may be written as

$$b|\beta\rangle_{\mathrm{g}} = \beta |\beta\rangle_{\mathrm{g}} . \tag{2.69}$$

From (2.68) it follows that

$$|\beta\rangle_{\mathrm{g}} = U|\beta\rangle \tag{2.70}$$

where $|\beta\rangle$ are the eigenstates of a.

The properties of $|\beta\rangle_{\mathrm{g}}$ may be proved to parallel those of the coherent states. The state $|\beta\rangle_{\mathrm{g}}$ may be obtained by operating on the vacuum

$$|\beta\rangle_{\mathrm{g}} = D_{\mathrm{g}}(\beta) |0\rangle_{\mathrm{g}} \tag{2.71}$$

with the displacement operator

$$D_g(\beta) = e^{\beta b^\dagger - \beta^* b} \tag{2.72}$$

and $|0\rangle_g = U|0\rangle$. The two-photon coherent states are complete

$$\int |\beta\rangle_g \,_g\langle\beta| \frac{d^2\beta}{\pi} = 1 \tag{2.73}$$

and their scalar product is

$$_g\langle\beta|\beta'\rangle_g = \exp\left(\beta^*\beta' - \frac{1}{2}|\beta|^2 - \frac{1}{2}|\beta'|^2\right) . \tag{2.74}$$

We now consider the relation between the two-photon coherent states and the squeezed states as previously defined. We first note that

$$U \equiv S(\varepsilon)$$

with $\mu = \cosh r$ and $v = e^{2i\phi}\sinh r$. Thus

$$|0\rangle_g \equiv |0, \varepsilon\rangle \tag{2.75}$$

with the above relations between (μ, v) and (r, θ). Using this result in (2.71) and rewriting the displacement operator, $D_g(\beta)$, in terms of a and $a^\dagger$ we find

$$|\beta\rangle_g = D(\alpha)S(\varepsilon)|0\rangle = |\alpha, \varepsilon| \tag{2.76}$$

where

$$\alpha = \mu\beta - v\beta^* .$$

Thus we have found the equivalent squeezed state for the given two-photon coherent state.

Finally, we note that the two-photon coherent state $|\beta\rangle_g$ may be written as

$$|\beta\rangle_g = S(\varepsilon)D(\beta)|0\rangle .$$

Thus the two-photon coherent state is generated by first displacing the vacuum state, then squeezing. This is the opposite procedure to that which defines the squeezed state $|\alpha, \varepsilon\rangle$. The two procedures yield the same state if the displacement parameters α and β are related as discussed above.

The completeness relation for the two-photon coherent states may be employed to derive the completeness relation for the squeezed states. Using the above results we have

$$\int \frac{d^2\beta}{\pi} |\beta\cosh r - \beta^* e^{2i\phi}\sinh r, \varepsilon\rangle\langle\beta\cosh r - \beta^* e^{2i\phi}\sinh r, \varepsilon| = 1 . \tag{2.77}$$

The change of variable

$$\alpha = \beta \cosh r - \beta^* e^{2i\phi} \sinh r \tag{2.78}$$

leaves the measure invariant, that is $d^2\alpha = d^2\beta$. Thus

$$\int \frac{d^2\alpha}{\pi} |\alpha, \, \varepsilon\rangle\langle\alpha,\varepsilon| = 1 \,. \tag{2.79}$$

2.6 Variance in the Electric Field

The electric field for a single mode may be written in terms of the operators X_1 and X_2 as

$$E(r,t) = \frac{1}{\sqrt{L^3}} \left(\frac{\hbar\omega}{2\varepsilon_0} \right)^{1/2} [X_1 \sin(\omega t - k \cdot r) - X_2 \cos(\omega t - k \cdot r)] \,. \tag{2.80}$$

The variance in the electric field is given by

$$V(E(r,t)) = K \{ V(X_1) \sin^2(\omega t - k \cdot r) + V(X_2) \cos^2(\omega t - k \cdot r)$$
$$- \sin[2(\omega t - k \cdot r)] V(X_1, X_2) \} \tag{2.81}$$

where

$$K = \frac{1}{L^3} \left(\frac{2\hbar\omega}{\varepsilon_0} \right),$$

$$V(X_1, X_2) = \frac{\langle (X_1 X_2) + (X_2 X_1) \rangle}{2} - \langle X_1 \rangle \langle X_2 \rangle.$$

For a minimum-uncertainty state

$$V(X_1, X_2) = 0 \,. \tag{2.82}$$

Hence (2.81) reduces to

$$V(E(r,t)) = K \left[V(X_1) \sin^2(\omega t - k \cdot r) + V(X_2) \cos^2(\omega t - k \cdot r) \right] \,. \tag{2.83}$$

The mean and uncertainty of the electric field is exhibited in Figs. 2.3a–c where the line is thickened about a mean sinusoidal curve to represent the uncertainty in the electric field.

The variance of the electric field for a coherent state is a constant with time (Fig. 2.3a). This is due to the fact that while the coherent-state-error circle rotates about the origin at frequency ω, it has a constant projection on the axis defining the electric field. Whereas for a squeezed state the rotation of the error ellipse leads to a variance that oscillates with frequency 2ω. In Fig. 2.3b the coherent excitation

Fig. 2.3 Plot of the electric field versus time showing schematically the uncertainty in phase and amplitude for (**a**) a coherent state, (**b**) a squeezed state with reduced amplitude fluctuations, and (**c**) a squeezed state with reduced phase fluctuations

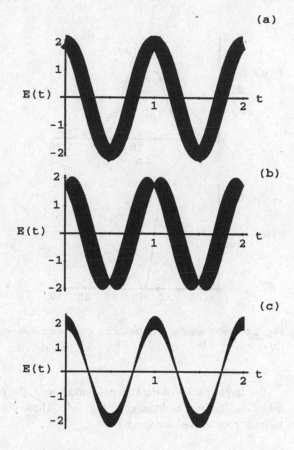

appears in the quadrature that has reduced noise. In Fig. 2.3c the coherent excitation appears in the quadrature with increased noise. This situation corresponds to the phase states discussed in [7] and in the final section of this chapter.

The squeezed state $|\alpha, r\rangle$ has the photon number distribution [6]

$$p(n) = (n!\cosh r)^{-1} \left[\frac{1}{2}\tanh r\right]^n \exp\left[-|\alpha|^2 - \frac{1}{2}\tanh r\left((\alpha^*)^2 e^{i\phi} + \alpha^2 e^{-i\phi}\right)\right]|H_n(z)|^2$$

(2.84)

where

$$z = \frac{\alpha + \alpha^* e^{i\phi}\tanh r}{\sqrt{2e^{i\phi}\tanh r}}.$$

The photon number distribution for a squeezed state may be broader or narrower than a Poissonian depending on whether the reduced fluctuations occur in the phase (X_2) or amplitude (X_1) component of the field. This is illustrated in Fig. 2.4a where we plot $P(n)$ for $r = 0$, $r > 0$, and $r < 0$. Note, a squeezed vacuum ($\alpha = 0$) contains only even numbers of photons since $H_n(0) = 0$ for n odd.

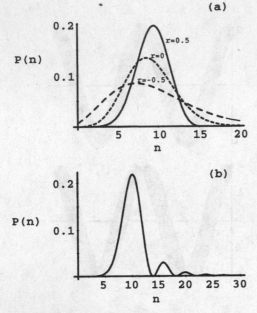

Fig. 2.4 Photon number distribution for a squeezed state $|\alpha, r\rangle$: **(a)** $\alpha = 3$, $r = 0$, 0.5, -0.5, **(b)** $\alpha = 3$, $r = 1.0$

For larger values of the squeeze parameter r, the photon number distribution exhibits oscillations, as depicted in Fig. 2.4b. These oscillations have been interpreted as interference in phase space [8].

2.7 Multimode Squeezed States

Multimode squeezed states are important since several devices produce light which is correlated at the two frequencies ω_+ and ω_-. Usually these frequencies are symmetrically placed either side of a carrier frequency. The squeezing exists not in the single modes but in the correlated state formed by the two modes.

A two-mode squeezed state may be defined by [9]

$$|\alpha_+, \alpha_-\rangle = D_+ (\alpha_+) D_- (\alpha_-) S(G) |0\rangle \qquad (2.85)$$

where the displacement operator is

$$D_\pm (\alpha) = \exp\left(\alpha a_\pm^\dagger - \alpha^* a_\pm\right), \qquad (2.86)$$

and the unitary two-mode squeeze operator is

$$S(G) = \exp\left(G^* a_+ a_- - G a_+^\dagger a_-^\dagger\right). \tag{2.87}$$

The squeezing operator transforms the annihilation operators as

$$S^\dagger(G) a_\pm \, S(G) = a_\pm \cosh r - a_\mp^\dagger \, e^{i\theta} \sinh r, \tag{2.88}$$

where $G = r e^{i\theta}$.

This gives for the following expectation values

$$\begin{aligned}
\langle a_\pm \rangle &= \alpha_\pm \\
\langle a_\pm a_\pm \rangle &= \alpha_\pm^2 \\
\langle a_+ a_- \rangle &= \alpha_+ \alpha_- - e^{i\theta} \sinh r \cosh r \\
\langle a_\pm^\dagger a_\pm \rangle &= |\alpha_\pm|^2 + \sinh^2 r.
\end{aligned} \tag{2.89}$$

The quadrature operator X is generalized in the two-mode case to

$$X = \frac{1}{\sqrt{2}}\left(a_+ + a_+^\dagger + a_- + a_-^\dagger\right). \tag{2.90}$$

As will be seen in Chap. 5, this definition is a particular case of a more general definition. It corresponds to the degenerate situation in which the frequencies of the two modes are equal.

The mean and variance of X in a two-mode squeezed state is

$$\begin{aligned}
\langle X \rangle &= 2(\mathrm{Re}\{\alpha_+\} + \mathrm{Re}\{\alpha_-\}), \\
V(X) &= \left(e^{-2r}\cos^2\frac{\theta}{2} + e^{2r}\sin^2\frac{\theta}{2}\right).
\end{aligned} \tag{2.91}$$

These results for two-mode squeezed states will be used in the analyses of nondegenerate parametric oscillation given in Chaps. 4 and 6.

2.8 Phase Properties of the Field

The definition of an Hermitian phase operator corresponding to the physical phase of the field has long been a problem. Initial attempts by P. Dirac led to a non-Hermitian operator with incorrect commutation relations. Many of these difficulties were made quite explicit in the work of *Susskind* and *Glogower* [10]. *Pegg* and *Barnett* [11] showed how to construct an Hermitian phase operator, the eigenstates of which, in an appropriate limit, generate the correct phase statistics for arbitrary states. We will first discuss the *Susskind–Glogower* (SG) phase operator.

Let a be the annihilation operator for a harmonic oscillator, representing a single field mode. In analogy with the classical polar decomposition of a complex amplitude we define the SG phase operator,

$$e^{i\phi} = (aa^\dagger)^{-1/2} a \,. \tag{2.92}$$

The operator $e^{i\phi}$ has the number state expansion

$$e^{i\phi} = \sum_{n=1}^{\infty} |n\rangle\langle n+1| \tag{2.93}$$

and eigenstates $|e^{i\phi}\rangle$ like

$$|e^{i\phi}\rangle = \sum_{n=1}^{\infty} e^{in\phi}|n\rangle \quad \text{for} \quad -\pi < \phi \leq \pi\,. \tag{2.94}$$

It is easy to see from (2.93) that $e^{i\phi}$ is not unitary,

$$\left[e^{i\phi}, \left(e^{i\phi} \right)^\dagger \right] = |0\rangle\langle 0|\,. \tag{2.95}$$

An equivalent statement is that the SG phase operator is not Hermitian. As an immediate consequence the eigenstates $|e^{i\phi}\rangle$ are not orthogonal. In many ways this is similar to the non-orthogonal eigenstates of the annihilation operator a, i.e. the coherent states. None-the-less these states do provide a resolution of identity

$$\int_{-\pi}^{\pi} d\phi \left| e^{i\phi}\rangle\langle e^{i\phi} \right| = 2\pi\,. \tag{2.96}$$

The phase distribution over the window $-\pi < \phi \leq \pi$ for any state $|\psi\rangle$ is then defined by

$$P(\phi) = \frac{1}{2\pi} |\langle e^{i\phi}|\psi\rangle|^2\,. \tag{2.97}$$

The normalisation integral is

$$\int_{-\pi}^{\pi} P(\phi)\,d\phi = 1\,. \tag{2.98}$$

The question arises; does this distribution correspond to the statistics of any physical phase measurement? At the present time there does not appear to be an answer. However, there are theoretical grounds [12] for believing that $P(\phi)$ is the correct distribution for optimal phase measurements. If this is accepted then the fact that the SG phase operator is not Hermitian is nothing to be concerned about. However, as we now show, one can define an Hermitian phase operator, the measurement statistics of which converge, in an appropriate limit, to the phase distribution of (2.97) [13].

Consider the state $|\phi_0\rangle$ defined on a finite subspace of the oscillator Hilbert space by

$$|\phi_0\rangle = (s+1)^{-1/2} \sum_{n=1}^{s} e^{in\phi_0} |n\rangle .$$ (2.99)

It is easy to demonstrate that the states $|\phi\rangle$ with the values of ϕ differing from ϕ_0 by integer multiples of $2\pi/(s+1)$ are orthogonal. Explicitly, these states are

$$|\phi_m\rangle = \exp\left(i\frac{a^\dagger a\, m\, 2\pi}{s+1}\right) |\phi_0\rangle; \quad m = 0, 1, \ldots, s ,$$ (2.100)

with

$$\phi_m = \phi_0 + \frac{2\pi m}{s+1} .$$

Thus $\phi_0 \le \phi_m < \phi_0 + 2\pi$. In fact, these states form a complete orthonormal set on the truncated $(s+1)$ dimensional Hilbert space. We now construct the *Pegg–Barnett* (PB) Hermitian phase operator

$$\phi = \sum_{m=1}^{s} \phi_m |\phi_m\rangle\langle\phi_m| .$$ (2.101)

For states restricted to the truncated Hilbert space the measurement statistics of ϕ are given by the discrete distribution

$$P_m = |\langle\phi_m|\psi\rangle_s|^2$$ (2.102)

where $|\psi\rangle_s$ is any vector of the truncated space.

It would seem natural now to take the limit $s \to \infty$ and recover an Hermitian phase operator on the full Hilbert space. However, in this limit the PB phase operator does not converge to an Hermitian phase operator, but the distribution in (2.102) does converge to the SG phase distribution in (2.97). To see this, choose $\phi_0 = 0$. Then

$$P_m = (s+1)^{-1} \left| \sum_{n=0}^{s} \exp\left(-i\frac{nm2\pi}{s+1}\right) \psi_n \right|^2$$ (2.103)

where $\psi_n = \langle n|\psi\rangle_s$.

As ϕ_m are uniformly distributed over 2π we define the probability density by

$$P(\phi) = \lim_{s\to\infty} \left[\left(\frac{2\pi}{s+1}\right)^{-1} P_m \right] = \frac{1}{2\pi} \left| \sum_{n=1}^{\infty} e^{in\phi} \psi_n \right|^2$$ (2.104)

where

$$\phi = \lim_{s\to\infty} \frac{2\pi m}{s+1} ,$$ (2.105)

and ψ_n is the number state coefficient for any Hilbert space state. This convergence in distribution ensures that the moments of the PB Hermitian phase operator converge, as $s \to \infty$, to the moments of the phase probability density.

The phase distribution provides a useful insight into the structure of fluctuations in quantum states. For example, in the number state $|n\rangle$, the mean and variance of the phase distribution are given by

$$\langle \phi \rangle = \phi_0 + \pi \,, \tag{2.106}$$

and

$$V(\phi) = \frac{2}{3}\pi \,, \tag{2.107}$$

respectively. These results are characteristic of a state with random phase. In the case of a coherent state $|re^{i\theta}\rangle$ with $r \gg 1$, we find

$$\langle \phi \rangle = \phi \,, \tag{2.108}$$

$$V(\phi) = \frac{1}{4\bar{n}} \,, \tag{2.109}$$

where $\bar{n} = \langle a^\dagger a \rangle = r^2$ is the mean photon number. Not surprisingly a coherent state has well defined phase in the limit of large amplitude.

Exercises

2.1 If $|X_1\rangle$ is an eigenstate for the operator X_1 find $\langle X_1 | \psi \rangle$ in the cases (a) $|\psi\rangle = |\alpha\rangle$; (b) $|\psi\rangle = |\alpha, r\rangle$.

2.2 Prove that if $|\psi\rangle$ is a minimum-uncertainty state for the operators X_1 and X_2, then $V(X_1, X_2) = 0$.

2.3 Show that the squeeze operator

$$S(r, \phi) = \exp\left[\frac{r}{2}\left(e^{-2i\phi}a^2 - e^{2i\phi}a^{\dagger 2}\right)\right]$$

may be put in the normally ordered form

$$S(r, \phi) = (\cosh r)^{-1/2} \exp\left(-\frac{\Gamma}{2}a^{\dagger 2}\right) \exp\left[-\ln(\cosh r)a^\dagger a\right] \exp\left(\frac{\Gamma^*}{2}a^2\right)$$

where $\Gamma = e^{2i\theta}\tanh r$.

2.4 Evaluate the mean and variance for the phase operator in the squeezed state $|\alpha, r\rangle$ with α real. Show that for $|r| \gg |\alpha|$ this state has either enhanced or diminished phase uncertainty compared to a coherent state.

References

1. E.A. Power: *Introductory Quantum Electrodynamics* (Longmans, London 1964)
2. R.J. Glauber: Phys. Rev. B **1**, 2766 (1963)
3. W.H. Louisell: *Statistical Properties of Radiation* (Wiley, New York 1973)

4. D.F. Walls: Nature **324**, 210 (1986)
5. C.M. Caves: Phys. Rev. D **23**, 1693 (1981)
6. H.P. Yuen: Phys. Rev. A **13**, 2226 (1976)
7. R. Loudon: *Quantum Theory of Light* (Oxford Univ. Press, Oxford 1973)
8. W. Schleich, J.A. Wheeler: Nature **326**, 574 (1987)
9. C.M. Caves, B.L. Schumaker: Phys. Rev. A **31**, 3068 (1985)
10. L. Susskind, J. Glogower: Physics **1**, 49 (1964)
11. D.T. Pegg, S.M. Barnett: Phys. Rev. A **39**, 1665 (1989)
12. J.H. Shapiro, S.R. Shepard: Phys. Rev. A **43**, 3795 (1990)
13. M.J.W. Hall: Quantum Optics **3**, 7 (1991)

Further Reading

Glauber, R.J.: In *Quantum Optics and Electronics*, ed. by C. de Witt, C. Blandin, C. Cohen Tannoudji Gordon and Breach, New York 1965)
Klauder, J.R., Sudarshan, E.C.G.: *Fundamentals of Quantum Optics* (Benjamin, New York 1968)
Loudon, R.: *Quantum Theory of Light* (Oxford Univ. Press, Oxford 1973)
Louisell, W.H.: *Quantum Statistical Properties of Radiation* (Wiley, New York 1973)
Meystre, P., M. Sargent, III: *Elements of Quantum Optics*, 2nd edn. (Springer, Berlin, Heidelberg 1991)
Nussenveig, H.M.: *Introduction to Quantum Optics* (Gordon and Breach, New York 1974)

Chapter 3
Coherence Properties
of the Electromagnetic Field

Abstract In this chapter correlation functions for the electromagnetic field are introduced from which a definition of optical coherence may be formulated. It is shown that the coherent states possess nth-order optical coherence. Photon-correlation measurements and the phenomena of photon bunching and antibunching are described. Phase-dependent correlation functions which are accessible via homodyne measurements are introduced. The theory of photon counting measurements is given.

3.1 Field-Correlation Functions

We shall now consider the detection of an electromagnetic field. A large-scale macroscopic device is complicated, hence, we shall study a simple device, an ideal photon counter. The most common devices in practice involve a transition where a photon is absorbed. This has important consequences since this type of counter is insensitive to spontaneous emission. A complete theory of detection of light requires a knowledge of the interaction of light with atoms. We shall postpone this until a study of the interaction of light with atoms is made in Chap. 10. At this stage we shall assume we have an ideal detector working on an absorption mechanism which is sensitive to the field $E^{(+)}(r,t)$ at the space-time point (r,t). We follow the treatment of *Glauber* [1].

The transition probability of the detector for absorbing a photon at position r and time t is proportional to

$$T_{if} = |\langle f|E^{(+)}(r,\,t)|i\rangle|^2 \tag{3.1}$$

if $|i\rangle$ and $|f\rangle$ are the initial and final states of the field. We do not, in fact, measure the final state of the field but only the total counting rate. To obtain the total count rate we must sum over all states of the field which may be reached from the initial state by an absorption process. We can extend the sum over a complete set of final states since the states which cannot be reached (e.g., states $|f\rangle$ which differ from $|i\rangle$

by two or more photons) will not contribute to the result since they are orthogonal to $E^{(+)}(r,t)|i\rangle$. The total counting rate or average field intensity is

$$I(r,\ t) = \sum_f T_{fi} = \sum_f \langle i|E^{(-)}(r,\ t)|f\rangle\langle f|E^{(+)(r,\ t)|i\rangle}$$

$$= \langle i|E^{(-)}(r,\ t)E^{(+)}(r,\ t)|i\rangle\ , \tag{3.2}$$

where we have used the completeness relation

$$\sum_f |f\rangle\langle f| = 1\ . \tag{3.3}$$

The above result assumes that the field is in a pure state $|i\rangle$. The result may be easily generalized to a statistical mixture state by averaging over initial states with the probability P_i, i.e.,

$$I(r,\ t) = \sum_i P_i\langle i|E^{(-)}(r,\ t)E^{(+)}(r,\ t)|i\rangle\ . \tag{3.4}$$

This may be written as

$$I(r,\ t) = \text{Tr}\ \{\rho E^{(-)}(r,\ t)E^{(+)}(r,\ t)\}\ , \tag{3.5}$$

where ρ is the density operator defined by

$$\rho = \sum_i P_i|i\rangle\langle i|\ . \tag{3.6}$$

If the field is initially in the vacuum state

$$\rho = |0\rangle\langle 0|\ , \tag{3.7}$$

then the intensity is

$$I(r,\ t) = \langle 0|E^{(-)}E^{(+)}|0\rangle = 0\ . \tag{3.8}$$

The normal ordering of the operators (that is, all annihilation operators are to the right of all creation operators) yields zero intensity for the vacuum. This is a consequence of our choice of an absorption mechanism for the detector. Had we chosen a detector working on a stimulated emission principle, problems would arise with vacuum fluctuations. More generally the correlation between the field at the space-time point $x = (r,t)$ and the field at the space-time point $x' = (r,t')$ may be written as the correlation function

$$G^{(1)}(x,\ x') = \text{Tr}\{\rho E^{(-)}(x)E^{(+)}(x')\}\ . \tag{3.9}$$

The first-order correlation function of the radiation field is sufficient to account for classical interference experiments. To describe experiments involving intensity correlations such as the Hanbury-Brown and Twiss experiment, it is necessary to define higher-order correlation functions. We define the nth-order correlation function of the electromagnetic field as

$$G^{(n)}(x_1 \ldots x_n, x_{n+1} \ldots x_{2n}) = \mathrm{Tr}\{\rho E^{(-)}(x_1) \ldots E^{(-)}(x_n)$$
$$\times E^{(+)}(x_{n+1}) \ldots E^{(+)}(x_{2n})\}. \qquad (3.10)$$

Such an expression follows from a consideration of an n-atom photon detector [1]. The n-fold delayed coincidence rate is

$$W^{(n)}(t_1 \ldots t_n) = s^n G^{(n)}(r_1 t_1 \ldots r_n t_n, r_n t_n \ldots r_1 t_1), \qquad (3.11)$$

where s is the sensitivity of the detector.

3.2 Properties of the Correlation Functions

A number of interesting inequalities can be derived from the general expression

$$\mathrm{Tr}\{\rho A^\dagger A\} \geq 0, \qquad (3.12)$$

which follows from the non-negative character of $A^\dagger A$ for any linear operator A.
   Thus choosing $A = E^{(+)}(x)$ gives

$$G^{(1)}(x, x) \geq 0. \qquad (3.13)$$

In general, taking

$$A = E^{(+)}(x_n) \ldots E^{(+)}(x_1) \qquad (3.14)$$

yields

$$G^{(n)}(x_1 \ldots x_n, x_n \ldots x_1) \geq 0 \qquad (3.15)$$

Choosing

$$A = \sum_{j=1}^{n} \lambda_j E^{(+)}(x_j), \qquad (3.16)$$

where λ_j are an arbitrary set of complex numbers gives

$$\sum_{ij} \lambda_i^* \lambda_j G^{(1)}(x_i, x_j) \geq 0. \qquad (3.17)$$

Thus the set of correlation functions $G^{(1)}(x_i, x_j)$ forms a matrix of coefficients for a positive definite quadratic form. Such a matrix has a positive determinant, i.e.,

$$\det[G^{(1)}(x_i, x_j)] \geq 0. \qquad (3.18)$$

For $n = 1$, this is simply (3.13). For $n = 2$ we find

$$G^{(1)}(x_1, x_1) G^{(1)}(x_2, x_2) \geq |G^{(1)}(x_1, x_2)|^2 \qquad (3.19)$$

which is a simple generalisation of the Schwarz inequality.

Choosing

$$A = \lambda_1 E^{(+)}(x_1) \; \dots \; E^{(+)}(x_n) + \lambda_2 E^{(+)}(x_{n+1}) \; \dots \; E^{(+)}(x_{2n}) \,, \tag{3.20}$$

we find the general relation

$$G^{(n)}(x_1 \; \dots \; x_n, x_n \; \dots x_1) G^{(n)}(x_{n+1} \; \dots \; x_{2n}, x_{2n} \; \dots x_{n+1})$$
$$\geq |G^{(n)}(x_1 \; \dots \; x_n, x_{n+1} \; \dots \; x_{2n})|^2 \,. \tag{3.21}$$

For two beams we may take

$$A = \lambda_1 E_1^{(+)}(x) E_1^{(+)}(x') + \lambda_2 E_2^{(+)}(x) E_2^{(+)}(x') \,, \tag{3.22}$$

with $x \equiv (r, 0)$ and $x' \equiv (r, t)$. The Cauchy–Schwartz inequality then becomes

$$G_{11}^{(2)}(0) G_{22}^{(2)}(0) \geq [G_{12}^{(2)}(t)]^2 \,, \tag{3.23}$$

where

$$G_{ij}^{(2)}(t) = \mathrm{Tr}\{\rho E_i^{(-)}(x) E_i^{(-)}(x') E_j^{(+)}(x') E_j^{(+)}(x)\} \,; \tag{3.24}$$

we have noted explicitly that $G_{ii}^{(2)}$ is time independent.

An inequality closely related to (3.23) may be derived by choosing

$$A = \lambda_1 E_1^{(-)}(x) E_1^{(+)}(x) + \lambda_2 E_2^{(-)}(x) E_2^{(+)}(x) \,. \tag{3.25}$$

This gives

$$|\langle E_1^{(-)}(x) E_1^{(+)}(x) E_2^{(+)}(x) E_2^{(-)}(x) \rangle|^2$$
$$\leq \langle [E_1^{(-)}(x) E_1^{(+)}(x)]^2 \rangle \langle [E_2^{(-)}(x) E_2^{(+)}(x)]^2 \rangle \,. \tag{3.26}$$

This inequality will be used in Chap. 5.

3.3 Correlation Functions and Optical Coherence

Classical optical interference experiments correspond to a measurement of the first-order correlation function. We shall consider Young's interference experiment as a measurement of the first-order correlation function of the field and show how a definition of first-order optical coherence arises from considerations of the fringe visibility.

A schematic sketch of Young's interference experiment is depicted in Fig. 3.1. The field incident on the screen at position r and time t is the superposition of the fields at the two pin holes

$$E^{(+)}(r, t) = E_1^{(+)}(r, t) + E_2^{(+)}(r, t) \tag{3.27}$$

Fig. 3.1 Schematic representation of Young's interference experiment

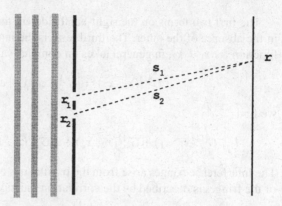

where $E_i^{(+)}(r,t)$ is the field produced by pinhole i at the screen with

$$E_i^{(+)}(r,t) = E_i^{(+)}\left(r_i,\, t - \frac{s_i}{c}\right)\left(\frac{1}{s_i}\right)\, e^{i(k-\frac{\omega}{c})\, s_i} \tag{3.28}$$

where $s_i = |r_i - r|$
and $E_i^{(+)}(r_i, t - s_i/c)$ is the field at the ith pinhole and for a spherical wave

$$k - \frac{\omega}{c} = 0 \,.$$

Therefore (3.27) becomes

$$E^{(+)}(r,\, t) = \frac{E_1^{(+)}\left(r_1,\, t - \dfrac{s_1}{c}\right)}{s_1} + \frac{E_2^{(+)}\left(r_2,\, t - \dfrac{s_2}{c}\right)}{s_2} \,. \tag{3.29}$$

For $s_1 \approx s_2 \approx R$, we have

$$E^{(+)}(r,\, t) = \frac{1}{R}[E_1^{(+)}(x_1) + E_2^{(+)}(x_2)] \tag{3.30}$$

where

$$x_1 = \left(r_1,\, t - \frac{s_1}{c}\right), \qquad x_2 = \left(r_2,\, t - \frac{s_2}{c}\right) \,.$$

The intensity observed on the screen is proportional to

$$I = \mathrm{Tr}\{\rho E^{(-)}(r,\, t) E^{(+)}(r,\, t)\} \,. \tag{3.31}$$

Using (3.27) we find

$$I = G^{(1)}(x_1,\, x_1) + G^{(1)}(x_2,\, x_2) + 2\mathrm{Re}\{G^{(1)}(x_1,\, x_2)\} \tag{3.32}$$

where the R^{-2} factor is absorbed into a normalisation constant.

The first two terms on the right-hand side are the intensities from each pinhole in the absence of the other. The third term is the interference term. The correlation function for $x_1 \neq x_2$, in general takes on complex values. Writing this as

$$G^{(1)}(x_1, x_2) = |G^{(1)}(x_1, x_2)|\, e^{i\Psi(x_1, x_2)} \,, \tag{3.33}$$

we find

$$I = G^{(1)}(x_1, x_1) + G^{(1)}(x_2, x_2) + 2|G^{(1)}(x_1, x_2)|\cos\Psi(x_1, x_2) \,. \tag{3.34}$$

The interference fringes arise from the oscillations of the cosine term. The envelope of the fringes is described by the correlation function $G^{(1)}(x_1, x_2)$.

3.4 First-Order Optical Coherence

The idea of coherence in optics was first associated with the possibility of producing interference fringes when two fields are superposed. The highest degree of optical coherence was associated with a field which exhibits fringes with maximum visibility. If $G^{(1)}(x_1, x_2)$ was zero there would be no fringes and the fields are then described as incoherent. Thus the larger $G^{(1)}(x_1, x_2)$ the more coherent the field. The magnitude of $|G^{(1)}(x_1, x_2)|$ is limited by the relation

$$|G^{(1)}(x_1, x_2)| \leq [G^{(1)}(x_1, x_1)G^{(1)}(x_2, x_2)]^{1/2} \,. \tag{3.35}$$

The best possible fringe contrast is given by the equality sign. Thus the necessary condition for full coherence is

$$|G^{(1)}(x_1, x_2)| = [G^{(1)}(x_1, x_1)G^{(1)}(x_2, x_2)]^{1/2} \,. \tag{3.36}$$

Introducing the normalized correlation function

$$g^{(1)}(x_1, x_2) = \frac{G^{(1)}(x_1,x_2)}{[G^{(1)}(x_1,x_1)G^{(1)}(x_2,x_2)]^{1/2}} \,, \tag{3.37}$$

the condition (3.36) becomes

$$|g^{(1)}(x_1, x_2)| = 1 \tag{3.38}$$

or

$$g^{(1)}(x_1, x_2) = e^{i\Psi(x_1, x_2)} \,.$$

The visibility of the fringes is given by

$$\upsilon = \frac{I_{\max} - I_{\min}}{I_{\max} - I_{\min}} \,. \tag{3.39}$$

Using (3.27 and 3.31) for the intensity we may write v as

$$v = \left| \frac{G^{(1)}(x_1, x_2)}{(G^{(1)}(x_1, x_1)G^{(1)}(x_2, x_2))^{1/2}} \right| \frac{2(I_1 I_2)^{1/2}}{I_1 + I_2}$$

$$= |g^{(1)}| \frac{2(I_1 I_2)^{1/2}}{I_1 + I_2} . \tag{3.40}$$

If the fields incident on each pinhole have equal intensities the fringe visibility is equal to $|g^{(1)}|$. Thus the condition for first-order optical coherence $|g^{(1)}| = 1$ corresponds to the condition for maximum fringe visibility.

A more general definition of first-order coherence of the field $E(x)$ is that the first-order correlation function factorizes

$$G^{(1)}(x_1, x_2) = \varepsilon^{(-)}(x_1)\varepsilon^{(+)}(x_2) . \tag{3.41}$$

It is readily seen that this is equivalent to the condition for first-order optical coherence given by (3.38). It is clear that for a field in a left eigenstate of the operator $E^{(+)}(x)$ this factorization holds. The coherent states are an example of such a field. It is precisely this coherence property of the coherent states which led to their names.

We may generalize (3.41) to give the condition for nth optical coherence. This requires that the nth order correlation function factorizes:

$$G^{(n)}(x_1 \ldots x_n, x_{n+1}, \ldots ,x_{2n}) = \varepsilon^{(-)}(x_1) \ldots \varepsilon^{(-)}(x_n)\varepsilon^{(+)}(x_{n+1}) \ldots \varepsilon^{(-)}(x_{2n}) . \tag{3.42}$$

Again the coherent states possess nth-order optical coherence.

Photon interference experiments of the kind typified by Young's interference experiment and Michelson's interferometer played a central role in early discussions of the dual wave and corpuscular nature of light. These experiments basically detect the interference pattern resulting from the superposition of two components of a light beam. Classical theory based on the wave nature of light readily explained the observed interference pattern. The quantum-mechanical explanation is based on the interference of the probability amplitudes for the photon to take either of two paths. We shall demonstrate how interference occurs even for a one photon field. For full details of the classical theory and experimental arrangements the reader is referred to the classic text of *Born* and *Wolf* [2].

We consider an interference experiment of the type performed by Young which consists of light from a monochromatic point source S incident on a screen possessing two pinholes P_1 and P_2 which are equidistant from S (see Fig. 3.1).

The pinholes act as secondary monochromatic point sources which are in phase and the beams from them are superimposed on a screen at position r and time t. In this region an interference pattern is formed.

To avoid calculating the diffraction pattern for the pinhole, we assume their dimensions are of the order of the wavelength of light in which case they effectively act as sources for single modes of spherical radiation in keeping with Huygen's principle. The appropriate mode functions for spherical radiation are

$$u_k(\mathbf{r}) = \sqrt{\frac{1}{4\pi L}} \frac{e^{i\mathbf{k}\cdot\mathbf{r}}}{r} \hat{e}_{\mathbf{k}} \,, \tag{3.43}$$

where L is the radius of the normalization volume, and $\hat{e}_k$ is the unit polarization vector.

The field detected on the screen at position $\mathbf{r}$ and time t is then the sum of the two spherical modes emitted by the two pinholes

$$E^{(+)}(\mathbf{r},\, t) = f(\mathbf{r},t)(a_1 e^{iks_1} + a_2 e^{iks_2}) \,, \tag{3.44}$$

with

$$f(\mathbf{r},\, t) = i\left(\frac{\hbar\omega}{2}\right)^{1/2} \frac{\hat{e}_k}{(4\pi L)^{1/2}} \frac{1}{R} e^{-i\omega t} \,,$$

where s_1 and s_2 are the distances of the pinholes P_1 and P_2 to the point on the screen, and we have set $s_1 \approx s_2 = R$ in the denominator of the mode functions. Substituting (3.43) into (3.2) for the intensity we find

$$I(\mathbf{r},\, t) = \eta\left[\mathrm{Tr}\{\rho a_1^\dagger a_1\} + \mathrm{Tr}\{\rho a_2^\dagger a_2\} + 2|\mathrm{Tr}\{\rho a_1^\dagger a_2\}|\cos\Phi\right]. \tag{3.45}$$

where

$$\mathrm{Tr}\{\rho a_1^\dagger a_2\} = |\mathrm{Tr}\{\rho a_1^\dagger a_2\}|e^{i\phi},$$
$$\eta = |f(\mathbf{r},\, t)|^2,$$
$$\Phi = k(s_1 - s_2) + \phi.$$

This expression exhibits the typical interference fringes with the maximum of intensity occurring at

$$k(s_1 - s_2) + \phi = n2\pi \,, \tag{3.46}$$

with n an integer.

The maximum intensity of the fringes falls off as one moves the point of observation further from the central line by the R^{-2} factor in $|f(\mathbf{r},t)|^2$.

We shall evaluate the intensity for fields which may be generated by a single-mode excitation and hence have first-order coherence. A general representation of such a field is

$$|\psi\rangle = f(b^\dagger)|0\rangle \,, \tag{3.47}$$

where $|0\rangle$ denotes the vacuum state of the radiation field and $b^\dagger$ is the creation operator for a single mode of the radiation field. The operator $b^\dagger$ may be expressed as a linear combination of $a_1^\dagger$ and $a_2^\dagger$ as follows

$$b^\dagger = -\frac{1}{\sqrt{2}}(a_1^\dagger + a_2^\dagger) \,, \tag{3.48}$$

where we have assumed equal intensities through each slit. We shall now consider as a special case the field with only one photon incident on the pinholes, i.e.,

$$|1\text{ photon}\rangle = b^\dagger|0\rangle = \frac{1}{\sqrt{2}}(|1,0\rangle + |0,1\rangle) \,, \tag{3.49}$$

where the notation used for the eigenkets $|n_1, n_2\rangle$ implies that there are n_1 photons present in mode k_1 and n_2 photons present in mode k_2. This state of the field reflects the fact that we don't know which pinhole the photon goes through.

From (3.45) this yields the following expression for the mean intensity on the screen

$$I(r, t) = \eta(1 + \cos \Phi) . \tag{3.50}$$

It is clear from this equation that an interference pattern may be built up from a succession of one-photon interference fringes.

The quantum explanation for the interference pattern was first put forward by *Dirac* [3] in his classic text on quantum mechanics. There he argued that the observed intensity pattern results from interference between the probability amplitudes of a single photon to take either of two possible paths. The crux of the quantum mechanical explanation is that the wavefunction gives information about the probability of one photon being in a particular place and not the probable number of photons in that place. *Dirac* pointed out that the interference between the two beams does not arise because photons of one beam sometimes annihilate photons from the other, and sometimes combine to produce four photons. "This would contradict the conservation of energy. The new theory which connects the wave functions with probabilities for one photon gets over the difficulty by making each photon go partly into each of two components. Each photon then interferes only with itself. Interference between two different photons never occurs". We stress that the above-quoted statement of *Dirac* was only intended to apply to experiments of the Young's type where the interference pattern is revealed by detecting single photons. It was not intended to apply to experiments of the type where correlations between different photons are measured.

A very early experiment to test if interference would result from a single photon was performed by *Taylor* [4] in 1905. In this experiment the intensity of the source was so low that on average only one photon was incident on the slits at a time. The photons were detected on a photographic plate so that the detection time was very large. Interference fringes were observed in this experiment. This experiment did not definitively show that the interference fringes resulted from a single photon since the statistical distribution of photons meant that sometimes two photons could be incident on the slits. A definitive experiment was conducted by *Grangier* et al. [5] using a two-photon cascade as a source. A coincidence technique which detected one photon of the pair enabled them to prepare a one photon source.

We now consider the interference patterns produced by other choices of a field.

3.5 Coherent Field

We consider a coherent field as generated by an ideal laser incident on the pinholes. The wavefunction for this coherent field is

$$|\text{coherent field}\rangle = |\alpha_1, \alpha_2\rangle = |\alpha_1\rangle|\alpha_2\rangle . \tag{3.51}$$

Since this wavefunction is a product state, it may well represent two independent light beams. This particular product may, however, be generated by a single-mode excitation in the following manner:

$$|\alpha_1\rangle|\alpha_2\rangle = \exp(\alpha b^\dagger - \alpha^* b)|0\rangle$$

$$= \exp\frac{1}{\sqrt{2}}(\alpha a_1^\dagger - \alpha^* a_1)\exp\frac{1}{\sqrt{2}}(\alpha a_2^\dagger - \alpha^* a_2)|0\rangle$$

$$= \left|\frac{\alpha}{\sqrt{2}}\right\rangle\left|\frac{\alpha}{\sqrt{2}}\right\rangle . \tag{3.52}$$

The intensity pattern produced by this coherent field is

$$I(r, t) = \eta(|\alpha|^2 + |\alpha|^2 \cos \phi) . \tag{3.53}$$

The above example demonstrates the possibility of obtaining interference between independent light beams. Experimentally, this requires that the phase relation between the two beams be slowly varying or else the fringe pattern will be washed out. Such experiments have been performed by *Pfleegor* and *Mandel* [6]. Interference between independent light beams is, however, only possible for certain states of the radiation field, for example, the coherent states as demonstrated above. Interference is not generally obtained from independent light beams, as we shall demonstrate in the following example. We consider the two independent light beams to be Fock states, that is, described by the wavefunction

$$|\psi\rangle = |n_1\rangle|n_2\rangle . \tag{3.54}$$

This yields a zero correlation function and consequently no fringes are obtained.

The analysis we performed leading to (3.50) bears out *Dirac's* argument that the interference fringes may be produced by a series of one photon experiments. However, Young's interference fringes may perfectly well be explained by the interference of classical waves. Experiments of this kind which measure the first-order correlation functions of the electromagnetic field do not distinguish between the quantum and classical theories of light.

3.6 Photon Correlation Measurements

The first experiment performed outside the domain of one photon optics was the intensity correlation experiment of *Hanbury-Brown* and *Twiss* [7]. Although the original experiment involved the analogue correlation of photo-currents, later experiments used photon counters and digital correlations and were truly photon correlation measurements. In essence these experiments measure the joint photocount probability of detecting the arrival of a photon at time t and another photon at time $t + \tau$. This may be written as an intensity or photon-number correlation function.

Using the quantum detection theory developed by *Glauber*, the measured quantity is the normally ordered correlation function

$$G^{(2)}(\tau) = \langle E^{(-)}(t)E^{(-)}(t+\tau)E^{(+)}(t+\tau)E^{(+)}(t)\rangle$$
$$= \langle : I(t)I(t+\tau) : \rangle$$
$$\propto \langle : n(t)n(t+\tau) : \rangle \tag{3.55}$$

where : : indicates normal ordering, $I(t)$ is the intensity for analogue measurements and $n(t)$ is the photon number in photon counting experiments. It is useful to introduce the normalized second-order correlation function defined by

$$g^{(2)}(\tau) = \frac{G^{(2)}(\tau)}{|G^{(1)}(0)|^2} . \tag{3.56}$$

We shall evaluate $g^{(2)}(\tau)$ for certain classes of field. For a field which possesses second-order coherence

$$G^{(2)}(\tau) = \varepsilon^{(-)}(t)\varepsilon^{(-)}(t+\tau)\varepsilon^{(+)}(t+\tau)\varepsilon^{(+)}(t) = [G^{(1)}(0)]^2 . \tag{3.57}$$

Hence $g^{(2)}(\tau) = 1$.

For a fluctuating classical field we may introduce a probability distribution $P(\varepsilon)$ describing the probability of the field $E^{(+)}(\varepsilon,t)$ having the amplitude ε where

$$E^{(+)}(\varepsilon, t) = -\left(i\frac{\hbar\omega}{2\varepsilon_0 V}\right)^{1/2} \varepsilon e^{-i\omega t} .$$

For a multimode field we have a multivariate probability distribution $P(\{\varepsilon_k\})$. The second-order correlation function $G^{(2)}(\tau)$ may be written as

$$G^{(2)}(\tau) = \int P(\{\varepsilon_k\})E^{(-)}(\varepsilon_k, t)E^{(-)}(\varepsilon_k, t+\tau)E^{(+)}(\varepsilon_k, t+\tau)E^{(+)}(\varepsilon_k, t)d^2\{\varepsilon_k\} . \tag{3.58}$$

For zero time delay $\tau = 0$ we may write for a single-mode field

$$g^{(2)}(0) = 1 + \frac{\int P(\varepsilon)(|\varepsilon|^2 - \langle|\varepsilon|^2\rangle)^2 d^2\varepsilon}{(\langle|\varepsilon|^2\rangle)^2} . \tag{3.59}$$

For classical fields the probability distribution $P(\varepsilon)$ is positive, hence $g^{(2)}(0) \geq 1$.

For a field obeying Gaussian statistics with zero mean amplitude

$$\langle E^{(-)}(\varepsilon, t)E^{(-)}(\varepsilon, t+\tau)E^{(+)}(\varepsilon, t)E^{(+)}(\varepsilon, t+\tau)\rangle$$
$$= \langle E^{(-)}(\varepsilon, t)E^{(-)}(\varepsilon, t+\tau)\rangle\langle E^{(+)}(\varepsilon, t+\tau)E^{(+)}(\varepsilon, t)\rangle$$
$$+ \langle E^{(-)}(\varepsilon, t)E^{(+)}(\varepsilon, t)\rangle\langle E^{(-)}(\varepsilon, t+\tau)E^{(+)}(\varepsilon, t+\tau)\rangle$$
$$+ \langle E^{(-)}(\varepsilon, t)E^{(+)}(\varepsilon, t+\tau)\rangle\langle E^{(-)}(\varepsilon, t+\tau)E^{(+)}(\varepsilon, t)\rangle . \tag{3.60}$$

For fields with no phase-dependent fluctuations the first term may be neglected. Then

$$G^{(2)}(\tau) = G^{(1)}(0)^2 + |G^{(1)}(\tau)|^2 \,. \tag{3.61}$$

Hence the normalized second-order correlation function is

$$g^{(2)}(\tau) = 1 + |g^{(1)}(\tau)|^2 \,. \tag{3.62}$$

Now $G^{(1)}(\tau)$ is the Fourier transform of the spectrum of the field

$$S(\omega) = \int\limits_{-\infty}^{\infty} d\tau \, e^{-i\omega\tau} G^{(1)}(\tau) \,. \tag{3.63}$$

Hence for a field with a Lorentzian spectrum

$$g^{(2)}(\tau) = 1 + e^{-\gamma\tau} \tag{3.64}$$

and for a field with a Gaussian spectrum

$$g^{(2)}(\tau) = 1 + e^{-\gamma^2\tau^2} \,, \tag{3.65}$$

where γ is the spectral linewidth.

For a values of $\tau \gg \tau_c$ the correlation time of the light, the correlation function factorizes and $g^{(2)}(\tau) \to 1$. The increased value of $g^{(2)}(\tau)$ for $\tau < \tau_c$ for chaotic light over coherent light $[g^{(2)}(0)_{\text{chaotic}} = 2g^{(2)}(0)_{\text{coh}}]$ is due to the increased intensity fluctuations in the chaotic light field. There is a high probability that the photon which triggers the counter occurs during a high intensity fluctuation and hence a high probability that a second photon will be detected arbitrarily soon. This effect known as photon bunching was first detected by *Hanbury-Brown* and *Twiss*. Later experiments [8] showed excellent agreement with the theoretical predictions for chaotic and coherent light (Fig. 3.2). We note that the above analysis does not rely on any quantisation of the electromagnetic field but may be deduced from a purely classical analysis of the electromagnetic field with fluctuating amplitudes for the modes.

Measurement of the second-order correlation function of light with Gaussian statistics has formed the basis of photon correlation spectroscopy [9]. Photon correlation spectroscopy may be used to measure very narrow linewidths $(1–10^8\,\text{Hz})$ which are outside the range of conventional spectrometers. The second-order correlation function $g^2(\tau)$ is measured using electronic correlators and the linewidth extracted using (3.64 or 3.65). This has found application, for example, in the measurement of the diffusion coefficient of macromolecules where the scattered light has Gaussian statistics. The linewidth of the scattered light contains information on the diffusion coefficient of the macromolecule. This technique has been applied to determine the size of biological molecules such as viruses as well as in studying turbulent flows.

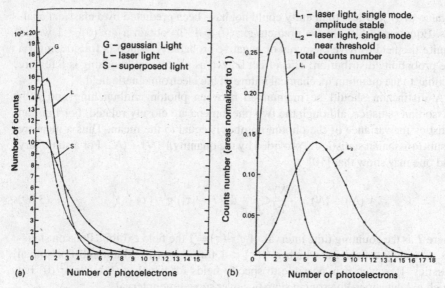

Fig. 3.2 Measured photo-count statistics for (**a**) Gaussian, laser and superposed fields. Measuring time of a single sample: 10 μs. Coherence time of the Gaussian field; 40 μs. (**b**) Two laser fields. Measuring time of a single sample: 10 μs

3.7 Quantum Mechanical Fields

We shall now evaluate the second-order correlation function for some quantum-mechanical fields. We shall restrict our attention to a single-mode field and calculate $g^{(2)}(0)$ and the variance in the photon number $V(n)$

$$g^{(2)}(0) = \frac{\langle a^{\dagger} a^{\dagger} a a \rangle}{\langle a^{\dagger} a \rangle^2} = 1 + \frac{V(n) - \bar{n}}{\bar{n}^2} , \qquad (3.66)$$

where $V(n) = \langle (a^{\dagger} a)^2 \rangle - \langle a^{\dagger} a \rangle^2$.
Coherent State
    For a coherent state

$$\rho = |\alpha\rangle\langle\alpha|, \quad g^{(2)}(0) = 1 \qquad (3.67)$$

and $V(n) = \bar{n}$ for a Poisson distribution in photon number.
    Number state

$$\rho = |n\rangle\langle n|, \qquad g^{(2)}(0) = 1 - \frac{1}{n}, \quad n > 2 . \qquad (3.68)$$

A number state has zero variance in the photon number ($V(n) = 0$). If $g^{(2)}(\tau) < g^{(2)}(0)$ there is a tendency for photons to arrive in pairs. This situation is referred to as *photon bunching*. The converse situation, $g^{(2)}(\tau) > g^{(2)}(0)$ is called *antibunching*. As noted above, however, $g^{(2)}(\tau) \rightarrow 1$ on a sufficiently long time scale. Thus a field for which $g^{(2)}(0) < 1$ will always exhibit antibunching on some time scale.

A value of $g^{(2)}(0)$ less than unity could not have been predicted by a classical analysis. Equation (3.59) always predicts $g^{(2)}(0) \geq 1$. To obtain a $g^{(2)}(0) < 1$ would require the field to have elements of negative probability, which is forbidden for a true probability distribution. This effect known as photon antibunching is a feature peculiar to the quantum mechanical nature of the electromagnetic field.

A distinction should be maintained between photon antibunching and sub-Poissonian statistics, although the two phenomena are closely related. For Poisson statistics the variance of the photon number is equal to the mean. Thus a measure of sub-Poissonian statistics is provided by the quantity $V(N) - \langle N \rangle$. For a stationary field one may show that [10].

$$V(N) - \langle N \rangle = \frac{\langle N \rangle^2}{T^2} \int_{-T}^{T} \mathrm{d}\tau (T - |\tau|)[g^{(2)}(\tau) - 1] \,, \tag{3.69}$$

where T is the counting time interval. If $g^{(2)}(\tau) = 1$ the field exhibits Poisson statistics. Certainly a field for which $g^{(2)}(\tau) < 1$ for all τ will exhibit sub-Poissonian statistics. However, it is possible to specify fields for which $g^{(2)}(\tau) > g^{(2)}(0)$ but which exhibit super-Poissonian statistics over some time interval.

3.7.1 Squeezed State

We consider a squeezed state $|\alpha, r\rangle$ with r defined as positive (Fig. 3.3). We align our axes such that the X_1 direction is parallel to the minor axis of the error ellipse. The direction (*1*) is referred to as the direction of squeezing and the direction (*2*) as the direction of coherent excitation. We then define α by $2\alpha = \langle X_1 \rangle + \mathrm{i}\langle X_2 \rangle$ with $\theta = \tan^{-1}(\langle X_2 \rangle / \langle X_1 \rangle)$. The variance in the photon number for this squeezed state is

$$\frac{V(n) - \bar{n}}{\bar{n}^2} = \frac{|\alpha|^2(\cosh 2r - \sinh 2r \cos 2\theta - 1) + \sinh^2 r \cosh 2r}{(|\alpha|^2 + \sinh^2 r)^2} \,. \tag{3.70}$$

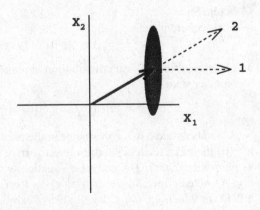

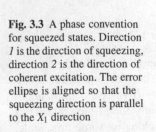

Fig. 3.3 A phase convention for squeezed states. Direction *1* is the direction of squeezing, direction *2* is the direction of coherent excitation. The error ellipse is aligned so that the squeezing direction is parallel to the X_1 direction

When $\theta = \pi/2$, that is the squeezing is out of phase with the complex amplitude

$$V(n) = |\alpha|^2 e^{2r} + 2\sinh^2 r\cosh^2 r. \qquad (3.71)$$

Thus this state with increased amplitude fluctuations has super-Poissonian statistics as expected.

When $\theta = 0$, that is the squeezing is in phase with the complex amplitude

$$V(n) = |\alpha|^2 e^{-2r} + 2\sinh^2 r\cosh^2 r. \qquad (3.72)$$

The first term corresponds to the reduction in number fluctuations in the original Poisson distribution. The second term is due to the fluctuations of the additional photons in the squeezed vacuum.

When $|\alpha|^2 \gg 2\sinh^2 r\cosh^2 r$ this is an amplitude squeezed state with sub-Poissonian photon statistics. The maximum reduction in photon number fluctuations one can get in an amplitude squeezed state may be estimated as follows: For $r \geq 1$

$$V(n) \approx |\alpha|^2 e^{-2r} + \frac{1}{8}e^{4r}. \qquad (3.73)$$

The minimum value of $V(n)$ occurs for $e^{6r} = 4|\alpha|^2$ which corresponds to $V_{\min}(n) \approx 0.94|\alpha|^{4/3}$. Diagrams depicting squeezed states with reduced amplitude and reduced phase fluctuations are shown in Fig. 3.4.

In Chap. 5 we will discuss a nonlinear interaction which produces a state with Poisson distribution in photon number, but can also exhibit amplitude squeezing.

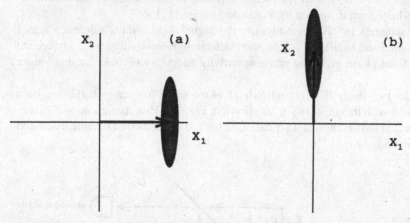

Fig. 3.4 Phase-space of amplitude and phase squeezed states. (a) The quadrature carrying the coherent excitation is squeezed ($\theta = 0$). (b) The quadrature out of phase with the coherent excitation is squeezed ($\theta = \pi/2$)

3.7.2 Squeezed Vacuum

For a squeezed vacuum $\alpha = 0$

$$V(n) = \bar{n}(1 + \cosh 2r) . \tag{3.74}$$

Hence a squeezed vacuum always exhibits super-Poissonian statistics.

We may compare the characteristics of a squeezed state with that of a number state. A number state has reduced photon number fluctuations but has complete uncertainty in phase. Thus a number state will not show any squeezing. For a number state

$$\Delta X_1^2 = \Delta X_2^2 = 2n + 1 . \tag{3.75}$$

A number state may be represented in an (X_1, X_2) phase space plot as an annulus with radius $\sqrt{n}$ and width $= 1$.

3.8 Phase-Dependent Correlation Functions

The even-ordered correlation functions such as the second-order correlation function $G^{(n,n)}(x)$ contain no phase information and are a measure of the fluctuations in the photon number. The odd-ordered correlation functions $G^{(n,m)}(x_1 \ldots x_n, x_{n+1} \cdots x_{n+m})$ with $n \neq m$ will contain information about the phase fluctuations of the electromagnetic field. The variances in the quadrature phases ΔX_1^2 and ΔX_2^2 are given by measurements of this type. A number of schemes to make quadrature phase measurements have been discussed by *Yuen* and *Shapiro* [11].

These schemes involve homodyning the signal field with a reference signal known as the local oscillator before photodetection. Homodyning with a reference signal of fixed phase gives the phase sensitivity necessary to yield the quadrature variances.

Consider two fields $E_1(r,t)$ and $E_2(r,t)$ of the same frequency, combined on a beam splitter with transmittivity η, as shown in Fig. 3.5. This configuration is essentially identical to the single field quadrature homodyne detection scheme discussed by *Yuen* and *Shapiro*.

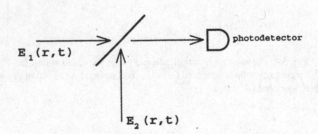

Fig. 3.5 Schematic representation of homodyne detection of squeezed states

We expand the two incident fields into the usual positive and negative frequency components

$$E_1(r, t) = i \left(\frac{\hbar \omega}{2V \varepsilon_0} \right)^{1/2} (a e^{i(k \cdot r - \omega t)} - a^\dagger e^{-i(k \cdot r - \omega t)}) , \qquad (3.76)$$

$$E_2(r, t) = i \left(\frac{\hbar \omega}{2V \varepsilon_0} \right)^{1/2} (b e^{i(k \cdot r - \omega t)} - b^\dagger e^{-i(k \cdot r - \omega t)}) , \qquad (3.77)$$

where a, b are boson operators which characterise the two modes E_1 and E_2, respectively. Both fields are taken to have the same sense of polarization, and are phase locked.

The total field after combination is given by

$$E_T(r, t) = i \left(\frac{\hbar \omega}{2V \varepsilon_0} \right)^{1/2} (c e^{i(k \cdot r - \omega t)} - c^\dagger e^{-i(k \cdot r - \omega t)}) , \qquad (3.78)$$

where

$$c = \sqrt{\eta} a + i \sqrt{1 - \eta} b . \qquad (3.79)$$

We have included a 90° phase shift between the reflected and transmitted beams at the beam splitter.

The photon detector, of course, responds to the moments of $c^\dagger c$. We thus define the number operator $\hat{N} = c^\dagger c$.

The mean photo-electron current in the detector is proportional to $\langle c^\dagger c \rangle$ which is given by

$$\langle c^\dagger c \rangle = \eta \langle a^\dagger a \rangle + (1 - \eta) \langle b^\dagger b \rangle - i \sqrt{\eta(1 - \eta)} (\langle a \rangle \langle b^\dagger \rangle - \langle a^\dagger \rangle \langle b \rangle) . \qquad (3.80)$$

Let us take the field E_2 to be the local oscillator and assume it to be in a coherent state of large amplitude β. Then we may neglect the first term in (3.80) and write $\langle c^\dagger c \rangle$ in the form

$$\langle c^\dagger c \rangle \approx (1 - \eta) |\beta|^2 + |\beta| \sqrt{\eta(1 - \eta)} \langle X_{\theta + \pi/2} \rangle , \qquad (3.81)$$

where

$$X_\theta \equiv a e^{-i\theta} + a^\dagger e^{i\theta} , \qquad (3.82)$$

and θ is the phase of β. We see that when the contribution from the reflected local-oscillator intensity level is subtracted, the mean photo-current in the detector is proportional to the mean quadrature phase amplitude of the signal field defined with respect to the local oscillator phase. If we change θ through $\pi/2$ we can determine the mean amplitude of the two canonically conjugate quadrature phase operators.

We now turn to a consideration of the fluctuations in the photo-current. The rms fluctuation current is determined by the variance of $c^\dagger c$. For an intense local oscillator in a coherent state this variance is

$$V(n_c) \approx (1 - \eta)^2 |\beta|^2 + |\beta|^2 \eta(1 - \eta) V(x_{\theta + \pi/2}) . \qquad (3.83)$$

The first term here represents reflected local oscillator intensity fluctuations. If this term is subtracted out, the photo-current fluctuations are determined by the variances in $X_{\theta+\pi/2}$, the measured quadrature phase operator. To subtract out the contribution of the reflected local oscillator field balanced homodyne detection may be used. In this scheme the output from both ports of the beam splitter is directed to a photodetector and the resulting currents combined with appropriate phase shifts before subsequent analysis. Balanced homodyne detection realises a direct measurement of the signal field quadrature phase operators [11]. Alternatively, the contribution from the local oscillator intensity fluctuations may be reduced by making the transmittivity $\eta \approx 1$, in which case the dominant contribution to $V(n_c)$ comes from the second term in (3.83).

3.9 Photon Counting Measurements

3.9.1 Classical Theory

Consider radiation of intensity $I(t)$ falling on a photo-electric counter. The probability that a count occurs in a time dt is given by

$$\Delta p(t) = \alpha I(t)dt . \tag{3.84}$$

The parameter α is a measure of the sensitivity of the detector, and depends on the area of the detector and the spectral range of the incident light. Suppose initially there are no random fluctuations in the intensity $I(t)$. Now $1 - \Delta p(t')$ represents the probability that no counts occur in the time interval dt' at t'. Then assuming the independence of photocounts in different time intervals the joint probability that no counts occur in an entire interval t to $t + T$ is given by the product

$$\prod_{t}^{t+T}[1 - \Delta p(t')] \approx \prod_{t}^{t+T} \exp[-\Delta p(t')]$$

$$= \exp\left[-\sum_{t}^{t+T} \Delta p(t')\right]$$

$$= \exp\left[-\int_{t}^{t+T} dp(t')\right] . \tag{3.85}$$

Thus the probability for no counts in the interval t to $t + T$ is

$$P_0(T+t,t) = \exp\left[-\alpha \int_{t}^{t+T} I(t')dt'\right] . \tag{3.86}$$

The probability $P_1(T+t,t)$ that one count occurs between t and $t+T$ is

$$\sum_{t''}^{t+T} dp(t'') \prod_t [1-\Delta p(t')] \rightarrow \int_t^{t+T} dp(t'') \exp\left[-\int_t^{t+T} dp(t')\right] . \qquad (3.87)$$

Hence

$$P_1(T+t, t) = \left[\alpha \int_t^{t+T} I(t')dt'\right] \exp\left[-\alpha \int_t^{t+T} I(t')dt'\right] . \qquad (3.88)$$

Following this reasoning the probability for n counts in the interval t to $t+T$ is

$$P_n(t, T) = \frac{1}{n!}[\alpha T \bar{I}(t, T)]^n \exp[-\alpha T \bar{I}(t, T)] , \qquad (3.89)$$

where

$$\bar{I}(t, T) = \frac{1}{T} \int_t^{t+T} I(t')dt'$$

is the mean intensity during the counting interval.

Now since $\bar{I}(t,T)$ may vary from one counting interval to the next, $P_n(T)$ is a time average of $P_n(t,T)$ over a large number of different starting times

$$P_n(T) = \langle P_n(t,T)\rangle$$

$$= \left\langle \frac{[\alpha \bar{I}(t, T)T]^n}{n!} \exp[-\alpha \bar{I}(t, T)T]\right\rangle . \qquad (3.90)$$

This formula was first derived by *Mandel* [12].

We note a useful generating function for the photon-counting distribution is

$$Q(\lambda, T) = \sum_{n=0}^{\infty} (1-\lambda)^n P_n(T) . \qquad (3.91)$$

The factorial moments of the photon counting distribution may be obtained as follows:

$$\overline{n(n-1)\dots(n-k)} = \sum_{n=0}^{\infty} n(n-1)\dots(n-k)P_n(T)$$

$$= (-1)^k \frac{\partial^k}{\partial \lambda^k} Q(\lambda, T)\bigg|_{\lambda=0} \qquad (3.92)$$

We shall now consider some important cases of the photon counting formula (3.89).

3.9.2 Constant Intensity

In the simplest case of a constant intensity $\bar{I}(t,T)$ is independent of t and T, hence

$$\bar{I}(t,T) = I . \tag{3.93}$$

In this case the averaging over a fluctuating intensity $I(t)$ is unnecessary and

$$P_n(T) = \frac{\bar{n}^n}{n!}\exp(-\bar{n}) , \tag{3.94}$$

where

$$\bar{n} = \alpha I T .$$

This is a Poisson distribution for which the variance $V(n) = \bar{n}$.

3.9.3 Fluctuating Intensity–Short-Time Limit

When the intensity is fluctuating, (3.89) can be simplified in the limit where the counting time T is short compared to the coherence time τ_c over which the intensity changes. If, during the interval $T, I(t)$ remains reasonably constant then

$$\bar{I}(t,T) = \bar{I}(t) . \tag{3.95}$$

With ergodic hypothesis for a stationary light source we may convert the time average in (3.90) into an ensemble average over the distribution $p(\bar{I}(t))$.

The photon counting formula may then be written

$$P_n(T) = \int\limits_0^\infty \frac{[\alpha\bar{I}(t)T]^n}{n!} e^{-\alpha\bar{I}(t)T} p(\bar{I}(t)) d\bar{I}(t) . \tag{3.96}$$

In the following we replace $\bar{I}(t)$ by the stochastic variable I for ease of notation. The mean photon count is

$$\bar{n} = \sum_{n=0}^\infty n P_n(T) = \int\limits_0^\infty \sum_{n=0}^\infty n \frac{(\alpha I T)^n}{n!} e^{-\alpha I T} p(I) dI$$

$$= \int\limits_0^\infty \alpha T I p(I) dI = \alpha T \langle I \rangle . \tag{3.97}$$

Defining moments of intensity as

$$\langle I^n \rangle = \int\limits_0^\infty I^n p(I) dI , \tag{3.98}$$

we find for the mean square count

$$\overline{n^2} = \sum_{n=0}^{\infty} n^2 P_n(T) = \int_0^{\infty} (\alpha^2 T^2 I^2 + \alpha T I) p(I) \mathrm{d}I$$

$$= \alpha^2 T^2 \langle I^2 \rangle + \alpha T \langle I \rangle . \tag{3.99}$$

Thus the variance is

$$V(n) = \overline{n^2} - \bar{n}^2 = \alpha T \langle I \rangle + \alpha^2 T^2 (\langle I^2 \rangle - \langle I \rangle^2) . \tag{3.100}$$

We note that this is always greater than the mean unless $p(I)$ is a Dirac delta function $\delta(I - I_0)$. This is true for classical fields. For certain quantum mechanical fields we shall see that it is possible to obtain $V(n) < \bar{n}$.

A thermal light field has the following probability distribution for its intensity

$$p(I) = \frac{1}{I_0} \exp\left(\frac{-I}{I_0}\right) , \tag{3.101}$$

with moments

$$\langle I^n \rangle = n! I_0^n .$$

The mean and variance of the photocount distribution are

$$\bar{n} = \alpha T I_0, \quad V(n) = \bar{n}(1 + \bar{n}) . \tag{3.102}$$

The photon-counting distribution is

$$P_n(T) = \frac{(\alpha T)^n}{I_0 n!} \int_0^{\infty} I^n \exp\left[-I\left(\alpha T + \frac{1}{I_0}\right)\right] \mathrm{d}I$$

$$= \frac{(\alpha T)^n}{I_0 n!} \left(\alpha T + \frac{1}{I_0}\right)^{-(n+1)} \int_0^{\infty} x^n e^{-x} \mathrm{d}x \, n!$$

$$= \frac{1}{(1 + \bar{n})} \left(\frac{\bar{n}}{1 + \bar{n}}\right)^n . \tag{3.103}$$

This power-law distribution for thermal light has been verified in photon-counting experiments. Experiments have also shown that the photon count distribution of highly stabilized lasers is approximated by a Poisson distribution [8, 13].

We conclude with a comment on the form assumed by $\bar{I}(t, T)$ if the depletion of the signal field by the detection process is taken into account. Then

$$I(t) = I_0 e^{-\lambda t} , \tag{3.104}$$

where λ is the rate of photon absorption. Then

$$\bar{I}(t, T) = \frac{I_0}{T} \int_t^{t+T} e^{-\lambda t'} \, dt' ,$$

(3.105)

Thus

$$\dot{I}(\tau, T) = \frac{I(t)}{\lambda T}(1 - e^{-\lambda T}) .$$

(3.106)

We note that for short counting times this has the same form as (3.95).

3.10 Quantum Mechanical Photon Count Distribution

The photon count distribution for a quantum mechanical field may be written in a formally similar way to the classical expression [14]

$$P_n(T) = \left\langle : \frac{[\alpha \bar{I}(T)T]^n}{n!} \exp[-\alpha \bar{I}(T)T] : \right\rangle$$

(3.107)

where

$$\bar{I}(T) = \frac{1}{T} \int_0^T I(t) dt$$

$$= \frac{1}{T} \int_0^T E^{(-)}(\boldsymbol{r}, t) E^{(+)}(\boldsymbol{r}, t) \, dt$$

(3.108)

and : : denotes normal ordering of the operators. We shall demonstrate the use of this formula for a single-mode field, in which case (3.107) may be written as

$$P_n(T) = \mathrm{Tr}\left(\rho : \frac{[\mu(T)a^\dagger a]^n}{n!} \exp[-a^\dagger a \mu(T)] : \right)$$

(3.109)

where $\mu(T)$ is the probability for detecting one photon in time T from a one photon field. The explicit form of $\mu(T)$ depends on the physical situation, e.g., $\mu(T) = \lambda T$ for an open system and $\mu(T) = (1 - e^{-\lambda T})$ for a closed system.

The photon count distribution may be related to the diagonal matrix elements $P_n = \langle n|\rho|n \rangle$ of ρ by

$$P_m(T) = \sum_n P_n \frac{[\mu(T)]^m}{m!} \left\langle n \left| \sum_{l=0}^\infty \frac{\mu(T)^l}{l!} a^{\dagger m+l} a^{m+l} \right| n \right\rangle .$$

(3.110)

This gives

$$P_m(T) = \sum_{n=m}^\infty P_n \sum_{l=0}^{n-m} (-1)^l \frac{\mu(T)^l}{l!} \frac{n!}{(n-m-l)!} .$$

(3.111)

The l summation is equivalent to a binomial expansion and we may write [15]

$$P_m(T) = \sum_{n=m}^{\infty} P_n \binom{n}{m} [\mu(T)]^m [1 - \mu(T)]^{n-m} \tag{3.112}$$

where

$$\binom{n}{m} = \frac{n!}{m!(n-m)!} .$$

This distribution is known as the *Bernoulli distribution*.

The photo-count distribution $P_n(T)$ is only the same as P_n in the case of unit quantum efficiency

$$P_m(T) = P_m, \qquad \mu(T) = 1 . \tag{3.113}$$

In practice, quantum efficiencies are less than unity and the photon-count distribution is only indirectly related to P_n.

The following results may be proved:

3.10.1 Coherent Light

$$P_n = \frac{\bar{n}^n}{n!} \exp(-\bar{n}) , \tag{3.114}$$

$$P_m(T) = \frac{[\mu(T)\bar{n}]^m}{m!} \exp[-\mu(T)\bar{n}] . \tag{3.115}$$

3.10.2 Chaotic Light

$$P_n = \frac{(\bar{n})^n}{(1+\bar{n})^{1+n}} , \tag{3.116}$$

$$P_m(T) = \frac{[\mu(T)\bar{n}]^m}{[1+\mu(T)\bar{n}]^{1+m}} . \tag{3.117}$$

These results agree with those obtained by semiclassical methods, see (3.94 and 3.103). In these cases P_n and $P_m(T)$ have the same mathematical form with the mean number $\bar{m}$ of counted photons related to the mean number $\bar{n}$ of photons in the mode by $\bar{m} = \mu(T)\bar{n}$. No such simple relation holds in general.

For example, for a photon number state, P_n is a delta function δ_{nn_0} but the photo-count distribution $P_m(T)$ is non zero for all $m \leq n_0$. However the normalized second order factorial moments are the same in all cases.

For a single-mode field

$$\sum m(m-1) \frac{P_m(T)}{\bar{m}^2} = \sum_n (n-1) \frac{P_n}{\bar{n}^2} = g^{(2)}(0) . \tag{3.118}$$

Thus the second-order correlation function $g^{(2)}(0)$ is directly obtainable from the photo-count distribution without any dependence on the quantum efficiency $\mu(T)$. For a multimode field a more complicated relation holds.

3.10.3 Photo-Electron Current Fluctuations

We now consider how the photon number statistics determines the statistics of the observed photo-electron current. Each individual photon detection produces a small current pulse, the observed current over a counting interval from $t - T$ to t is then due to the accumulated electrical pulses over this interval. Thus we write

$$i(t) = \int_{t-T}^{t} F(t') dn(t') . \tag{3.119}$$

Here $F(t')$ is a response function which determines the current resulting from each photon detection event. We assume $F(t')$ is *flat*, i.e. independent of t,

$$F(t') = \frac{Ge}{T} , \tag{3.120}$$

where e is the electronic charge and G is a gain factor. Then the photo-electron current is given by

$$i(t) = \frac{Ge}{T} n , \tag{3.121}$$

where n is the total number of photon detection events over the counting interval. The mean current is then given by

$$\overline{i(t)} = \frac{Ge}{T} \sum_{n=0}^{\infty} n P_n(T; t) , \tag{3.122}$$

where $P_n(T,t)$ is given by (3.89) with

$$\bar{I}(t, T) = \frac{1}{T} \int_{t-T}^{t} dt' \, E^{(-)}(t') E^{(+)}(t') . \tag{3.123}$$

Thus

$$\overline{i(t)} = (\alpha Ge) \langle : \bar{I}(t, T) : \rangle . \tag{3.124}$$

The current power spectrum is directly related to the statistical properties of the current by

$$S(\omega) = \frac{1}{\pi} \int_{0}^{\infty} d\tau \, \cos(\omega t) \, \overline{i(0)i(\tau)} . \tag{3.125}$$

The two-time correlation function is determined by joint emission probabilities for photo-electrons which are generalisations of the single photon result in (3.84). Explicit expressions were given by *Carmichael* [16]. The result is, with the definitions of (3.120).

$$\overline{i(0)i(\tau)} = (\alpha G e)^2 [\langle : \bar{I}(T, 0)\bar{I}(T, \tau) : \rangle + \theta(T - \tau)\langle : \bar{I}(\tau - T, 0) : \rangle] \qquad (3.126)$$

where $\theta(x)$ is zero for $x \le 0$ and unity otherwise. For multiple time correlations :: also signifies time ordering (time arguments increasing to the left in products of annihilation operators). In the case of constant intensity

$$\begin{aligned}
\overline{i(0)i(\tau)} =&(\alpha \zeta \, G e)^2 [\langle a^\dagger(0)a^\dagger(\tau)a(\tau)a(0)\rangle] \\
&+ (Ge)^2 \alpha \zeta \left[\theta(T - \tau)\frac{(T - \tau)}{T^2} \langle a^\dagger(0)a(0)\rangle \right]
\end{aligned} \qquad (3.127)$$

where ζ is a scale factor that converts the intensity operator into a photon-flux operator. For plane waves it is given by

$$\zeta = \frac{\varepsilon_0 \, c A}{\hbar \omega c} \qquad (3.128)$$

where A is the transverse area of the field over which the field is measured, and ω_c the frequency of the field. Using the following result for the delta function

$$\int\limits_0^\infty dt' f(t')\delta(t') = \frac{1}{2}f(0) , \qquad (3.129)$$

one may show that

$$\lim_{T \to 0} \frac{\theta(T - t)}{T^2}(T - t) = \delta(t) . \qquad (3.130)$$

Then in the limit of broad-band detector response $(T \to 0)$

$$\overline{i(0)i(\tau)} = (\alpha G e \zeta)^2 \langle a^\dagger(0)a^\dagger(\tau)a(\tau)a(0)\rangle + \alpha \zeta (Ge)^2 \langle a^\dagger(0)a(0)\rangle \delta(\tau) \qquad (3.131)$$

The last term in this expression is the *shot noise* contribution to the current.

It is more convenient to write this expression directly in terms of the normally-ordered correlation function

$$\langle : I(0), I(\tau) : \rangle \equiv \zeta^2 [\langle a^\dagger(0)a^\dagger(\tau)a(\tau)a(0)\rangle - \langle a^\dagger(0)a(0)\rangle^2] . \qquad (3.132)$$

Then

$$\begin{aligned}
\overline{i(0)i(\tau)} =& (\alpha G e \zeta)^2 \langle a^\dagger(0)a(0)\rangle^2 + \alpha \zeta (Ge)^2 \langle a^\dagger(0)a(0)\rangle \delta(\tau) \\
&+ (\alpha G e)^2 \langle : I(0), I(\tau) : \rangle .
\end{aligned} \qquad (3.133)$$

The first term is a DC term and does not contribute to the spectrum. The second term is the shot-noise contribution. The final term represents intensity fluctuations, which for a coherent field is zero.

Exercises

3.1 Calculate the mean intensity at the screen when the two slits of a Young's interference experiment are illuminated by the two photon state $(b^\dagger)^2|0\rangle/\sqrt{2}$ where $b = (a_1 + a_2)/\sqrt{2}$ and a_i is the annihilation operator for the mode radiated by slit i.

3.2 In balanced homodyne detection the measured photocurrent is determined by the moments of the photon number difference at the two output ports of the beam splitter. Show that the variance of the photon-number difference for a 50/50 beam splitter is

$$V(n_-) = |\beta|^2 V(X_{\theta+\pi/2})$$

where $|\beta|^2$ is the intensity of the local oscillator. Thus the local oscillator intensity fluctuations do not contribute.

3.3 Show that the probability to detect m photons with unit quantum efficiency in a field which has been transmitted by a beam splitter of transmitivity μ, is given by

$$P_m(\mu) = \sum_{n=m}^{\infty} P_n \binom{n}{m} \mu^m (1-\mu)^{n-m}$$

where P_n is the photon number distribution for the field before passing through the beam splitter.

3.4 A beam splitter transforms incoming mode operators a_i, b_i to the outgoing operators a_0, b_0 where

$$a_0 = \sqrt{\eta}\,a_i - i\sqrt{1-\eta}\,b_i, \qquad b_0 = \sqrt{\eta}\,b_i - i\sqrt{1-\eta}\,a_i .$$

(a) Show that such a transformation may be generated by the unitary operator

$$T = \exp[-i\theta(a^\dagger b + ab^\dagger)], \qquad \eta = \cos^2\Theta .$$

(b) Thus show that if the incoming state is a coherent state $|\alpha_i\rangle \otimes |\beta_i\rangle$, the outgoing state is also a coherent state with

$$\alpha_0 = \sqrt{\eta}\,\alpha_i - i\sqrt{1-\eta}\,\beta_i, \qquad \beta_0 = \sqrt{\eta}\,\beta_i - i\sqrt{1-\eta}\,\alpha_i .$$

(c) Show that, if the incoming state is the product number state $|1\rangle \otimes |1\rangle$, the outgoing state is

$$(2\eta - 1)|1\rangle|1\rangle + i\sqrt{2\eta(1-\eta)}(|2\rangle|0\rangle + |0\rangle|2\rangle).$$

Note that when $\eta = 1/2$ the 'coincidence' term $|1\rangle|1\rangle$ does not appear, a result known as Hong-Ou-Mandel interference.

References

1. R.J. Glauber: Phys. Rev. **130**, 2529 (1963)
2. M. Born, E. Wolf: *Principles of Optics*, 3rd edn. (Pergamon, London 1965)
3. P.A.M. Dirac: *The Principles of Quantum Mechanics*, 3rd edn. (Clarendon, London 1947)
4. G.I. Taylor: Proc. Cam. Phil. Soc. Math. Phys. Sci. **15**, 114 (1909)
5. P. Grangier, G. Roger, A. Aspect: Europhys. Lett. **1**, 173 (1986)
6. R.F. Pfleegor, L. Mandel: Phys. Rev. **159**, 1084 (1967)
7. R. Hanbury-Brown, R.W. Twiss: Nature **177**, 27 (1956)
8. F.T. Arecchi, E. Gatti, A. Sona: Phys. Rev. Lett. **20**, 27 (1966); F.T. Arecchi, Phys. Lett. **16**, 32 (1966)
9. W. Demtroeder: *Laser Spectroscopy*, 2nd edn. (Springer, Berlin, Heidelberg 1993)
10. L. Mandel: Phys. Rev. Lett. **49**, 136 (1982)
11. H.P. Yuen, J.H. Shapiro: IEEE Trans. IT **26**, 78 (1980)
12. L. Mandel: *Progress in Optics* **2**, 183 (North-Holland, Amsterdam 1963)
13. E.R. Pike: In *Quantum Optics*, ed. by R.J. Glauber (Academic, New York 1970)
14. P.L. Kelley, W.H. Kleiner: Phys. Rev. **136**, 316 (1964)
15. M.O. Scully, W.E. Lamb: Phys. Rev. **179**, 368 (1969)
16. H.J. Carmichael: J. Opt. Soc. Am. B **4**, 1588 (1987)
17. D. Mettzer, L. Mandel: IEEE J. QE-6, 661 (1970)

Further Reading

Meystre, P., M. Sargent III: *Elements of Quantum Optics*, 2nd edn. (Springer, Berlin, Heidelberg 1991)
Perina, J.: *Coherence of Light* (van Nostrand Reinhold, London 1972)

Chapter 4
Representations of the Electromagnetic Field

Abstract A full description of the electromagnetic field requires a quantum statistical treatment. The electromagnetic field has an infinite number of modes and each mode requires a statistical description in terms of its allowed quantum states. However, as the modes are described by independent Hilbert spaces, we may form the statistical description of the entire field as the product of the distribution function for each mode. This enables us to confine our description to a single mode without loss of generality.

In this chapter we introduce a number of possible representations for the density operator of the electromagnetic field. One representation is to expand the density operator in terms of the number states. Alternatively the coherent states allow a number of possible representations via the P function, the Wigner function and the Q function.

4.1 Expansion in Number States

The number or Fock states form a complete set, hence a general expansion of ρ is

$$\rho = \sum C_{nm} |n\rangle \langle m| . \tag{4.1}$$

The expansion coefficients C_{nm} are complex and there is an infinite number of them. This makes the general expansion rather less useful, particularly for problems where the phase-dependent properties of the electromagnetic field are important and hence the full expansion is necessary. However, in certain cases where only the photon number distribution is of interest the reduced expansion

$$\rho = \sum P_n |n\rangle \langle n| , \tag{4.2}$$

may be used. Here P_n is a probability distribution giving the probability of having n photons in the mode. This is not a general representation for all fields but may prove useful for certain fields. For example, a chaotic field, which has no phase information, has the distribution

$$P_n = \frac{1}{(1+\bar{n})} \left(\frac{\bar{n}}{1+\bar{n}} \right)^n , \tag{4.3}$$

where $\bar{n}$ is the mean number of photons. This is derived by maximising the entropy

$$S = -\text{TR}\{\rho \ln \rho\} , \tag{4.4}$$

subject to the constraint $\text{Tr}\{\rho a^{\dagger} a\} = \bar{n}$, and is just the usual Planck distribution for black-body radiation with

$$\bar{n} = \frac{1}{e^{\hbar \omega / kT} - 1} . \tag{4.5}$$

The second-order correlation function $g^2(0)$ may be written according to (3.66)

$$g^2(0) = 1 + \frac{V(n) - \bar{n}}{\bar{n}^2} , \tag{4.6}$$

where $V(n)$ is the variance of the distribution function P_n. Hence, for the power-law distribution $V(n) = \bar{n}^2 + \bar{n}$ we find $g^{(2)}(0) = 2$. For a field with a Poisson distribution of photons

$$P_n = \frac{e^{-\bar{n}}}{n!} \bar{n}^n \tag{4.7}$$

the variance $V(n) = \bar{n}$, hence $g^{(2)}(0) = 1$.

A coherent state has a Poisson distribution of photons. However, a measurement of $g^{(2)}(0)$ would not distinguish between a coherent state and a field prepared from an incoherent mixture with a Poisson distribution. In order to distinguish between these two fields a phase-dependent measurement such as a measurement of ΔX_1, ΔX_2 would need to be made.

4.2 Expansion in Coherent States

4.2.1 P Representation

The coherent states $|\alpha\rangle$ form a complete set of states, in fact, an overcomplete set of states. They may therefore be used as a basis set despite the fact that they are non-orthogonal. The following diagonal representation in terms of coherent states was introduced independently by *Glauber* [1] and *Sudarshan* [2]

$$\rho = \int P(\alpha) |\alpha\rangle \langle \alpha| d^2 \alpha , \tag{4.8}$$

where $d^2 \alpha = d(\text{Re}\{\alpha\}) d(\text{Im}\{\alpha\})$. It has found wide-spread application in quantum optics.

Now it might be imagined that the function $P(\alpha)$ is analogous to a probability distribution for the values of α. However, in general this is not the case since the projection operator $|\alpha\rangle\langle\alpha|$ is onto non-orthogonal states, and hence $P(\alpha)$ cannot be interpreted as a genuine probability distribution. We may note that the coherent states $|\alpha\rangle$ and $|\alpha'\rangle$ are approximately orthogonal for $|\alpha - \alpha'| \gg 1$, see (2.41). Hence, if $P(\alpha)$ is slowly varying over such large ranges of the parameter there is an approximate sense in which $P(\alpha)$ may be interpreted with a classical description. There are, however, certain quantum states of the radiation field where $P(\alpha)$ may take on negative values or become highly singular. For these fields there is no classical description and $P(\alpha)$ clearly cannot be interpreted as a classical probability distribution. Let us now consider examples of fields which may be described by the P representation:

(i) Coherent state

If

$$\rho = |\alpha_0\rangle\langle\alpha_0| ,\tag{4.9}$$

then

$$P(\alpha) = \delta^{(2)}(\alpha - \alpha_0) .\tag{4.10}$$

(ii) Chaotic state

For a chaotic state it follows from the central limit theorem that $P(\alpha)$ is a Gaussian

$$P(\alpha) = \frac{1}{\pi\bar{n}}e^{-|\alpha|^2/\bar{n}} .\tag{4.11}$$

That this is equivalent to the result for P_n is clear if we take matrix elements

$$P_n = \langle n|\rho|n\rangle = \int P(\alpha)|\langle n|\alpha\rangle|^2 d^2\alpha = \frac{1}{\pi\bar{n}} \int e^{-|\alpha|^2/\bar{n}} \frac{|\alpha|^{2n}}{n!}e^{-|\alpha|^2}d^2\alpha .\tag{4.12}$$

Using the identity

$$\pi^{-1}(l!m!)^{-1/2} \int \exp(-C|\alpha|^2)\alpha^l(\alpha^*)^m d^2\alpha = \delta_{lm}C^{-(m+1)} ,\tag{4.13}$$

which holds for $C > 0$ and choosing

$$C = \frac{1+\bar{n}}{\bar{n}}$$

we find

$$P_n = \frac{1}{1+\bar{n}}\left(\frac{\bar{n}}{1+\bar{n}}\right)^n .\tag{4.14}$$

For a mixture of a coherent and a chaotic state the P function is

$$P(\alpha) = \frac{1}{\pi\bar{n}}e^{-|\alpha-\alpha_0|^2/\bar{n}} ,\tag{4.15}$$

which may be derived using the following convolution property of $P(\alpha)$. Consider a field produced by two independent sources. The first source acting constructs the field

$$\rho_1 = \int P_1(\alpha_1)|\alpha_1\rangle\langle\alpha_1|d^2\alpha_1 . \tag{4.16}$$

Acting alone the second source would produce the field

$$\rho_2 = \int P_2(\alpha_2)|\alpha_2\rangle\langle\alpha_2|d^2\alpha_2 = \int P_2(\alpha_2)D(\alpha_2)|0\rangle\langle0|D^{-1}(\alpha_2)d^2\alpha_2 . \tag{4.17}$$

The second source acting after the first field generates the field

$$\rho = \int P_2(\alpha_2)D(\alpha_2)\rho_1 D^{-1}(\alpha_2)d^2\alpha_2$$
$$= \int P_2(\alpha_2)P_1(\alpha_1)|\alpha_1 + \alpha_2\rangle\langle\alpha_1 + \alpha_2|d^2\alpha_1 \, d^2\alpha_2 . \tag{4.18}$$

The weight function $P(\alpha)$ for the superposed excitations is therefore

$$P(\alpha) = \int \delta^2(\alpha - \alpha_1 - \alpha_2)P_1(\alpha_1)P_2(\alpha_2)d^2\alpha_1 \, d^2\alpha_2$$
$$= \int P_1(\alpha - \alpha')P_2(\alpha')d^2\alpha' . \tag{4.19}$$

We see that the distribution function for the superposition of two fields is the convolution of the distribution functions for each field.

a) Correlation Functions

The $P(\alpha)$ representation is convenient for evaluating normally-ordered products of operators, for example

$$\langle a^{\dagger n} a^m \rangle = \int P(\alpha)\alpha^{*n}\alpha^m d^2\alpha . \tag{4.20}$$

This reduces the taking of quantum mechanical expectation values to a form similar to classical averaging.

Let us express the second-order correlation function in terms of $P(\alpha)$

$$g^{(2)}(0) = 1 + \frac{\int P(\alpha)[(|\alpha|^2) - \langle|\alpha|^2\rangle]^2 \, d^2\alpha}{[\int P(\alpha)|\alpha|^2 \, d^2\alpha]^2} . \tag{4.21}$$

This looks functionally identical to the expression for classical fields. However, the argument that $g^{(2)}(0)$ must be greater than or equal to unity no longer holds since for certain fields as we have mentioned $P(\alpha)$ may take on negative values and allow for a $g^{(2)}(0) < 1$, that is, photon antibunching.

A similar result may be derived for the squeezing. We may write the variances in X_1 and X_2 as

$$\Delta X_1^2 = \{1 + \int P(\alpha)[(\alpha + \alpha^*) - (\langle\alpha\rangle + \langle\alpha^*\rangle)]^2 \, d^2\alpha\},$$
$$\Delta X_2^2 = \left\{1 + \int P(\alpha)\left[\left(\frac{\alpha - \alpha^*}{i}\right) - \left(\frac{\langle\alpha\rangle - \langle\alpha^*\rangle}{i}\right)\right]^2 d^2\alpha\right\} . \tag{4.22}$$

The condition for squeezing $\Delta X_1^2 < 1$, requires that $P(\alpha)$ takes on a negative value along either the real or imaginary axis in the complex plane, but not both simultaneously. Thus squeezing and antibunching are phenomena which are the exclusive property of quantum fields and may not be generated by classical fields. Some ambiguity may arise in the case of squeezing which only has significance for quantum fields. If a classical field is assumed from the outset arbitrary squeezing may occur in either quadrature or both simultaneously.

Quantised fields for which $P(\alpha)$ is a positive function do not exhibit quantum properties such as photon antibunching and squeezing. Such fields may be simulated by a classical description which treats the complex field amplitude ε as a stochastic random variable with the probability distribution $P(\varepsilon)$ and hence may be considered as quasiclassical. Coherent and chaotic fields are familiar examples of fields with a positive P representation. Quantum fields exhibiting antibunching and/or squeezing cannot be described in classical terms. For such fields the P representation may be negative and highly singular. The coherent state has a P representation which is a delta function, defining the boundary between quantum and classical behaviour. For fields exhibiting quantum behaviour such as a number state $|n\rangle$ or squeezed state $|\alpha,\varepsilon\rangle$ no representation for $P(\alpha)$ in terms of tempered distributions exists. Though representations in terms of generalised functions do exist [3], such representations are highly singular, for example, derivatives of delta functions. We shall therefore look for alternative representations to describe such quantum fields.

b) *Covariance Matrix*

Gaussian processes which arise, for example, in linearized fluctuation theory may be characterized by a covariance matrix. A covariance matrix may be defined by

$$C(a,a^\dagger) = \begin{pmatrix} \langle a^2 \rangle - \langle a \rangle^2 & \frac{1}{2}\langle aa^\dagger + a^\dagger a \rangle - \langle a^\dagger \rangle \langle a \rangle \\ \frac{1}{2}\langle aa^\dagger + a^\dagger a \rangle - \langle a^\dagger \rangle \langle a \rangle & \langle a^{\dagger 2} \rangle - \langle a^\dagger \rangle^2 \end{pmatrix} . \tag{4.23}$$

One may also introduce a correlation matrix $C(X_1, X_2)$ for the quadrature phase operators X_1 and X_2:

$$C(X_1,X_2)_{p,q} = \frac{1}{2}\langle X_p X_q + X_q X_p \rangle - \langle X_p \rangle \langle X_q \rangle \ (p,q = 1,2) . \tag{4.24}$$

These two correlation matrices are related by

$$C(X_1,X_2) = \Omega C(a,a^\dagger)\Omega^T \tag{4.25}$$

where

$$\Omega = \begin{pmatrix} 1 & 1 \\ -i & i \end{pmatrix} .$$

The covariance matrix $C_p(\alpha, \alpha^*)$ defined by the moments of α and α^* over $P(\alpha, \alpha^*)$ is related to the covariance matrix $C(a, a^\dagger)$ by

$$C(a,a^\dagger) = C_p(\alpha,\alpha^*) + \frac{1}{2}\begin{pmatrix} 0 & 1 \\ 1 & 0 \end{pmatrix} . \tag{4.26}$$

The distribution can be written in terms of the real variables

$$x_1 = \alpha + \alpha^*, \quad x_2 = \frac{1}{i}(\alpha - \alpha^*)$$

for which the covariance matrix relation is

$$C(X_1, X_2) = C_p(x_1, x_2) + I. \tag{4.27}$$

c) *Characteristic Function*

In practice, it proves useful to evaluate the P function through a characteristic function.

The density operator ρ is uniquely determined by its characteristic function

$$\chi(\eta) = \text{Tr}\{\rho e^{\eta a^\dagger - \eta^* a}\}.$$

We may also define normally and antinormally ordered characteristic functions

$$\chi_N(\eta) = \text{Tr}\{\rho e^{\eta a^\dagger} e^{-\eta^* a}\}, \tag{4.28}$$

$$\chi_A(\eta) = \text{Tr}\{\rho e^{-\eta^* a} e^{\eta a^\dagger}\}. \tag{4.29}$$

Using the relation (2.25) the characteristic functions are related by

$$\chi(\eta) = \chi_N(\eta) \exp\left(-\frac{1}{2}|\eta|^2\right). \tag{4.30}$$

If the density operator ρ has a P representation, then $\chi_N(\eta)$ is given by

$$\chi_N(\eta) = \int \langle \alpha | e^{\eta a^\dagger} e^{-\eta^* a} | \alpha \rangle P(\alpha) d^2\alpha = \int e^{\eta \alpha^* - \eta^* \alpha} P(\alpha) d^2\alpha. \tag{4.31}$$

Writing η and α in terms of their real and imaginary parts we find that (4.31) expresses $\chi_N(\eta)$ as a two-dimensional Fourier transform of $P(\alpha)$. The solution for $P(\alpha)$ is the inverse Fourier transform

$$P(\alpha) = \frac{1}{\pi^2} \int e^{\alpha \eta^* - \alpha^* \eta} \chi_N(\eta) d^2\eta. \tag{4.32}$$

Thus the criterion for the existence of a P representation is the existence of a Fourier transform for the normally-ordered characteristic function $\chi_N(\eta)$.

4.2.2 Wigner's Phase-Space Density

The first quasi-probability distribution was introduced into quantum mechanics by *Wigner* [4]. The Wigner function may be defined as the Fourier transform of the symmetrically ordered characteristic function $\chi(\eta)$

$$W(\alpha) = \frac{1}{\pi^2} \int \exp(\eta^*\alpha - \eta\alpha^*)\chi(\eta)\mathrm{d}^2\eta \ . \qquad (4.33)$$

The Wigner distribution always exists but is not necessarily positive.

The relationship between the Wigner distribution and the $P(\alpha)$ distribution may be obtained via the characteristic functions. Using (4.30) we may express the Wigner function as

$$\begin{aligned} W(\alpha) &= \frac{1}{\pi^2} \int \exp(\eta^*\alpha - \eta\alpha^*)\chi_N(\eta)\mathrm{e}^{-1/2|\eta|^2}\mathrm{d}^2\eta \\ &= \frac{1}{\pi^2} \int \mathrm{Tr}\{\rho\mathrm{e}^{\eta(a^\dagger - \alpha^*)}\mathrm{e}^{-\eta^*(a-\alpha)}\}\mathrm{e}^{-1/2|\eta|^2}\mathrm{d}^2\eta \\ &= \frac{1}{\pi^2} \int P(\beta)\exp[\eta(\beta^* - \alpha^*) - \eta^*(\beta - \alpha) - \frac{1}{2}|\eta|^2]\mathrm{d}^2\eta\ \mathrm{d}^2\beta \ . \qquad (4.34) \end{aligned}$$

Substituting $\varepsilon = \eta/\sqrt{2}$ leads to

$$W(\alpha) = \frac{2}{\pi^2} \int P(\beta)\exp[\sqrt{2}\varepsilon(\beta^* - \alpha^*) - \sqrt{2}\varepsilon^*(\beta - \alpha) - |\varepsilon|^2]\mathrm{d}^2\varepsilon\ \mathrm{d}^2\beta \ . \qquad (4.35)$$

The integral may be evaluated using the identity

$$\frac{1}{\pi} \int \mathrm{d}^2\eta \exp(-\lambda|\eta|^2 + \mu\eta + \nu\eta^*) = \frac{1}{\lambda} \exp\left(\frac{\mu\nu}{\lambda}\right) \qquad (4.36)$$

which holds for $\mathrm{Re}\{\lambda\} > 0$ and arbitrary μ, ν. This gives

$$W(\alpha) = \frac{2}{\pi} \int P(\beta)\exp(-2|\beta - \alpha|^2)\mathrm{d}^2\beta \ . \qquad (4.37)$$

That is, the Wigner function is a Gaussian convolution of the P function.

The covariance matrix $C_w(\alpha, \alpha^*)$ defined by the moments of α and α^* over $W(\alpha, \alpha^*)$ is related to the covariance matrix $C(a, a^\dagger)$ defined by (4.23) by

$$C(a, a^\dagger) = C_w(\alpha, \alpha^*) \ . \qquad (4.38)$$

The error areas discussed in Chap. 2 may rigorously be derived as contours of the Wigner function. We shall now study the Wigner functions for several states of this radiation field and their corresponding contours.

As it is defined in (4.33) the Wigner function is normalised as

$$\int \mathrm{d}^2\alpha W(\alpha) = 1 \qquad (4.39)$$

We have defined the quadrature phase operators X_i as (2.56)

$$a = \frac{1}{2}(X_1 + \mathrm{i}X_2) \qquad (4.40)$$

If we denote the eigenstates of X_i as $|x_i\rangle$, we can write the Wigner function as a function of x_i using

$$W(x_1,x_2) = \frac{1}{4}W(\alpha)\big|_{\alpha=(x_1+ix_2)/2} \qquad (4.41)$$

so that the marginal distributions of W,

$$P(x_k) = \int_{-\infty}^{\infty} dx_{\bar{k}} W(x_1,x_2) \qquad (4.42)$$

are given by $P(x_i) = \langle x_i|\rho|x_i\rangle$, where $\bar{k} = k - (-1)^k$.

For certain states of the radiation field the Wigner function may be written in the Gaussian form

$$W(x_1,x_2) = N\exp\left(-\frac{1}{2}Q\right) \qquad (4.43)$$

where Q is the quadratic form

$$Q = (x-a)^{\mathrm{T}} A^{-1}(x-a) \qquad (4.44)$$

and N is the normalization. A contour of the Wigner function is the curve $Q = 1$. We choose to work in the phase space where x_1 and x_2 are the c-number variables corresponding to the quadrature phase amplitudes X_1 and X_2.

a) Coherent State

For a coherent state $|\alpha\rangle = |\frac{1}{2}(X_1 + iX_2)\rangle$ the Wigner function is

$$W(x_1,x_2) = \frac{1}{2\pi}\exp\left[-\frac{1}{2}(x_1'^2 + x_2'^2)\right] \qquad (4.45)$$

where $x_i' = x_i - X_i$. The contour of the Wigner function given by $Q = 1$ is

$$x_1'^2 + x_2'^2 = 1 . \qquad (4.46)$$

Thus the error area is a circle with radius 1 centred on the point (X_1, X_2) (Fig. 2.1a).

b) Squeezed State

The Wigner function for a squeezed state is

$$W(x_1,x_2) = \frac{1}{2\pi}\exp\left[-\frac{1}{2}(x_1'^2 e^{2r} + x_2'^2 e^{-2r})\right] . \qquad (4.47)$$

The contour of the Wigner function given by $Q = 1$ is

$$\frac{x_1'^2}{e^{-2r}} + \frac{x_1'^2}{e^{2r}} = 1 \qquad (4.48)$$

which is an ellipse with the length of the major and minor axes given by e^r and e^{-r}, respectively (Fig. 2.1b).

c) *Number State*

The Wigner function for a number state $|n\rangle$ is

$$W(x_1,x_2) = \frac{2}{\pi}(-1)^n L_n(4r^2)e^{-2r^2}$$
(4.49)

where $r^2 = x_1^2 + x_2^2$, and $L_n(x)$ is the Laguerre polynomial. This Wigner function is clearly negative.

The Wigner function gives direct symmetrically-ordered moments such as those arising in the calculation of the variances of quadrature phases.

4.2.3 Q Function

An alternative function is the diagonal matrix elements of the density operator in a pure coherent state

$$Q(\alpha) = \frac{\langle\alpha|\rho|\alpha\rangle}{\pi} \geq 0 .$$
(4.50)

This is clearly a non-negative function since the density operator is a positive operator. It is also a bounded function

$$Q(\alpha) < \frac{1}{\pi} .$$

Writing the distribution in terms of the real variables

$$x_1 = \alpha + \alpha^*, \quad x_2 = -i(\alpha - \alpha^*)$$

the covariance matrix relation is

$$C(X_1,X_2) = C_Q(x_1,x_2) - I .$$

The Q function may be expressed as the Fourier transform of the antinormally-ordered characteristic function $\chi_A(\eta)$

$$\chi_A(\eta) = \text{Tr}\{\rho e^{-\eta^* a}e^{\eta a^\dagger}\} = \int \frac{d^2\alpha}{\pi} \langle\alpha|e^{\eta a^\dagger}\rho e^{-\eta^* a}|\alpha\rangle = \int e^{\eta\alpha^* - \eta^*\alpha}Q(\alpha)d^2\alpha .$$
(4.51)

Thus $Q(\alpha)$ is the inverse Fourier transform

$$Q(\alpha) = \frac{1}{\pi^2}\int e^{\alpha\eta^* - \alpha^*\eta}\chi_A(\eta)d^2\eta .$$
(4.52)

The relation between the $P(\alpha)$ and the $Q(\alpha)$ follows from

$$Q(\alpha) = \frac{\langle\alpha|\rho|\alpha\rangle}{\pi} = \frac{1}{\pi}\int P(\beta)|\langle\alpha|\beta\rangle|^2 d^2\beta = \frac{1}{\pi}\int P(\beta)e^{-|\alpha-\beta|^2}d^2\beta .$$
(4.53)

That is, the Q function like the Wigner function is a Gaussian convolution of the P function. However, it is convoluted with a Gaussian which has $\sqrt{2}$ times the width of the Wigner function which accounts for the rather more well-behaved properties.

The Q function is convenient for evaluating anti-normally-ordered moments

$$\langle a^n a^{\dagger m}\rangle = \int \alpha^n \alpha^{*m} Q(\alpha, \alpha^*) \mathrm{d}^2\alpha. \tag{4.54}$$

The covariance matrix $C_Q(\alpha,\ \alpha^*)$ defined by the moments of α and α^* over $Q(\alpha,\ \alpha^*)$ is related to the covariance matrix $C(a,\ a^\dagger)$ defined by (4.23) by

$$C(a,a^\dagger) = C_Q(\alpha,\alpha^*) - \frac{1}{2}\begin{pmatrix} 0 & 1 \\ 1 & 0 \end{pmatrix}. \tag{4.55}$$

The Q function has the advantage of existing for states where no P function exists and unlike the Wigner or P function is always positive. The Q functions for a coherent state and a number state are easily obtained.

For a coherent state $|\beta\rangle$ the Q function is

$$Q(\alpha) = \frac{|\langle\alpha|\beta\rangle|^2}{\pi} = \frac{\mathrm{e}^{-|\alpha-\beta|^2}}{\pi}. \tag{4.56}$$

For a number state $|n\rangle$ the Q function is

$$Q(\alpha) = \frac{|\langle\alpha|n\rangle|^2}{\pi} = \frac{|\alpha|^{2n}\mathrm{e}^{-|\alpha|^2}}{\pi n!}. \tag{4.57}$$

The Q function for a squeezed state $|\alpha,\ r\rangle$ is defined as

$$Q(\beta,\beta^*) = \frac{1}{\pi}|\langle\beta|D(\alpha)S(r)|0\rangle|^2. \tag{4.58}$$

This is a multivariate Gaussian distribution and may be written in terms of the quadrature phase variables x_1 and x_2 as

$$Q(x_1,x_2) = \frac{1}{4\pi^2\cosh r}\exp\left[-\frac{1}{2}(x-x_0)^{\mathrm{T}}C^{-1}(x-x_0)\right] \tag{4.59}$$

where

$$x_0 = 2(\mathrm{Re}\{\alpha\},\ \mathrm{Im}\{\alpha\}),$$
$$x = (x_1,x_2),$$
$$C = \begin{pmatrix} \mathrm{e}^{-2r}+1 & 0 \\ 0 & \mathrm{e}^{2r}+1 \end{pmatrix}.$$

The Q function for a squeezed state is shown in Fig. 4.1.

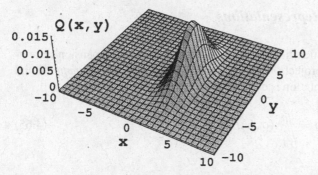

Fig. 4.1 Q function for a squeezed state with coherent amplitude $\alpha = 2.0$, $r = 1.0$

4.2.4 R Representation

Any density operator ρ may be represented in a unique way by means of a function of two complex variables $R(\alpha^*, \beta)$ which is analytic throughout the finite α^* and β planes. The function R is given explicitly as

$$R(\alpha^*, \beta) = \langle \alpha | \rho | \beta \rangle \exp[(|\alpha|^2 + |\beta|^2)/2] \,. \tag{4.60}$$

We may express the density operator in terms of $R(\alpha^*, \beta)$

$$\rho = \frac{1}{\pi^2} \int |\alpha\rangle R(\alpha^*, \beta) \langle \beta | e^{-(|\alpha|^2 + |\beta|^2)/2} d^2\alpha \, d^2\beta \,. \tag{4.61}$$

The normalization condition

$$\mathrm{Tr}\{\rho\} = 1$$

implies

$$\int \langle \gamma | \alpha \rangle \langle \alpha | \rho | \beta \rangle \langle \beta | \gamma \rangle \frac{d^2\alpha}{\pi} \frac{d^2\beta}{\pi} \frac{d^2\gamma}{\pi} = 1 \,. \tag{4.62}$$

Interchanging the scalar products and performing the integrations over β and γ we arrive at the result

$$\frac{1}{\pi} \int \langle \alpha | \rho | \alpha \rangle d^2\alpha = 1 \tag{4.63}$$

which gives the normalization condition on R

$$\frac{1}{\pi} \int R(\alpha^*, \alpha) e^{-|\alpha|^2} d^2\alpha = 1 \,. \tag{4.64}$$

The function $R(\alpha^*, \beta)$ is analytic in α^* and β (and therefore non-singular) and is by definition non-positive. It has a normalization that includes a Gaussian weight factor. For these reasons it cannot have a Fokker–Planck equation or any direct interpretation as a quasiprobability. Nevertheless, the existence of this representation does demonstrate that a calculation of normally-ordered observables for any ρ is possible with a non-singular representation.

4.2.5 Generalized P Representations

Another representation which like the R representation uses an expansion in non-diagonal coherent state projection operators was suggested by *Drummond* and *Gardiner* [5]. The representation is defined as follows

$$\rho = \int_D \Lambda(\alpha, \beta) P(\alpha, \beta) \mathrm{d}\mu(\alpha, \beta) \tag{4.65}$$

where

$$\Lambda(\alpha, \beta) = \frac{|\alpha\rangle\langle\beta^*|}{\langle\beta^*|\alpha\rangle}$$

and $\mathrm{d}\mu(\alpha, \beta)$ is the integration measure which may be chosen to define different classes of possible representations and D is the domain of integration. The projection operator $\Lambda(\alpha, \beta)$ is analytic in α and β. It is clear that the normalization condition on ρ leads to the following normalization condition on $P(\alpha, \beta)$

$$\int_D P(\alpha, \beta) \mathrm{d}\mu(\alpha, \beta) = 1 . \tag{4.66}$$

Thus, the $P(\alpha, \beta)$ is normalisable and we shall see in Chap. 6 that it gives rise to Fokker–Planck equations. The definition given by (4.65) leads to different representations depending on the integration measure.

Useful choices of integration measure are

1. Glauber–Sudarshan P Representation

$$\mathrm{d}\mu(\alpha, \beta) = \delta^2(\alpha^* - \beta) \mathrm{d}^2\alpha\, \mathrm{d}^2\beta . \tag{4.67}$$

This measure corresponds to the diagonal Glauber–Sudarshan P representation defined in (4.8).

2. Complex P Representation

$$\mathrm{d}\mu(\alpha, \beta) = \mathrm{d}\alpha\, \mathrm{d}\beta . \tag{4.68}$$

Here (α, β) are treated as complex variables which are to be integrated on individual contours C, C'. The conditions for the existence of this representation are discussed in the appendix. This particular representation may take on complex values so in no sense can it have any physical interpretation as a probability distribution. However, as we shall see it is an extremely useful representation giving exact results for certain problems and physical observables such as all the single time correlation functions.

We shall now give some examples of the complex P representation.

(a) Coherent State $|\gamma_0\rangle$

Consider a density operator with an expansion in coherent states as

$$\rho = \iint_{DD'} \rho(\alpha,\beta)|\alpha\rangle\langle\beta^*|d^2\alpha\,d^2\beta\ . \tag{4.69}$$

Using the residue theorem

$$\rho = -\frac{1}{4\pi^2}\iint_{DD'} \rho(\alpha,\beta)\langle\beta^*|\alpha\rangle\left[\oiint_{CC'}\frac{\Lambda(\alpha',\beta')d\alpha'\,d\beta'}{(\alpha-\alpha')(\beta-\beta')}\right]d^2\alpha\,d^2\beta\ . \tag{4.70}$$

Exchanging the order of integration we see the complex P function is

$$P(\alpha,\beta) = -\frac{1}{4\pi^2}\iint_{DD'}\rho(\alpha',\beta')\langle\beta'^*|\alpha'\rangle\frac{d^2\alpha'd^2\beta'}{(\alpha-\alpha')(\beta-\beta')}\ . \tag{4.71}$$

Thus for a coherent state $|\gamma_0\rangle$

$$P(\alpha,\beta) = -\frac{1}{4\pi^2(\alpha-\gamma_0)(\beta-\gamma_0^*)}\ . \tag{4.72}$$

Examples of complex P functions for nonclassical fields where the Glauber–Sudarshan P function would be highly singular are given below.

(b) Number State $|n\rangle$

$$P(\alpha,\beta) = -\frac{1}{4\pi^2}e^{\alpha\beta}\frac{n!}{(\alpha\beta)^{n+1}}\ . \tag{4.73}$$

This may be proved as follows. Using

$$\langle\alpha|\beta\rangle = e^{\alpha^*\beta-|\alpha|^2/2-|\beta|^2/2} \tag{4.74}$$

and

$$|\alpha\rangle = \sum\frac{e^{-|\alpha|^2/2}\alpha^n|n\rangle}{n!^{1/2}}\ , \tag{4.75}$$

we may write ρ as

$$\rho = \int P(\alpha,\beta)\sum\frac{|n'\rangle\langle m'|}{(n'!)^{1/2}(m'!)^{1/2}}e^{-\alpha\beta}(\alpha^{n'}\beta^{m'})d\alpha\,d\beta\ . \tag{4.76}$$

Substituting (4.73) for $P(\alpha,\beta)$

$$\rho = -\frac{1}{4\pi^2}\sum\frac{(n!)^2}{(n'!)^{1/2}(m'!)^{1/2}}\int\alpha^{-(n+1-n')}\beta^{-(n+1-m')}|n'\rangle\langle m'|d\alpha\,d\beta\ . \tag{4.77}$$

Choosing any contour of integration encircling the origin and using Cauchy's theorem

$$\frac{1}{2\pi i} \oint dz\, z^n = 0 \quad \text{if } n \geq 0,$$
$$= 1 \quad \text{if } n = -1,$$
$$= 0 \quad \text{if } n < -1. \tag{4.78}$$

We find

$$\rho = |n\rangle\langle n| .$$

(c) Squeezed State $|\gamma, r\rangle$

The complex P representation for a squeezed state is

$$P(\alpha, \beta) = N \exp\{(\alpha - \gamma)(\beta - \gamma^*) + \coth r[(\alpha - \gamma)^2 + (\beta - \gamma^*)^2]\} . \tag{4.79}$$

This may be normalized by integrating along the imaginary axis for r real. The resulting normalization for this choice of contour is

$$N = -\frac{1}{2\pi \sinh r} .$$

As an example of the use of the complex P representation we shall consider the photon counting formula given by (3.107). Using the diagonal coherent-state representation for ρ we may write the photon counting probability $P_m(T)$ as

$$P_m(T) = \int d^2 z P(z) \frac{(|z|^2 \mu(T))^m}{m!} \exp[-|z|^2 \mu(T)] . \tag{4.80}$$

An appealing feature of this equation is that $P_m(T)$ is given by an averaged Poisson distribution with $P(z)$ in the role of a probability distribution over the complex field amplitude. It is a close analogue of the classical expression (3.96). We know however that $P(z)$ is not a true probability distribution and may take on negative values, and this may cause some anxiety over the validity of (4.80). In such cases we may consider a simple generalization of (4.80) by using the complex P representation for ρ. The photocount probability is then given by

$$P_m(T) = \int_{CC'} dz\, dz'\, P(z, z') \frac{(zz' \mu(T))^m}{m!} \exp[-zz' \mu(T)] . \tag{4.81}$$

We shall demonstrate the use of this formula to calculate $P_m(T)$ for states for which no well behaved diagonal P distribution exists.

a) Number State

For a number state with density operator $\rho = |n\rangle\langle n|$ we have

$$P(z, z') = -\frac{1}{4\pi^2} \exp(zz') n! (zz')^{-n-1} \tag{4.82}$$

and the contours C and C' enclose the origin. Substituting (4.81) for $m > n$ the integrand contains no poles and $P_m(T) = 0$, while for $m < n$ poles of order $n - m + 1$ contribute in each integration and we obtain the result of (3.112)

$$P_m(T) = \sum_{n=m}^{\infty} \binom{n}{m} [\mu(T)]^m [1 - \mu(T)]^{n-m} . \qquad (4.83)$$

b) Squeezed State

For a squeezed state with density operator $\rho = |\gamma, r\rangle\langle\gamma, r|$ where γ and r are taken to be real, we have

$$P(z,z') = -\frac{1}{2\pi}(\sinh r)^{-1} \exp\{(z - \gamma)(z' - \gamma) + \coth r[(z - \gamma)^2$$
$$+ (z' - \gamma)^2]\} \qquad (4.84)$$

and the contours C and C' are along the imaginary axes in z and z' space, respectively. Performing the integration in (4.81) gives the formula, see (3.112),

$$P_m(T) = \sum_{n=m}^{\infty} \binom{n}{m} [\mu(T)]^m [1 - \mu(T)]^{n-m} P_n \qquad (4.85)$$

with

$$P_n = (n!\ \cosh r)^{-1} \exp[-\gamma^2 e^{2r}(1 + \tanh r)](\tanh r)^n H_n^2 \left(\frac{\gamma e^r}{\sqrt{\sinh 2r}} \right)$$

where the $H_n(x)$ are Hermite polynomials. This agrees with the result of a derivation using the number state representation (3.109) when we recognise that $P_n = |\langle n|\gamma, r\rangle|^2$.

4.2.6 Positive P Representation

The integration measure is chosen as

$$d\mu(\alpha, \beta) = d^2\alpha\, d^2\beta . \qquad (4.86)$$

This representation allows α, β to vary independently over the whole complex plane. It was proved in [5] that $P(\alpha, \beta)$ always exists for a physical density operator and can always be chosen positive. For this reason we call it the positive P representation. $P(\alpha, \beta)$ has all the mathematical properties of a genuine probability. It may also have an interpretation as a probability distribution [6]. It proves a most useful representation, in particular, for problems where the Fokker–Planck equation in other representations may have a non-positive definite diffusion matrix. It may be shown that provided any Fokker–Planck equation exists for the time development in the Glauber–Sudarshan representation, a corresponding Fokker–Planck equation exists with a positive semidefinite diffusion coefficient for the positive P representation.

Exercises

4.1 Show that if a field with the P representation $P_i(\alpha)$ is incident on a 50/50 beam splitter the output field has a P representation given by $P_0(\alpha) = 2P_i(\sqrt{2}\alpha)$.

4.2 Show that the Wigner function may be written, in terms of the matrix elements of ρ in the eigenstates of X_1, as

$$W(x_1, x_2) = \frac{1}{2\pi} \int_{-\infty}^{\infty} dx\, e^{-ixx_2} \langle x_1 + x | \rho | x_1 - x \rangle$$

where $x_1 = \alpha + \alpha^*$ and $x_2 = -i(\alpha - \alpha^*)$.

4.3 The complex P representation for a number state $|n\rangle$ is

$$P(\alpha, \beta) = -\frac{1}{4\pi^2} e^{\alpha\beta} \frac{n!}{(\alpha\beta)^{n+1}} .$$

Show that

$$\langle a^\dagger a \rangle = \oint d\alpha\, d\beta\, \alpha\beta\, P(\alpha, \beta) = n .$$

4.4 Use the complex P representation for a squeezed state $|\gamma, r\rangle$ with γ and r both real, to show that the photon number distribution for such a state is

$$P_n = (n!\, \cosh r)^{-1} \exp[-\gamma^2 e^{2r}(1 + \tanh r)](\tanh r)^n H_n^2 \left(\frac{\gamma e^r}{\sqrt{\sinh 2r}} \right) .$$

References

1. R.J. Glauber: Phys. Rev. **131**, 2766 (1963)
2. E.C.G. Sudarshan: Phys. Rev. Lett. **10**, 277 (1963)
3. J.R. Klauder, E.C.G. Sudarshan: *Fundamentals of Quantum Optics* (Benjamin, New York 1968)
4. E.P. Wigner: Phys. Rev. **40**, 749 (1932)
5. P.D. Drummond, C.W. Gardiner: J. Phys. A **13**, 2353 (1980)
6. S.L. Braunstein, C.M. Caves and G.J. Milburn: Phys. Rev. A **43**, 1153 (1991)

Chapter 5
Quantum Phenomena in Simple Systems in Nonlinear Optics

Abstract In this chapter we will analyse some simple processes in nonlinear optics where analytic solutions are possible. This will serve to illustrate how the formalism developed in the preceding chapters may be applied. In addition, the simple examples chosen illustrate many of the quantum phenomena studied in more complex systems in later chapters.

This chapter will serve as an introduction to how quantum phenomena such as photon antibunching, squeezing and violation of certain classical inequalities may occur in nonlinear optical systems. In addition, we include an introduction to quantum limits to amplification.

5.1 Single-Mode Quantum Statistics

A single-mode field is the simplest example of a quantum field. However, a number of quantum features such as photon antibunching and squeezing may occur in a single-mode field. To illustrate these phenomena we consider the degenerate parametric amplifier which displays interesting quantum behaviour.

5.1.1 Degenerate Parametric Amplifier

One of the simplest interactions in nonlinear optics is where a photon of frequency 2ω splits into two photons each with frequency ω. This process known as parametric down conversion may occur in a medium with a second-order nonlinear susceptibility $\chi^{(2)}$. A detailed discussion on nonlinear optical interactions is left until Chap. 9.

We shall make use of the process of parametric down conversion to describe a parametric amplifier. In a parametric amplifier a signal at frequency ω is amplified by pumping a crystal with a $\chi^{(2)}$ nonlinearity at frequency 2ω. We consider a simple model where the pump mode at frequency 2ω is classical and the signal mode at frequency ω is described by the annihilation operator a. The Hamiltonian describing the interaction is

$$\mathcal{H} = \hbar\omega a^\dagger a - i\hbar\frac{\chi}{2}\left(a^2 e^{2i\omega t} - a^{\dagger 2} e^{-2i\omega t}\right) , \tag{5.1}$$

where χ is a constant proportional to the second-order nonlinear susceptibility and the amplitude of the pump. If we work in the interaction picture we have the time-independent Hamiltonian

$$\mathcal{H}_I = -i\hbar\frac{\chi}{2}\left(a^2 - a^{\dagger 2}\right) . \tag{5.2}$$

The Heisenberg equations of motion are

$$\frac{da}{dt} = \frac{1}{i\hbar}[a,\mathcal{H}_I] = \chi a^\dagger, \quad \frac{da^\dagger}{dt} = \frac{1}{i\hbar}\left[a^\dagger,\mathcal{H}_I\right] = \chi a . \tag{5.3}$$

The interaction picture can be viewed equivalently as transforming to a frame rotating at frequency ω.

These equations have the solution

$$a(t) = a(0)\cosh\chi t + a^\dagger(0)\sinh\chi t , \tag{5.4}$$

which has the form of a generator of the squeezing transformation, see (2.60). As such we expect the light produced by parametric amplification to be squeezed. This can immediately be seen by introducing the two quadrature phase amplitudes

$$X_1 = a + a^\dagger, \quad X_2 = \frac{a - a^\dagger}{i} \tag{5.5, 5.6}$$

which diagonalize (5.2 and 5.3)

$$\frac{dX_1}{dt} = +\chi X_1, \quad \frac{dX_2}{dt} = -\chi X_2. \tag{5.7, 5.8}$$

These equations demonstrate that the parametric amplifier is a phase-sensitive amplifier which amplifies one quadrature and attenuates the other:

$$X_1(t) = e^{\chi t} X_1(0) , \tag{5.9}$$

$$X_2(t) = e^{-\chi t} X_2(0) . \tag{5.10}$$

The parametric amplifier also reduces the noise in the X_2 quadrature and increases the noise in the X_1 quadrature. The variances $V(X_i,t)$ satisfy the relations

$$V(X_1,t) = e^{2\chi t} V(X_1,0) , \tag{5.11}$$

$$V(X_2,t) = e^{-2\chi t} V(X_2,0) . \tag{5.12}$$

For initial vacuum or coherent states $V(X_i, 0) = 1$, hence

$$V(X_1, t) = e^{2\chi t},$$
$$V(X_2, t) = e^{-2\chi t}, \tag{5.13}$$

and the product of the variances satisfies the minimum uncertainty relation $V(X_1)$ $V(X_2) = 1$. Thus the deamplified quadrature has less quantum noise than the vacuum level. The amount of squeezing or noise reduction is proportional to the strength of the nonlinearity, the amplitude of the pump and the interaction time.

5.1.2 Photon Statistics

We shall next consider the photon statistics of the light produced by the parametric amplifier. First we analyse the light produced from an initial vacuum state. The intensity correlation function $g^{(2)}(0)$ in this case is

$$g^{(2)}(0) = \frac{\langle a^\dagger(t) a^\dagger(t) a(t) a(t) \rangle}{\langle a^\dagger(t) a(t) \rangle^2}$$
$$= 1 + \frac{\cosh 2\chi t}{\sinh^2 \chi t}. \tag{5.14}$$

This indicates that the squeezed light generated from an initial vacuum exhibits photon bunching ($g^{(2)}(0) > 1$). This is expected for a squeezed vacuum which must contain correlated pairs of photons.

For an initial coherent state $|\alpha\rangle$ we find the mean photon number

$$\langle a^\dagger(t) a(t) \rangle = |\alpha|^2 (\cosh 2\chi t + \cos 2\theta \sinh 2\chi t) + \sinh^2 \chi t, \tag{5.15}$$

where we have used $\alpha = |\alpha| e^{i\theta}$, and the intensity correlation function

$$g^{(2)}(0) \approx 1 + \frac{1}{|\alpha|^2 e^{-2\chi t}} \left(e^{-2\chi t} - 1 \right), \qquad \theta = \frac{\pi}{4}, \tag{5.16}$$

where $|\alpha|^2$ is large compared with $\sinh^2 \chi t$ and $\sinh \chi t \cosh \chi t$.

Thus under these conditions the photon statistics of the output light is anti-bunched. We see that a parametric amplifier evolving from an initial coherent state $|i|\alpha|\rangle$ evolves towards an amplitude squeezed state with a coherent amplitude of $|\alpha| e^{-\chi t}$. This reduction in amplitude is due to the dynamic contraction in the X_2 direction described by (5.10) (Fig. 5.1).

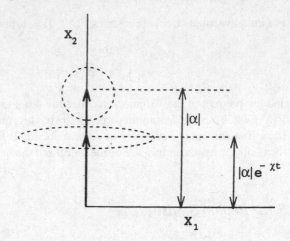

5.1.3 *Wigner Function*

The full photon statistics of the light generated in parametric amplification may be
calculated via a quasi-probability distribution. While we could choose to calculate
the P function we would find that it would become singular due to the quantum
correlations which build up during the amplification process. Therefore we shall
calculate the Wigner distribution which is a nonsingular positive function for this
problem.

The Wigner function describing the state of the parametric oscillator at any time
t may now be calculated via the symmetrically ordered characteristic function,

$$\chi(\eta,t) = \mathrm{Tr}\left\{\rho(0)\,e^{\eta a^\dagger(t) - \eta^* a(t)}\right\}\,. \tag{5.17}$$

Let us take the initial state to be the coherent state $\rho(0) = |\alpha_0\rangle\langle\alpha_0|$. Then substitut-
ing (5.4) into (5.17) we find

$$\chi(\eta,t) = \exp\left[\eta\,\alpha_0^*(t) - \eta^*\alpha_0(t) - \frac{|\eta|^2}{2}\cosh 2\chi t + \tfrac{1}{4}\left(\eta^2 + \eta^{*2}\right)\sinh 2\chi t\right], \tag{5.18}$$

where

$$\alpha_0(t) = \alpha_0\cosh\chi t + \alpha_0^*\sinh\chi t\,. \tag{5.19}$$

This may be written as

$$\chi(\eta,t) = \exp\left[\eta^{\mathrm{T}}\cdot\alpha_0^*(t) + \tfrac{1}{2}\eta^{\mathrm{T}}\Lambda\eta\right]\,, \tag{5.20}$$

$$\eta^{\mathrm{T}} = (\eta, -\eta^*)\,, \tag{5.21}$$

$$\alpha_0^{\mathrm{T}}(t) = (\alpha_0(t), \alpha_0^*(t)) \tag{5.22}$$

and

$$\Lambda = \frac{1}{2} \begin{pmatrix} \sinh 2\chi t & \cosh 2\chi t \\ \cosh 2\chi t & \sinh 2\chi t \end{pmatrix} . \tag{5.23}$$

The Wigner function is then given by the Fourier transform of $\chi(\eta, t)$, see (4.33). Using (4.36) the result is

$$W(\alpha, t) = \frac{2}{\pi} \exp\left\{ \frac{1}{2} [\alpha - \alpha_0(t)]^{\mathrm{T}} C_\alpha^{-1} [\alpha - \alpha_0(t)] \right\} , \tag{5.24}$$

where $\alpha^{\mathrm{T}} = (\alpha, \alpha^*)$. This is a two variable Gaussian with mean $\alpha_0(t)$ and covariance matrix $C_\alpha = \Lambda$. In terms of the real variables $x_1 = \alpha + \alpha^*$. $x_2 = -\mathrm{i}(\alpha - \alpha^*)$ (corresponding to the quadrature phase operators), the Wigner function becomes

$$W(x_1, x_2) = \frac{1}{2\pi} \exp\left\{ -\frac{1}{2} [x - x_0(t)] C_x^{-1} [x - x_0(t)] \right\} , \tag{5.25}$$

where

$$C_x = \begin{pmatrix} \mathrm{e}^{2\chi t} & 0 \\ 0 & \mathrm{e}^{-2\chi t} \end{pmatrix} . \tag{5.26}$$

Thus the Wigner function is a two-dimensional Gaussian with the variance in fluctuations in the quadratures X_1 and X_2 given by the major and minor axes of the elliptic contours.

5.2 Two-Mode Quantum Correlations

In two-mode systems there are a richer variety of quantum phenomena since there exists the possibility of quantum correlations between the modes. These correlations may give rise to two mode squeezing such, as described by (2.85). There may also exist intensity and phase correlations between the modes. A simple system which displays many of the above features is the non-degenerate parametric amplifier [1].

5.2.1 Non-degenerate Parametric Amplifier

The non-degenerate parametric amplifier is a simple generalization of the degenerate parametric amplifier considered in the previous section. In this case the classical pump mode at frequency $2\omega_1$ interacts in a nonlinear optical medium with two modes at frequency ω_1 and ω_2. These frequencies sum to the pump frequency, $2\omega = \omega_1 + \omega_2$. It is conventional to designate one mode as the signal and the other as the idler.

The Hamiltonian describing this system is

$$\mathscr{H} = \hbar\omega_1 a_1^\dagger a_1 + \hbar\omega_2 a_2^\dagger a_2 + \mathrm{i}\hbar\chi \left(a_1^\dagger a_2^\dagger \mathrm{e}^{-2\mathrm{i}\omega t} - a_1 a_2 \mathrm{e}^{2\mathrm{i}\omega t} \right) , \tag{5.27}$$

where $a_1 (a_2)$ is the annihilation operator for the signal (idler) mode. The coupling constant χ is proportional to the second-order susceptibility of the medium and to the amplitude of the pump.

The Heisenberg equations of motion in the interaction picture are

$$\frac{da_1}{dt} = \chi a_2^\dagger , \tag{5.28}$$

$$\frac{da_2^\dagger}{dt} = \chi a_1 . \tag{5.29}$$

The solutions to these equations are

$$a_1 (t) = a_1 \cosh \chi t + a_2^\dagger \sinh \chi t , \tag{5.30}$$

$$a_2 (t) = a_2 \cosh \chi t + a_1^\dagger \sinh \chi t . \tag{5.31}$$

If the system starts in an initial coherent stat $|\alpha_1\rangle, |\alpha_2\rangle$, the mean photon number in mode one after time t is

$$\langle n_1 (t) \rangle = \langle \alpha_1, \alpha_2 | a_1^\dagger (t) a_1 (t) | \alpha_1, \alpha_2 \rangle$$

$$= |\alpha_1 \cosh \chi t + \alpha_2^* \sinh \chi t|^2 + \sinh^2 \chi t . \tag{5.32}$$

The last term in this equation represents the amplification of vacuum fluctuations since if the system initially starts in the vacuum ($\alpha_1 = \alpha_2 = 0$) a number of photons given by $\sinh^2 \chi t$ will be generated after a time t.

The intensity correlation functions of this system exhibit interesting quantum features. With a two-mode system we may consider cross correlations between the two modes. We shall show that quantum correlations may exist which violate classical inequalities.

Consider the moment $\langle a_1^\dagger a_1 a_2^\dagger a_2 \rangle$. We may express this moment in terms of the Glauber–Sudarshan P function as follows:

$$\langle a_1^\dagger a_1 a_2^\dagger a_2 \rangle = \int d^2 \alpha_1 \int d^2 \alpha_2 |\alpha_1|^2 |\alpha_2|^2 P(\alpha_1, \alpha_2) . \tag{5.33}$$

If a positive P function exists the right-hand side of this equation is the classical intensity correlation function for two fields with the fluctuating complex amplitudes α_1 and α_2. It follows from the Hölder inequality that

$$\int d^2 \alpha_1 d^2 \alpha_2 |\alpha_1|^2 |\alpha_2|^2 P(\alpha_1, \alpha_2) \leq \left[\int d^2 \alpha_1 d^2 \alpha_2 |\alpha_1|^4 P(\alpha_1, \alpha_2) \right]^{1/2}$$

$$\times \left[\int d^2 \alpha_1 d^2 \alpha_2 |\alpha_2|^4 P(\alpha_1, \alpha_2) \right]^{1/2} . \tag{5.34}$$

Re-expressed in terms of operators this inequality implies

$$\langle a_1^\dagger a_1 a_2^\dagger a_2 \rangle \leq [\langle (a_1^\dagger)^2 a_1^2 \rangle \langle (a_2^\dagger)^2 a_2^2 \rangle]^{1/2} , \tag{5.35}$$

a result known as the *Cauchy–Schwarz inequality*. If the two modes are symmetric as for the non-degenerate parametric amplifier this inequality implies

$$\langle a_1^\dagger a_1 a_2^\dagger a_2 \rangle \le \langle (a_1^\dagger)^2 a_1^2 \rangle \,. \tag{5.36}$$

Because we have assumed a positive P function this is a weak inequality and there exists certain quantum fields which will violate it.

It is more usual to express the Cauchy–Schwarz inequality in terms of the second-order intensity correlation functions defined for a single-mode field in (3.63). The two-mode intensity correlation function is defined by

$$g_{12}^{(2)}(0) = \frac{\langle a_1^\dagger a_1 a_2^\dagger a_2 \rangle}{\langle a_1^\dagger a_1 \rangle \langle a_2^\dagger a_2 \rangle} \,. \tag{5.37}$$

This definition together with

$$g_i^{(2)}(0) = \frac{\langle a_i^\dagger a_i^\dagger a_i a_i \rangle}{\langle a_i^\dagger a_i \rangle^2} \tag{5.38}$$

enables one to write the Cauchy–Schwarz inequality as

$$[g_{12}^{(2)}(0)]^2 \le g_1^{(2)}(0) g_2^{(2)}(0) \,. \tag{5.39}$$

A stronger inequality may be derived for quantum fields when a Glauber–Sudarshan P representation does not exist. The appropriate inequality for two non-commuting operators is, see (3.26),

$$\langle a_1^\dagger a_1 a_2^\dagger a_2 \rangle^2 \le \langle (a_1^\dagger a_1)^2 \rangle \langle (a_2^\dagger a_2)^2 \rangle \,. \tag{5.40}$$

For symmetrical systems this implies

$$\langle a_1^\dagger a_1 a_2^\dagger a_2 \rangle \le \langle (a_1^\dagger)^2 a_1^2 \rangle + \langle a_1^\dagger a_1 \rangle \tag{5.41}$$

or

$$g_{12}^{(2)}(0) \le g_1^{(2)}(0) + \frac{1}{\langle a_1^\dagger a_1 \rangle} \,. \tag{5.42}$$

We now show that the non-degenerate parametric amplifier if initially in the ground state leads to a maximum violation of the Cauchy–Schwarz inequality (5.39), as is consistent with the inequality (5.42). That is, the correlations built up in the parametric amplifier are the maximum allowed by quantum mechanics.

In this system the following conservation law is easily seen to hold,

$$n_1(t) - n_2(t) = n_1(0) - n_2(0) \,, \tag{5.43}$$

where $n_i(t) \equiv a_i^\dagger(t) a_i(t)$. This conservation law has been exploited to give squeezing in the photon number difference in a parametric amplifier as will be described in

Chap. 8. Using this relation the intensity correlation function may be written

$$\langle n_1(t) n_2(t) \rangle = \langle n_1(t)^2 \rangle + \langle n_1(t) [n_2(0) - n_1(0)] \rangle . \tag{5.44}$$

If the system is initially in the vacuum state the last term is zero, thus

$$\langle n_1(t) n_2(t) \rangle = \langle a_1^\dagger(t) a_1^\dagger(t) a_1(t) a_1(t) \rangle + \langle a_1^\dagger(t) a_1(t) \rangle , \tag{5.45}$$

which corresponds to the maximum violation of the Cauchy–Schwarz inequality allowed by quantum mechanics.

Thus the non-degenerate parametric amplifier exhibits quantum mechanical correlations which violate certain classical inequalities. These quantum correlations may be further exploited to give squeezing and states similar to those discussed in the EPR paradox, as will be described in the following subsections.

5.2.2 Squeezing

In the interaction picture, the unitary operator for time evolution of the non-degenerate parametric amplifier is

$$U(t) = \exp \left[\chi t \left(a_1^\dagger a_2 - a_1 a_2 \right) \right] \tag{5.46}$$

Comparison with (2.87) shows that $U(t)$ is the unitary two-mode squeezing operator, $S(G)$ with $G = -\chi t$. We will define the squeezing with respect to the quadrature phase amplitudes of the field at the local oscillator frequency ω_{LO} and phase reference θ [2].

To see this explicitly consider the positive frequency components of the field that results for a super-position of the signal and a local oscillator field in a coherent state with $\langle E_{LO}^{(-)}(t) \rangle = |\alpha| e^{-i(\omega_{LO}t + \theta)}$. The positive frequency components of the superposed field is then well approximated by

$$E_T^{(-)}(t) = E^{(-)}(t) + |\alpha| e^{-i(\omega_{LO}t + \theta)} \tag{5.47}$$

The photo-current when such a field is directed to a detector is then proportional to $i(t) \propto \langle E_T^{(-)}(t) E_T^{(+)}(t) \rangle$. We now define the average homodyne detection signal by subtracting off the known local oscillator intensity and normalising by $|\alpha|$,

$$s(t) = \langle E^{(-)}(t) e^{i(\theta + \omega_{LO}t)} + E^{(+)}(t) e^{-i(\theta + \omega_{LO}t)} \rangle \tag{5.48}$$

The noise in the signal will then be determined by the variance in the operator $\hat{s}(t) = E^{(-)}(t) e^{i(\theta + \omega_{LO}t)} + E^{(+)}(t) e^{-i(\theta + \omega_{LO}t)}$.

We can now make a change of variable for the frequencies of the signal and idler fields by writing $\omega_1 = \omega - \varepsilon$, $\omega_2 = \omega + \varepsilon$ with $\varepsilon > 0$. This change of variable anticipates a homodyne detection scheme. If we mix these two modes with a local oscillator at half the pump frequency, i.e. at $\omega_{LO} = \omega$, the resulting signal will have Fourier components at frequencies $\pm\varepsilon$.

In the new frequency variables, the total field of the signal plus idler in the Heisenberg picture is the sum of two modes $\omega \pm \varepsilon$ symmetrically displaced about the local oscillator,

$$E(t) = \frac{1}{\sqrt{2}} \left[a_1(t) e^{-i(\omega+\varepsilon)t} + a_2(t) e^{-i(\omega+\varepsilon)t} + \text{h.c.} \right]$$

where $a_i(t)$ is the solution given in (5.30 and 5.31), h.c means hermitian conjugate, and the factor $1/\sqrt{2}$ has been inserted to give a convenient definition of the vacuum. This may be written as

$$E(t) = X_\theta(t,\varepsilon) \cos(\omega t + \theta) - X_{\theta+\pi/2}(t,\varepsilon) \sin(\omega t + \theta) \tag{5.49}$$

and the quadrature phase operators are defined as

$$X_\theta(t,\varepsilon) = \frac{1}{\sqrt{2}} \left[\left(a_1(t) e^{i\theta} + a_2^\dagger(t) e^{-i\theta} \right) e^{i\varepsilon t} + \text{h.c.} \right]$$

$$X_{\theta+\pi/2}(t,\varepsilon) = \frac{i}{\sqrt{2}} \left[\left(a_1(t) e^{i\theta} - a_2^\dagger(t) e^{-i\theta} \right) e^{i\varepsilon t} + \text{h.c.} \right]$$

In this form, as $\varepsilon > 0$, we can distinguish the positive and negative frequency components of the quadrature phase operators with respect to the local oscillator frequency.

If the system starts in the vacuum state, the homodyne detection signal at $\varepsilon = 0$ (the DC signal) will have a variance given by

$$X_\theta(t,\varepsilon = 0) = \cosh 2\chi t + \cos 2\theta \sinh 2\chi t \tag{5.50}$$

Thus for $\theta = 0$, we find that

$$V(X_0(t,\varepsilon = 0)) = e^{2\chi t} \tag{5.51}$$

$$V(X_{\pi/2}(t,\varepsilon = 0)) = e^{-2\chi t} \tag{5.52}$$

Changing the phase of the local oscillator by $\pi/2$ enables one to move from enhanced to diminished noise in the homodyne signal. We note that the squeezing in the non-degenerate parametric amplifier is due to the development of quantum correlations between the signal and idler mode. The individual signal and idler modes are not squeezed as is easily verified.

5.2.3 Quadrature Correlations and the Einstein–Podolsky–Rosen Paradox

The non-degenerate parametric amplifier can also be used to prepare states similar to those discussed in the Einstein–Podolsky–Rosen (EPR) paradox [3]. In the original treatment two systems are prepared in a correlated state. One of two canonically conjugate variables is measured on one system and the correlation is such that the value for a physical variable in the second system may be inferred with certainty.

To see how this behaviour is manifested in the non-degenerate parametric amplifier we first define two sets, one for each mode, of canonically conjugate variables, i.e.,

$$X_i^\theta = a_i e^{i\theta} + a_i^\dagger e^{-i\theta} \quad (i = 1, 2) . \tag{5.53}$$

The variables X_i^θ and $X_i^{\theta + \pi/2}$ obey the commutation relation

$$\left[X_i^\theta, X_i^{\theta + \pi/2} \right] = -2i \tag{5.54}$$

and are thus directly analogous to the position and momentum operators discussed in the original EPR paper.

To measure the degree of correlation between the two modes in terms of these operators, we consider the quantity

$$V(\theta, \phi) \equiv \tfrac{1}{2} \langle (X_1^\theta - X_2^\phi)^2 \rangle . \tag{5.55}$$

If $V(\theta, \phi) = 0$ then X_1^θ is perfectly correlated with X_2^ϕ. This means a measurement of X_1^θ can be used to infer a value of X_2^ϕ with certainty. To appreciate why such a correlation should occur in the non-degenerate parametric amplifier, we can write the interaction Hamiltonian directly in terms of the defined canonical variables,

$$\mathcal{H}_I = -2\hbar\chi \sin(\theta + \phi) \left(X_1^\theta X_2^\phi - X_1^{\theta + \pi/2} X_2^{\phi + \pi/2} \right)$$
$$- 2\hbar\chi \cos(\theta + \phi) \left(X_1^{\theta + \pi/2} X_2^\phi + X_1^\theta X_2^{\phi + \pi/2} \right) . \tag{5.56}$$

The Heisenberg equation of motion for X_1^θ is then

$$\dot{X}_1^\theta = -4\chi \left[X_2^\phi \cos(\theta + \phi) - X_2^{\phi + \pi/2} \sin(\theta + \phi) \right] \tag{5.57}$$

and we see that X_1^θ is coupled solely to X_2^ϕ when $\theta + \phi = 0$.

Direct calculation of $V(\theta, \phi)$ using the solutions in (5.30 and 5.31) gives

$$V(\theta, \phi) = \cosh 2\chi t - \sinh 2\chi t \cos(\theta + \phi) . \tag{5.58}$$

When $\theta + \phi = 0$, $V(\theta, \phi) = e^{-2\chi t}$ and, for long times, $V(\theta, \phi)$ becomes increasingly small reflecting the build up of correlation between the two variables for this case. Initially, of course, the two systems are uncorrelated and $V(\theta,\phi) = 1$. As $V(\theta,\phi)$ tends to zero the system becomes correlated in the sense of the EPR paradox. As time proceeds a measurement of X_1^θ yields an increasingly certain value of X_2^ϕ. However one could equally have measured $X_1^{\theta-\pi/2}$. Thus certain values for two noncommuting observables X_2^ϕ, $X_2^{\phi+\pi/2}$ may be obtained without in anyway disturbing system 2. This outcome constitutes the centre of the *EPR argument*.

Of course, in reality no measurement enables a perfect inference to be made. To quantify the extent of the apparent paradox, we can define the variances $V_{\text{inf}}(X_2^\phi)$ and $V_{\text{inf}}(X_2^{\phi+\pi/2})$ which determine the error in inferring X_2^ϕ and $X_1^{\phi+\pi/2}$ from direct measurements on X_2^ϕ and $X_1^{\theta-\pi/2}$. In the case of direct measurements made on $(X_2^\phi, X_2^{\phi+\pi/2})$ quantum mechanics would suggest

$$V\left(X_2^\phi\right) V\left(X_2^{\phi+\pi/2}\right) \geq 4 .$$

However the variances in the inferred values are not constrained. Thus whenever $V_{\text{inf}}(X_2^\phi)V_{\text{inf}}(X_2^{\phi+\pi/2}) < 4$, we can claim an EPR correlation paradoxically less than expected by direct measurement on the same state. This result seems to contradict the uncertainty principle. That this is not the case is seen as follows. In the standard uncertainty principle the variances are calculated with respect to the same state. However in the inference uncertainty product the variances are not calculated in the same state. That is to say $V_{\text{inf}}(X_2^{\phi+\pi/2})$ is calculated on the conditional state given a result for a measurement of $X_2^{\phi+\pi/2}$, however $V_{\text{inf}}(X_2^\phi)$ is calculated on the different conditional state given a result for the measurement of X_2^ϕ.

Ou et al. [4] performed an experimental test of these for the parametric amplifier. Using their quadrature normalization, the inferred variances indicate a paradoxical result if

$$V_{\text{inf}}\left(X_1^\phi\right) V_{\text{inf}}\left(X_1^{\phi+\pi/2}\right) \leq 2 .$$

The experimental result for the lowest value of the product was 0.7 ± 0.01.

5.2.4 Wigner Function

The full quantum correlations present in the parametric amplifier may be represented using a quasi-probability distribution. If both modes of the amplifier are initially in the vacuum state no Glauber P function for the total system exists at any time. However, a Wigner function may be obtained. We shall proceed to derive the Wigner function for the parametric amplifier.

We may define a two mode characteristic function by a simple generalization of the single-mode definition. For both modes initially in the vacuum state this may be expressed as

$$\chi(\eta_2,\eta_2,t) = \langle 0| \exp\left[\eta_1 a_1^\dagger(t) - \eta_1^* a_1(t)\right] \exp\left[\eta_2 a_2^\dagger(t) - \eta_2^* a_2(t)\right] |0\rangle$$

$$= \exp\left[-\frac{1}{2}\left[\eta_1(t)|^2 - \frac{1}{2}\left[\eta_2(t)|^2\right]\right] \right. \quad . \tag{5.59}$$

where

$$\eta_1(t) = \eta_1 \cosh \chi t - \eta_2^* \sinh \chi t,$$
$$\eta_2(t) = \eta_2 \cosh \chi t - \eta_1^* \sinh \chi t,$$

The Wigner function is then given by

$$W(\alpha_1,\alpha_2,t) = \frac{1}{\pi^4} \int d^2\eta_1 \int d^2\eta_2 \exp\left(\eta_1^* \alpha_1 - \eta_1 \alpha_1^*\right) \exp\left(\eta_2^* \alpha_2 - \eta_2 \alpha_2^*\right) \chi(\eta_1,\eta_2,t)$$

$$= \frac{4}{\pi^2} \exp\left(-2|\alpha_1 \cosh \chi t - \alpha_2^* \sinh \chi t|^2 \right.$$

$$\left. - 2|\alpha_2 \cosh \chi t - \alpha_1^* \sinh \chi t|^2\right) . \tag{5.60}$$

This distribution may be written in terms of the uncoupled c-number variables

$$\gamma_1 = \alpha_1 + \alpha_2^*,$$
$$\gamma_2 = \alpha_1 - \alpha_2^*.$$

In these new variables the Wigner function is

$$W(\gamma_1,\gamma_2) = \frac{4}{\pi^2} \exp\left[-\frac{1}{2}\left(\frac{|\gamma_1|^2}{e^{2\chi t}} + \frac{|\gamma_2|^2}{e^{-2\chi t}}\right)\right] , \tag{5.61}$$

in which form it is particularly easy to see that squeezing occurs in a linear combination of the two modes. The variances in the two quadratures being given by $e^{-2\chi t}$ and $e^{2\chi t}$, respectively. It is interesting to note that even though the state produced contains non-classical correlations the Wigner function always remains positive.

5.2.5 *Reduced Density Operator*

When a two component system is in a pure state the reduced state of each component system, determined by a partial trace operation, will be a mixed state. An interesting feature of the non-degenerate parametric amplifier is that the reduced state of each mode is a thermal state, if each mode starts from the vacuum.

To demonstrate this result we first show the high degree of correlation between the photon number in each mode. The state of the total system at time t is

$$|\psi(t)\rangle = \exp\left[\chi t \left(a_1^\dagger a_2^\dagger - a_1 a_2\right)\right] |0\rangle \tag{5.62}$$

We now make use of the disentangling theorem [5]

$$e^{\theta\left(a_1^\dagger a_2^\dagger - a_1 a_2\right)} = e^{\Gamma a_1^\dagger a_2^\dagger} e^{-g\left(a_1^\dagger a_1 + a_2^\dagger a_2 + 1\right)} e^{-\Gamma a_1 a_2} , \qquad (5.63)$$

where

$$\Gamma = \tanh\theta,$$
$$g = \ln(\cosh\theta).$$

Thus

$$|\psi(t)\rangle = e^{-g} e^{\Gamma a_1^\dagger a_2^\dagger} |0\rangle = (\cosh\chi t)^{-1} \sum_{n=1}^{\infty} (\tanh\chi t)^n |n,n\rangle , \qquad (5.64)$$

where $|n,n\rangle \equiv |n\rangle_1 \otimes |n\rangle_2$. As photons are created in pairs there is perfect correlation between the photon number in each mode. The reduced state of either mode is then easily seen to be

$$\rho_i(t) = Tr_j\{|\Psi(t)\rangle\langle\Psi(t)|\} = (\cosh\chi t)^{-2} \sum_{n=0}^{\infty} (\tanh\chi t)^{2n} |n\rangle i \langle n| \quad i \neq j . \quad (5.65)$$

This is a thermal state with mean $\bar{n} = \sinh^2\chi t$, having strong analogies with the Hawking effect associated with the thermal evaporation of black holes.

Suppose, however, that a photodetector with quantum efficiency μ has counted m photons in mode b. What is the state of mode a conditioned on this result? Such a conditional state for mode a is referred to as the *selected state* as it is selected from an ensemble of systems each with different values for the number of photons counted in mode b. We shall now describe how the conditional state of mode a may be calculated.

In Chap. 3 we saw that the probability to detect m photons from a field with the photon-number distribution $P(n)$ and detector efficiency μ is

$$P_\mu(m) = \sum_{n=m}^{\infty} \binom{n}{m} (1-\mu)^{n-m} \mu^m P_1(n) , \qquad (5.66)$$

where $P_1(n)$ is the photon number distribution for the field. This equation may be written as

$$P_\mu(m) \approx Tr\{\rho \Upsilon_\mu^\dagger(m) \Upsilon_\mu(m)\} \qquad (5.67)$$

and the operator Υ on mode b is defined by

$$\Upsilon_\mu(m) = \sum_{n=m}^{\infty} \binom{n}{m}^{1/2} (1-\mu)^{(n-m)/2} \mu^{m/2} |n-m\rangle_b \langle n| . \qquad (5.68)$$

Note that when $\mu \to 1$ this operator approaches the projection operator $|0\rangle_b \langle m|$. This is quite different to the projection operator $|m\rangle_b \langle m|$ that a naive application of the von Neumann projection postulate would indicate for photon counting measurements, and reflects the fact that real photon-counting measurements are destructive,

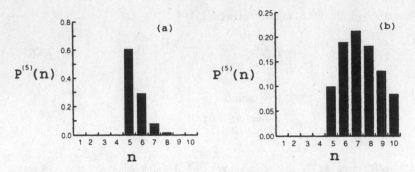

Fig. 5.2 Photon number distribution for mode a given that 5 photons are counted in mode b of a parametric amplifier. (a) $\mu = 0.9$. (b) $\mu = 0.6$

i.e. photons are absorbed upon detection. The conditional state of mode a is then given by

$$\rho^{(m)} = \left(P_\mu(m) \right)^{-1} \mathrm{Tr}_b \left\{ \Upsilon_\mu(m)_\rho \Upsilon_m^\dagger(m) \right\} . \tag{5.69}$$

This equation is a generalisation of the usual projection postulate. In the case of the correlated two mode state in (5.65), the conditional state of mode a becomes

$$\rho^{(m)} = \left(P_\mu(m) \right)^{-1} \sum_{n=m}^{\infty} \binom{n}{m} \mu^m (1-\mu)^{n-m} |n\rangle \langle n| , \tag{5.70}$$

with

$$P_\mu(m) = (1+\bar{n})^{-1} (\lambda \mu)^m \left[1 - \lambda (1-\mu) \right]^{(m+1)} , \tag{5.71}$$

where $\lambda = \tanh^2 \chi t$, $\bar{n} = \sinh^2 \chi t$. Equation (5.70) represents a state with *at least* m quanta. In Fig. 5.2 we show the photon number distribution $P^{(m)}(n)$ for this conditional state. As one would expect, when $\mu \to 1$, this approaches a number state $|m\rangle$. It should be noted, however, that the conditional state computed above refers to a situation in which the counting is done *after* the interaction which produces the correlated state, is turned off. In a cavity configuration, however, it is likely that photon counting is proceeding simultaneously with the process of parametric amplification. In that case one must proceed a little differently, however the overall result is much the same, i.e., mode a is left with at least m quanta. The details of this more complicated calculation will be found in the paper by *Holmes* et al. [6].

5.3 Quantum Limits to Amplification

The non-degenerate parametric amplifier exemplifies many features of general linear amplification. One such feature is the limit placed on the amplifier gain if the output is to be squeezed. To see how this limit arises, and to see how it might be overcome, we write the solutions (5.30 and 5.31) in the form

$$X_{1,\text{OUT}}^{\theta} = G^{1/2}X_{1,\text{IN}}^{\theta} + (G-1)^{1/2}X_{2,\text{IN}}^{\theta} , \qquad (5.72)$$

where $G = \cosh^2 \chi t$ and the quadrature phase operators are defined in (5.53). The subscript IN denotes operators defined at $t = 0$ and the subscripts 1 and 2 refer to the signal and idler modes, respectively. In (5.72) the first term describes the amplification of the quadrature and the second term the noise added by the amplifier. The variances obey the equation

$$V\left(X_{1,\text{OUT}}^{\theta}\right) = GV\left(X_{1,\text{IN}}^{\theta}\right) + (G-1)V\left(X_{2,\text{IN}}^{\theta}\right) . \qquad (5.73)$$

The maximum gain consistent with any squeezing at the output is

$$G_{\text{MAX}} = \frac{1 + V\left(X_{2,\text{IN}}^{\theta}\right)}{V\left(X_{1,\text{IN}}^{\theta}\right) + V\left(X_{2,\text{IN}}^{\theta}\right)} . \qquad (5.74)$$

If the idler mode is in the vacuum state, $V(X_{2,\text{IN}}^{\theta}) = 1$ then

$$G_{\text{MAX}} = \frac{2}{1 + V\left(X_{1,\text{IN}}^{\theta}\right)} , \qquad (5.75)$$

which gives a maximum gain of 2 for a highly squeezed state at the signal input. For higher values of the gain the squeezing at the signal output is lost due to contamination from the amplification of vacuum fluctuation in the idler input.

Greater gains may be achieved while still retaining the squeezing in the output signal if the input to the idler mode, is squeezed $(V(X_{1,\text{IN}}^{\theta}) < 1)$.

If we define the total noise in the signal as the sum of the noise in the two quadratures

$$N = \text{Var}\left\{X_1^{\theta}\right\} + \text{Var}\left\{X_1^{\theta+\pi/2}\right\} \qquad (5.76)$$

then

$$N_{\text{OUT}} = G\left(N_{\text{IN}} + A\right) , \qquad (5.77)$$

where

$$A = \left(1 - \frac{1}{G}\right)\left(\text{Var}\left\{X_{2,\text{IN}}^{\theta}\right\} + \text{Var}\left\{X_{2,\text{IN}}^{\theta+\pi/2}\right\}\right)$$

$$\leq 2\left(1 - \frac{1}{G}\right) . \qquad (5.78)$$

This is in agreement with a general theorem for the noise added by a linear amplifier [7]. The minimum added noise $A = 2(1 - 1/G)$ occurs when $\text{Var}(X_{2,\text{IN}}^{\theta}) = \text{Var}(X_{2,\text{IN}}^{\theta+\pi/2}) = 1$, that is, when the idler is in a coherent or vacuum state.

5.4 Amplitude Squeezed State with Poisson Photon Number Statistics

Finally we consider a simple nonlinear optical model which produces an amplitude squeezed state which has Poissonian photon number statistics [8, 9]. The model describes a quantised field undergoing a self-interaction via the Kerr effect. The Kerr effect is a nonlinear process involving the third-order nonlinear polarisability of a nonlinear medium. The field undergoes an intensity dependent phase shift, and thus we regard the medium as having a refractive index proportional to the intensity of the field.

Quantum mechanically the Kerr effect may be described by the effective Hamiltonian

$$\mathcal{H} = \hbar \frac{\chi}{2} \left(a^\dagger \right)^2 a^2 , \qquad (5.79)$$

where χ is proportional to the third-order nonlinear susceptibility. The Heisenberg equation of motion for the annihilation operator is

$$\frac{da}{dt} = -i\chi a^\dagger a\, a . \qquad (5.80)$$

As $a^\dagger a$, the photon number operator, is a constant of motion the photon number statistics is time invariant. The solution is then

$$a(t) = e^{-i\chi t a^\dagger a} a(0) . \qquad (5.81)$$

Assume the initial state is a coherent state with real amplitude α. The mean amplitude at a later time is then

$$\langle a(\theta) \rangle = \alpha \exp\left[-\alpha^2 (1 - \cos\theta) - i\alpha^2 \sin\theta \right] , \qquad (5.82)$$

where we have defined $\theta = \chi t$. Typically $\theta \ll 1$ and then

$$\langle a(\theta) \rangle \approx \alpha e^{-i\alpha^2\theta - \alpha^2\theta^2/2} . \qquad (5.83)$$

This result displays two effects. Firstly, there is a rotation of the mean amplitude by $\alpha^2\theta$; the expected nonlinear phase shift. Secondly, there is a decay of the amplitude which goes quadratically with time. This decay is due to the fact that the Kerr effect transforms intensity fluctuations in the initial coherent state into phase fluctuations (Fig. 5.3). In effect, the initial coherent state error circle undergoes a rotational shearing while the area remains constant.

Inspection of Fig. 5.3 suggests that, at least for short times, this system is likely to produce a squeezed state with reduced amplitude fluctuations. This is indeed the case. For short times ($\theta \ll 1$) and large intensities ($\alpha^2 \gg 1$) one finds the minimum variance of the in-phase quadrature approaches the value

$$V(X_1)_{\min} = 0.4 . \qquad (5.84)$$

Fig. 5.3 Contour of the
Q-function (at a height of
0.3) for the state of a single
mode, prepared in a coherent
state with $\alpha = 2.0$, evolved
with a Kerr nonlinearity for
$\theta = 0.25$. The equivalent contour for the initial coherent
state is shown as *dashed*

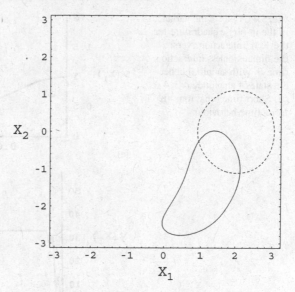

This occurs at the value

$$\theta_{\min} \approx \pm \frac{0.55}{\alpha^2} . \tag{5.85}$$

This short time behaviour is evident in Fig. 5.4a. Thus even though the photon statistics is at all times Poissonian, for short times the field is amplitude squeezed.

We now consider mixing the output of the nonlinear process with a coherent field
on a beam splitter of low reflectivity. The output field is now given by

$$a_0 = \sqrt{T} e^{-i\theta a^\dagger a} a + \sqrt{R} \beta , \tag{5.86}$$

where β is the coherent amplitude of the mixing field, and T and R are, respectively, the transmitivity and reflectivity of the beam splitter. We assume $T \to 1$ with
$\sqrt{R}\beta \to \xi$, that is the mixing field is very strong. In this limit we have

$$a_0 = e^{-i\theta a^\dagger a} a + \xi . \tag{5.87}$$

We now can choose ξ so as to minimise the photon number noise at the output.
This requires ξ to be $-\pi/2$ out-of-phase with the coherent excitation of the input.
As θ increases, the ratio of the number variance to number mean decreases to a
minimum at $\theta = 1/2\langle n_0 \rangle^{-2/3}$ (for optimal ξ), and then increases. The minimum
photon number variance is [10]

$$V(n_0) = \langle n_0 \rangle^{1/3} , \tag{5.88}$$

where $\langle n_0 \rangle = \langle n \rangle + |\xi|^2$. This is smaller than the similar result for a squeezed state,
which has a minimum value of $\langle n \rangle^{2/3}$.

Fig. 5.4 A plot of the variance in the in-phase quadrature for the Kerr interaction, versus the dimensionless interaction time θ, with an initial coherent state of amplitude $\alpha = 4.0$. (**a**) Short time behaviour. (**b**) Long time behaviour

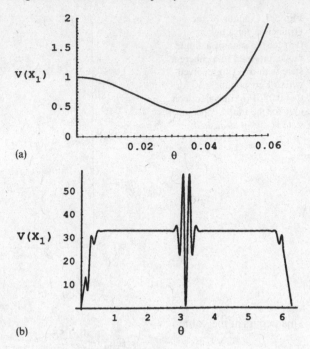

Were the rotational shearing to continue (as one might expect from a classical model) the variance in the in-phase and out-of-phase quadratures would saturate at the value $2\alpha^2 + 1$. This would be the variance for a number state with the photon number equal to α^2. That this does not happen is evident in Fig. 5.4b. Indeed, from (5.82) it is clear that for $\theta = 2\pi$ the mean amplitude returns to the initial value. A similar result holds for the variances (Exercise 5.6). This is an example of a quantum recurrence and arises from the discrete nature of the photon number distribution for a quantised field. The details are left for Exercise 5.6. In fact at $\theta = \pi$ the system evolves to a coherent superposition of coherent states:

$$|\psi(\theta = \pi)\rangle| = \frac{1}{\sqrt{2}}\left(e^{i\pi/4}|i\alpha\rangle + e^{-i\pi/4}|-i\alpha\rangle\right) . \qquad (5.89)$$

This phenomenon would be very difficult to observe experimentally as typical values of χ would require absurdly large interaction times, which in practice means extremely large interaction lengths. In Chap. 15 we will show that dissipation also makes the observation of such a coherent superposition state unlikely in a Kerr medium.

Exercises

5.1 Derive the Wigner and P functions for the reduced density operator of the signal mode for the non-degenerate parametric amplifier.

5.2 Show that, $n_1 - n_2$, the difference in the number of photons in the signal and idler mode is a constant for the parametric amplifier.

5.3 The Hamiltonian for the frequency up-converter is

$$\mathcal{H} = \hbar\omega_1 a_1^\dagger a_1 + \hbar\omega_2 a_2^\dagger a_2 + \hbar\kappa \left(e^{i\omega t} a_1^\dagger a_2 + e^{-i\omega t} a_1 a_2^\dagger \right),$$

where $\omega = \omega_2 - \omega_1$. Show that $n_1 + n_2$, the sum of the number of photons in the signal and idler modes, is a constant.

5.4 Show that the process of parametric frquency upconversion is noiseless, that is a coherent state remains coherent.

5.5 Take the initial state for the frequency upconverter to be $|N,N\rangle$. Express the density operator at time t as the tensor product of number states. *Hint: Use the disentangling theorem*, see (5.63). What is the reduced density operator for a single mode?

5.6(a) If the initial state for the Kerr-effect model is a coherent state with real mean amplitude, calculate the variances for the in-phase and out-of-phase quadratures. Show that at $\chi t = \pi$ the field exhibits amplitude squeezing for small values of the amplitude.

(b) Show that at $\chi t = \pi$ the state may be written in the form

$$\frac{1}{\sqrt{2}} \left(e^{i\pi/4}|-i\alpha\rangle + e^{-i\pi/4}|i\alpha\rangle \right).$$

References

1. B.R. Mollow, R.J. Glauber: Phys. Rev. **160**, 1097 (1967); ibid. **162**, 1256 (1967)
2. B.L. Schumaker, C.M. Caves: Phys. Rev. A **31**, 3093 (1985)
3. M.D. Reid, P.D. Drummond: Phys. Rev. Lett. A **41**, 3930 (1990)
4. Z.Y. Ou, L. Mandel: Phys. Rev. Lett. **61**, 50 (1988)
5. M.J. Collett: Phys. Rev. A **38**, 2233 (1988)
6. C.A. Holmes, G.J. Milburn, D.F. Walls: Phys. Rev. A **39**, 2493 (1989)
7. C.M. Caves: Phys. Rev. D **26**, 1817 (1982)
8. G.J. Milburn: Phys. Rev. A **33**, 674 (1985)
9. B. Yurke, D. Stoler: Phys. Rev. Lett. **57**, 13 (1985)
10. M. Kitagawa, Y. Yamamoto: Phys. Rev. A **34**, 345 (1986)

Chapter 6
Stochastic Methods

Abstract In all physical processes there is an associated loss mechanism. In this chapter we shall consider how losses may be included in the quantum mechanical equations of motion. There are several ways in which a quantum theory of damping may be developed. We shall adopt the following approach: We consider the system of interest coupled to a heat bath or reservoir. We first derive an operator master equation for the density operator of the system in the Schrödinger or interaction picture. Equations of motion for the expectation values of system operators may directly be derived from the operator master equation. Using the quasi-probability representations for the density operator discussed in Chap. 4, the operator master equation may be converted to a c-number Fokker–Planck equation. For linear problems a time-dependent solution to the Fokker–Planck equation may be found. In certain nonlinear problems with an appropriate choice of representation the steady-state solution for the quasi-probability distribution may be found from which moments may be calculated.

Using methods familiar in stochastic processes the Fokker–Planck equation may be converted into an equivalent set of stochastic differential equations. These stochastic differential equations of which the Langevin equations are one example are convenient when linearization is necessary. We begin then with a derivation of the master equation. We follow the method of *Haake* [1].

6.1 Master Equation

We consider a system described by the Hamiltonian $\mathcal{H}_S$ coupled to a reservoir described by the Hamiltonian $\mathcal{H}_R$. The reservoir may be considered to be a large number of harmonic oscillators as, for example, the modes of the free electromagnetic field or phonon modes in a solid. In some cases the reservoir may be more appropriately modelled as a set of atomic energy levels. The derivation of the master equation is not dependent on the specific reservoir model. There is a weak interaction between the system and the reservoir given by the Hamiltonian V. Thus the total Hamiltonian is

$$\mathcal{H} = \mathcal{H}_S + \mathcal{H}_R + V \tag{6.1}$$

Let $w(t)$ be the total density operator of the system plus reservoir in the interaction picture. The equation of motion in the interaction picture is

$$\frac{dw(t)}{dt} = -\frac{i}{\hbar}[V(t), w(t)] \tag{6.2}$$

The reduced density operator for the system is defined by

$$\rho(t) = \text{Tr}_R\{w(t)\} \tag{6.3}$$

where Tr_R indicates a trace over reservoir variables. We assume that initially the system and reservoir are uncorrelated so that

$$w(0) = \rho(0) \otimes \rho_R \tag{6.4}$$

where ρ_R is the density operator for the reservoir.

Integrating (6.2) we obtain

$$w(t) = w(0) - \frac{i}{\hbar} \int_0^t dt_1 [V(t_1), w(t_1)] . \tag{6.5}$$

Iterating this solution we find

$$w(t) = w(0) + \sum_{n=1}^{\infty} \left(-\frac{i}{\hbar}\right)^n \int_0^t dt_1 \int_0^{t_1} dt_2 \cdots$$

$$\times \int_0^{t_{n-1}} dt_n [V(t_1), [V(t_2), \ldots [V(t_n), w(0)]]] . \tag{6.6}$$

Performing the trace over reservoir variables

$$\rho(t) = \rho(0) + \sum_{n=1}^{\infty} \left(-\frac{i}{\hbar}\right)^n \int_0^t dt_1 \int_0^{t_1} dt_2 \cdots \int_0^{t_{n-1}} dt_n$$

$$\times \text{Tr}_R\{[V(t_1), [V(t_2), \ldots [V(t_n), \rho_R \otimes \rho(0)]]]\}$$

$$\equiv (1 + U_1(t) + U_2(t) + \cdots)\rho(0) \tag{6.7}$$

$$\equiv U(t)\rho(0)$$

where

$$U_n(t) = \left(-\frac{i}{\hbar}\right)^n \text{Tr}_R \int_0^t dt_1 \int_0^{t_1} dt_2 \cdots$$

$$\times \int_0^{t_{n-1}} dt_n [V(t_1), [V(t_2), \ldots [V(t_n), \rho_R \otimes (\cdot)]] \ldots] . \tag{6.8}$$

Thus

$$\frac{d\rho}{dt} = [\dot{U}_1(t) + \dot{U}_2(t) + \cdots]U(t)^{-1}\rho(t)$$

$$\equiv l(t)\rho(t) \tag{6.9}$$

where the generator of time development is

$$l(t) = [\dot{U}_1(t) + \dot{U}_2(t) + \cdots]U(t)^{-1} . \tag{6.10}$$

We now assume that $V(t)$ is such that

$$\text{Tr}_R(V(t)\rho_R) = 0 . \tag{6.11}$$

This ensures that $U_1(t) = 0$. If the perturbation is weak we may drop terms from $l(t)$ of order higher than two. Thus

$$l(t) = \dot{U}_2(t)$$

$$= -\frac{1}{\hbar^2}\int_0^t dt_1 \, \text{Tr}_R[V(t),[V(t_1),\rho_R \otimes (\cdot)]] . \tag{6.12}$$

Thus to second order in the perturbation

$$\frac{d\rho}{dt} = -\frac{1}{\hbar^2}\int_0^t dt_1 \, \text{Tr}_R[V(t),[V(t_1),\rho_R \otimes \rho(t)]] . \tag{6.13}$$

The next-order correction is at least quartic in the coupling and thus we expect (6.13) to be a good approximation.

Let us now consider the case of a damped simple harmonic oscillator. In this case

$$V(t) = \hbar(a^\dagger \Gamma(t)e^{i\omega_0 t} + a\Gamma^\dagger(t)e^{-i\omega_0 t}) \tag{6.14}$$

where

$$\Gamma(t) = \sum_j g_j b_j e^{-i\omega_j t} , \tag{6.15}$$

and

$$[b_j, b_k^\dagger] = \delta_{jk} . \tag{6.16}$$

Substituting (6.14) into (6.13) we find that the following integrals are required

$$I_1 = \int_0^t dt_1 \langle \Gamma(t)\Gamma(t_1)\rangle e^{i\omega_0(t+t_1)} , \tag{6.17}$$

$$I_2 = \int_0^t dt_1 \langle \Gamma^\dagger(t)\Gamma^\dagger(t_1)\rangle e^{-i\omega_0(t+t_1)} , \tag{6.18}$$

$$I_3 = \int_0^t dt_1 \langle \Gamma(t) \Gamma^\dagger(t_1) \rangle e^{i\omega_0(t-t_1)} , \qquad (6.19)$$

$$I_4 = \int_0^t dt_1 \langle \Gamma^\dagger(t) \Gamma(t_1) \rangle e^{-i\omega_0(t-t_1)} \qquad (6.20)$$

which we now evaluate.

Using the definition of $\Gamma(t)$ we have

$$I_1 = \int_0^t dt_1 \sum_{i,j} g_i g_j \langle b_i b_j \rangle_R e^{-i(\omega_i t + \omega_j t_1)} e^{i\omega_0(t+t_1)} . \qquad (6.21)$$

Converting the sum over modes to a frequency-space integral

$$I_1 = \int_0^t dt_1 \int_0^\infty \frac{d\omega_1}{2\pi} \rho(\omega_1) \int_0^\infty \frac{d\omega_2}{2\pi} \rho(\omega_2) g(\omega_1) g(\omega_2) \langle b(\omega_1) b(\omega_2) \rangle_R$$
$$\times e^{-i(\omega_1 t + \omega_2 t_1) + i\omega_0(t+t_1)} \qquad (6.22)$$

where $\rho(\omega)$ is the density of states function.

For a thermal bath the phase dependent correlation function $\langle b(\omega_1) b(\omega_2) \rangle_R = 0$. However, certain specially prepared reservoirs such as squeezed reservoirs may have phase-dependent correlations.

In order to include these we now assume that

$$\langle b(\omega_1) b(\omega_2) \rangle = 2\pi M(\omega_1) \delta(2\omega_0 - \omega_1 - \omega_2) \qquad (6.23)$$

which corresponds to a multimode squeezed vacuum state with the carrier frequency equal to the cavity resonance frequency. Thus

$$I_1 = \int_0^t dt_1 \int_0^\infty \frac{d\omega}{2\pi} \rho(\omega) \rho(2\omega_0 - \omega) g(\omega) g(2\omega_0 - \omega) M(\omega) e^{i(\omega_0 - \omega)(t - t_1)} . \quad (6.24)$$

Note that the time integral depends only on $t - t_1$. This suggests the change of variable $\tau = t - t_1$ and thus

$$I_1 = \int_0^t d\tau \int_0^\infty \frac{d\omega}{2\pi} \rho(\omega) \rho(2\omega_0 - \omega) g(\omega) g(2\omega_0 - \omega) M(\omega) e^{i(\omega_0 - \omega)\tau} . \qquad (6.25)$$

We now make the first Markov approximation by assuming $\rho(\omega)$, $g(\omega)$ and $M(\omega)$ are slowly varying functions around $\omega = \omega_0$, where ω_0 is very large. Thus it is convenient to make the change of variable $\varepsilon = \omega - \omega_0$ and write

$$I_1 \approx \int_0^t d\tau \int_{-\infty}^\infty \frac{d\varepsilon}{2\pi} \rho^2(\varepsilon + \omega_0) g^2(\varepsilon + \omega_0) M(\varepsilon + \omega_0) e^{-i\varepsilon\tau} , \qquad (6.26)$$

assuming a symmetry around ω_0. Since the integral over frequency is assumed to be a rapidly decaying function of time we have extended the upper limit of the time integration to infinity. Interchanging the order of the time and frequency integral this term becomes

$$I_1 \approx \int_{-\infty}^{\infty} \frac{d\varepsilon}{2\pi} \rho^2(\varepsilon + \omega_0) g^2(\varepsilon + \omega_0) M(\varepsilon + \omega_0) \left[\pi\delta(\varepsilon) - i \, \text{PV}\left(\frac{1}{\varepsilon}\right) \right] \qquad (6.27)$$

where we have used

$$\int_{0}^{\infty} d\tau \, e^{\pm i\varepsilon\tau} = \pi\delta(\varepsilon) \pm i \, \text{PV}\left(\frac{1}{\varepsilon}\right) \qquad (6.28)$$

with PV being the Cauchy principal value part defined by

$$\text{PV} \int_{-a}^{b} \frac{f(\omega)}{\omega} = \lim_{\varepsilon \to 0} \left(\int_{-a}^{-\varepsilon} \frac{f(\omega)}{\omega} \, d\omega + \int_{\varepsilon}^{b} \frac{f(\omega)}{\omega} \, d\omega \right) . \qquad (6.29)$$

If we now define the damping rate γ by

$$\gamma \equiv \rho^2(\omega_0) g^2(\omega_0) \qquad (6.30)$$

and a term

$$\bar{\Delta} = \text{PV} \int_{-\infty}^{\infty} \frac{d\varepsilon}{2\pi} \frac{1}{\varepsilon} \rho^2(\varepsilon + \omega_0) g^2(\varepsilon + \omega_0) M(\varepsilon + \omega_0) \qquad (6.31)$$

then

$$I_1 = \frac{\gamma}{2} M(\omega_0) + i\bar{\Delta} . \qquad (6.32)$$

Proceeding in a similar way we find

$$I_2 = \frac{\gamma}{2} M^*(\omega_0) - i\bar{\Delta} , \qquad (6.33)$$

$$I_3 = \frac{\gamma}{2} (N(\omega_0) + 1) - i\Delta , \qquad (6.34)$$

$$I_4 = \frac{\gamma}{2} N(\omega_0) - i\Delta' , \qquad (6.35)$$

where the function $N(\omega)$ is defined by

$$\langle b^{\dagger}(\omega) b(\omega') \rangle = 2\pi N(\omega) \delta(\omega - \omega')$$

and is thus proportional to the intensity spectrum of the reservoir. The term Δ is defined by

$$\Delta = \text{PV} \int_{-\infty}^{\infty} \frac{d\varepsilon}{2\pi} \frac{1}{\varepsilon} \rho^2(\omega_0 + \varepsilon) g^2(\omega_0 + \varepsilon)(N(\omega_0 + \varepsilon) + 1) . \qquad (6.36)$$

This term represents a small shift in the frequency of the oscillator. In the case where the system is a two-level atom, this term contributes to the Lamb shift which we discuss in Chap. 10. In what follows we ignore the effects of Δ and $\bar{\Delta}$.

Substituting the above results into (6.13) we find that the evolution of the system's density operator in the interaction picture is described by the master equation

$$\frac{d\rho}{dt} \doteq \frac{\gamma}{2}(N+1)(2a\rho a^\dagger - a^\dagger a\rho - \rho a^\dagger a) + \frac{\gamma}{2}N(2a^\dagger \rho a - aa^\dagger \rho - \rho aa^\dagger)$$
$$+ \frac{\gamma}{2}M(2a^\dagger \rho a^\dagger - a^\dagger a^\dagger \rho - \rho a^\dagger a^\dagger) + \frac{\gamma}{2}M^*(2a\rho a - aa\rho - \rho aa). \qquad (6.37)$$

(For convenience, we have suppressed the functional dependence of $N(\omega_0)$ and $M(\omega_0)$).

If the bath is in thermal equilibrium at temperature T, $M = 0$ and

$$N(\omega_0) = (e^{\hbar\omega_0/kT} - 1)^{-1} \qquad (6.38)$$

which is just the mean number of bath quanta at frequency ω_0. In this case the master equation considerably simplifies. If the bath temperature is zero, $N = 0$, and the master equation simplifies further. In general the positivity of the density operator requires, $|M|^2 \leq N(N+1)$.

Equations of motion for the expectation values of system operators may be directly derived from the master equation, (6.37). For example, the mean amplitude of the simple harmonic oscillator is given by

$$\frac{d\langle a \rangle}{dt} = \text{Tr}\left\{ a\frac{d\hat{\rho}}{dt} \right\} = -\frac{\gamma}{2}\langle a \rangle \qquad (6.39)$$

which has the solution

$$\langle a(t) \rangle = \langle a(0) \rangle e^{-\gamma t/2} . \qquad (6.40)$$

(Note in the Schrödinger picture the mean amplitude evolves as $\langle a(t) \rangle = \langle a(0) \rangle e^{-i\omega t} e^{-\gamma t/2}$). Thus the mean amplitude decays at a rate $\gamma/2$. The mean number of quanta $\langle n \rangle = \langle a^\dagger a \rangle$ obeys the equation

$$\frac{d\langle a^\dagger a \rangle}{dt} = -\gamma\langle a^\dagger a \rangle + \gamma N . \qquad (6.41)$$

The solution to this equation is

$$\langle n(t) \rangle = \langle n(0) \rangle e^{-\gamma t} + N(1 - e^{-\gamma t}) . \qquad (6.42)$$

In the steady state $\langle n(t) \rangle \rightarrow N$ and the mean number of quanta in the oscillator is equal to the mean number of quanta in the reservoir at that temperature. The role of the terms multiplied by M can be seen by evaluating the equation of motion for $\langle a^2 \rangle$,

$$\frac{d}{dt}\langle a^2 \rangle = -\gamma\langle a^2 \rangle + \gamma M . \qquad (6.43)$$

Thus these terms lead to a driving force on the second-order phase dependent moments.

The master equation (6.37) applies to a free damped harmonic oscillator in the interaction picture. If the harmonic oscillator is perturbed by an additional interaction $\mathcal{H}_1$, the master equation in the interaction picture becomes

$$
\begin{aligned}
\frac{d\rho}{dt} = &-\frac{1}{\hbar}[\mathcal{H}_1,\rho] + \frac{\gamma}{2}(N+1)(2a\rho a^\dagger - a^\dagger a\rho - \rho a^\dagger a) \\
&+ \frac{\gamma}{2}N(2a^\dagger\rho a - aa^\dagger\rho - \rho aa^\dagger) \\
&+ \frac{\gamma}{2}M(2a^\dagger\rho a^\dagger - (a^\dagger)^2\rho - \rho(a^\dagger)^2) + \frac{\gamma}{2}M^*(2a\rho a - a^2\rho - \rho a^2) .
\end{aligned}
\tag{6.44}
$$

In general, the equations of motion for the mean amplitude, mean quantum number etc. are not as easily obtained from (6.44), as they were for the free damped harmonic oscillator. To proceed in such situations, it is desirable to convert the master equation to an equivalent c-number partial differential equation. We now discuss various ways this may be done.

6.2 Equivalent c-Number Equations

6.2.1 Photon Number Representation

The operator master equation may be converted into an equation for the matrix elements of ρ in the number state basis:

$$
\begin{aligned}
\frac{\partial \rho_{mn}}{\partial t} = &\gamma N[(nm)^{1/2}\rho_{m-1,\,n-1} - \frac{1}{2}(m+n+2)\rho_{mn}]. \\
&+ \gamma(N+1)\{[(m+1)(n+1)]^{1/2}\rho_{m+1,\,n+1} - \frac{1}{2}(m+n)\rho_{mn}\} \\
&- \frac{\gamma}{2}M\left\{2[m(n+1)]^{1/2}\rho_{m-1,\,n+1} - \sqrt{(n+1)(n+2)}\rho_{m,n+2}\right. \\
&\left. - \sqrt{m(m-1)}\rho_{m-2,\,n}\right\} - \frac{\gamma}{2}M^*\left\{2[n(m+1)]^{1/2}\rho_{m+1,n-1}\right. \\
&\left. - \sqrt{(m+1)(m+2)}\rho_{m+2,\,n} - \sqrt{n(n-1)}\rho_{m,n-2}\right\}
\end{aligned}
\tag{6.45}
$$

where $\rho_{m,\,n} \equiv \langle m|\rho|n\rangle$. This gives an infinite hierarchy of coupled equations for the off-diagonal matrix elements. When $M=0$ the diagonal elements $\rho_{m,\,m}$ are coupled only amongst themselves and not coupled to the off-diagonal elements. In this case the diagonal elements satisfy

$$
\frac{dP(n)}{dt} = t_+(n-1)P(n-1) + t_-(n+1)P(n+1) - [t_+(n)+t_-(n)]P(n)
\tag{6.46}
$$

where we have set $P(n) = \langle n|\rho|n \rangle$ and defined the *transition probabilities*

$$t_+(n) \equiv \gamma N(n+1) , \tag{6.47}$$

$$t_-(n) \equiv \gamma(N+1)n . \tag{6.48}$$

In the steady state the detailed balance condition holds:

$$t_-(n)P(n) = t_+(n-1)P(n-1) \tag{6.49}$$

and the steady state solution is found by iteration

$$P_{ss}(n) = P(0) \prod_{k=1}^{n} \frac{t_+(k-1)}{t_-(k)} . \tag{6.50}$$

Thus the steady state solution for the ordinary damped harmonic oscillator $(M = 0)$ is

$$P_{ss}(n) = \frac{1}{1+N} \left(\frac{N}{1+N} \right)^n . \tag{6.51}$$

An optical cavity damped into a reservoir with phase-independent correlation functions has a power law photon number distribution of thermal light.

In the more general case $M \neq 0$, or when there are additional terms in the master equation such as linear driving with the Hamiltonian $\mathcal{H}_0 = \hbar[\varepsilon(t)a^\dagger + \varepsilon^*(t)a]$, the coupling of diagonal and off-diagonal matrix elements makes the photon number representation less convenient for determining $\rho(t)$.

6.2.2 P Representation

An operator master equation may be transformed to a c-number equation using the Glauber–Sudarshan representation for ρ. It is necessary to first establish the rules for converting operators to an equivalent c-number form. We know the relations

$$a|\alpha\rangle = \alpha|\alpha\rangle , \tag{6.52}$$

$$\langle\alpha|a^\dagger = \alpha^*\langle\alpha| . \tag{6.53}$$

To derive other relations it is convenient to use the Bargmann state $\| \alpha \rangle$ defined by

$$\| \alpha \rangle = e^{1/2|\alpha|^2}|\alpha\rangle \tag{6.54}$$

so that

$$a^\dagger \| \alpha \rangle = \sum_n \frac{\alpha^n}{\sqrt{n!}} \sqrt{n+1}|n+1\rangle$$

$$= \frac{\partial}{\partial\alpha} \| \alpha \rangle . \tag{6.55}$$

Similarly

$$\langle \alpha \| a = \frac{\partial}{\partial \alpha^*} \langle \alpha \| . \tag{6.56}$$

Hence, given the P representation

$$\rho = \int d^2\alpha \| \alpha \rangle \langle \alpha \| e^{-|\alpha|^2} P(\alpha) \tag{6.57}$$

we find

$$a^\dagger \rho = \int d^2\alpha \frac{\partial}{\partial \alpha}(\| \alpha \rangle \langle \alpha \|) e^{-\alpha\alpha^*} P(\alpha) \tag{6.58}$$

and integrating by parts

$$a^\dagger \rho = \int d^2\alpha \| \alpha \rangle \langle \alpha \| e^{-|\alpha|^2} \left(\alpha^* - \frac{\partial}{\partial \alpha} \right) P(\alpha) . \tag{6.59}$$

We can thus make an operator correspondence between $a^\dagger$ and $\alpha^* - \partial/\partial\alpha$. A similar formula holds for a. Summarizing we have the following operator correspondences:

$$a\rho \leftrightarrow \alpha P(\alpha),$$

$$a^\dagger \rho \leftrightarrow \left(\alpha^* - \frac{\partial}{\partial \alpha} \right) P(\alpha),$$

$$\rho a \leftrightarrow \left(\alpha - \frac{\partial}{\partial \alpha^*} \right) P(\alpha),$$

$$\rho a^\dagger \leftrightarrow \alpha^* P(\alpha) . \tag{6.60}$$

Consider the correspondences for operator products

$$a^\dagger a \rho \to \left(\alpha^* - \frac{\partial}{\partial \alpha} \right) \alpha P , \tag{6.61a}$$

$$\rho a^\dagger a \to \left(\alpha - \frac{\partial}{\partial \alpha^*} \right) \alpha^* P . \tag{6.61b}$$

Notice that the order of the operators in (6.61b) reverses, since acting on ρ, they operate from the right, whereas on P, they operate from the left.

Note that α and α^* are not independent variables. In terms of real variables we may write

$$\alpha = x + iy,$$

$$\alpha^* = x - iy,$$

$$\frac{\partial}{\partial \alpha} = \frac{1}{2} \left(\frac{\partial}{\partial x} - i \frac{\partial}{\partial y} \right),$$

$$\frac{\partial}{\partial \alpha^*} = \frac{1}{2} \left(\frac{\partial}{\partial x} + i \frac{\partial}{\partial y} \right) . \tag{6.62}$$

To obtain a c-number equation we substitute the P representation for ρ into the master equation and use the operator correspondences. This leads to the equation

$$\frac{\partial P(\alpha)}{\partial t} = \left[\frac{1}{2}\gamma \left(\frac{\partial}{\partial \alpha}\alpha + \frac{\partial}{\partial \alpha^*}\alpha^* \right) + \frac{\gamma}{2} \left(M^* \frac{\partial^2}{\partial \alpha^{*2}} + M \frac{\partial^2}{\partial \alpha^2} \right) + \gamma N \frac{\partial^2}{\partial \alpha \partial \alpha^*} \right] P(\alpha) .$$

$$(6.63)$$

When $M > N$ this equation has "non positive-definite" diffusion, hence the P representation is unable to describe the system in terms of a classical stochastic process. Alternative representations will be discussed later in this chapter. When $M \leq N$ (6.63) has the form of a Fokker–Planck equation. We shall discuss some useful properties of Fokker–Planck equations below.

6.2.3 Properties of Fokker–Planck Equations

A general Fokker–Planck equation in n variables may be written in the form

$$\frac{\partial}{\partial t}P(x) = \left[-\frac{\partial}{\partial x_j}A_j(x) + \frac{1}{2}\frac{\partial}{\partial x_i}\frac{\partial}{\partial x_j}D_{ij}(x) \right] P(x) .$$

$$(6.64)$$

The first derivative term determines the mean or deterministic motion and is called the *drift term*, while the second derivative term, provided its coefficient is positive definite, will cause a broadening or diffusion of $P(x, t)$ and is called the *diffusion term*. $A = (A_j)$ is the *drift vector* and $D = (D_{ij})$ is the *diffusion matrix*. The different role of the two terms may be seen in the equations of motion for $\langle x_k \rangle$ and $\langle x_k x_l \rangle$.

$$\frac{d\langle x_k \rangle}{dt} = \langle A_k \rangle ,$$

$$(6.65)$$

$$\frac{d\langle x_k x_l \rangle}{dt} = \langle x_k A_l \rangle + \langle x_l A_k \rangle + \frac{1}{2}\langle D_{kl} + D_{lk} \rangle .$$

$$(6.66)$$

We see that A_k determines the motion of the mean amplitude whereas D_{lk} enters into the equation for correlations.

Thus, for the damped harmonic oscillator described by (6.63)

$$\frac{d\langle \alpha \rangle_P}{dt} = -\frac{\gamma}{2}\langle \alpha \rangle_P ,$$

$$(6.67)$$

$$\frac{d\langle \alpha^* \alpha \rangle_P}{dt} = -\gamma \langle \alpha^* \alpha \rangle_P + \gamma N ,$$

$$(6.68)$$

which are equivalent to (6.39 and 6.41) derived directly from the master equation (6.37). Note that the expectation values $\langle \rangle_P$ are defined by integrals over $P(\alpha, t)$.

6.2.4 Steady State Solutions – Potential Conditions

For many problems in nonlinear optics it is sufficient to know the steady state solution. That is the solution after all transients have died out. We shall therefore seek a steady state solution to (6.64).

In the steady state we set the time derivatives to zero which gives

$$\frac{\partial}{\partial x_i}\left[-A_i(x)P(x) + \frac{1}{2}\frac{\partial}{\partial x_j}D_{ij}(x)P(x)\right] = 0 . \qquad (6.69)$$

As a first attempt consider

$$A_i(x)P(x) = \frac{1}{2}\frac{\partial}{\partial x_j}[D_{ij}(x)P(x)] \qquad (6.70)$$

which implies

$$D_{ij}\frac{\partial \ln P}{\partial x_j} = 2A_i(x) - \frac{\partial D_{ij}}{\partial x_j}(x) . \qquad (6.71)$$

Denoting $P(x) = \exp[-\phi(x)]$ we wish to solve

$$-\frac{\partial \phi(x)}{\partial x_i} = 2(D^{-1})_{ij}\left[A_j(x) - \frac{1}{2}\frac{\partial D_{jk}}{\partial x_k}\right] \equiv F_i(x) . \qquad (6.72)$$

If we consider $F_j(x)$ as a generalized force, $\phi(x)$ corresponds to a potential. The system of equations (6.72) can be solved by integration if the so called *potential conditions* are satisfied

$$\frac{-\partial^2 \phi}{\partial x_i \partial x_j} = \frac{\partial F_j}{\partial x_i} = \frac{\partial F_i}{\partial x_j} = \frac{-\partial^2 \phi}{\partial x_j \partial x_i} . \qquad (6.73)$$

These conditions say that the function ϕ is well behaved and that the multivariable integral is independent of the path of integration. The potential conditions are always satisfied in the one dimensional case.

Provided the potential conditions are satisfied a steady state solution of the form

$$P(x) = N\exp[-\phi(x)] \qquad (6.74)$$

exists where

$$\phi(x) = \int_0^x 2[D(x)^{-1}]_{ij}\left[-A_j(x) + \frac{1}{2}\frac{\partial D_{jk}}{\partial x_k}\right] dx_i .$$

The turning points of the potential ϕ correspond to the values x such that for each $j = 1, \ldots, n$

$$\left(A_j(x) - \frac{1}{2}\frac{\partial D_{jk}(x)}{\partial x_k}\right) = 0 . \qquad (6.75)$$

In systems where the diffusion matrix is diagonal and constant $(D_{ij} = W \delta_{ij})$ (6.72) become

$$-\frac{\partial \phi(x)}{\partial x_i} = \frac{2A_i(x)}{W} \, . \tag{6.76}$$

Hence the turning points of the potential ϕ correspond exactly to the deterministic steady state solutions, that is the steady state solutions to the first-order moment equations.

6.2.5 Time Dependent Solution

In the case where the drift term is linear in the variable (x) and the diffusion coefficient is a constant, a solution to the Fokker–Planck equation may be found using the method of *Wang* and *Uhlenbeck* [2]. We consider the Fokker–Planck equation

$$\frac{\partial P}{\partial t} = -\sum a_i \frac{\partial}{\partial x_i}(x_i P) + \frac{1}{2} d_{ij} \frac{\partial^2 P}{\partial x_i \partial x_j} \, . \tag{6.77}$$

The Greens function solution to this equation given by the initial condition

$$P(x_i, 0) = \delta(x_i - x_i^0)$$

is

$$P(x_i, x_i^0, t) = \frac{1}{\pi^{n/2} [\det \sigma_{ij}(t)]^{1/2}} \exp\left(-\sum_{ij} \sigma_{ij}(t)^{-1} \{ [x_i - x_i^0 \exp(a_i t)] \right.$$
$$\left. \times [x_j - x_j^0 \exp(a_j t)] \} \right) \tag{6.78}$$

where

$$\sigma_{ij}(t) = \frac{-2d_{ij}}{a_i + a_j} \{ 1 - \exp[(a_i + a_j)t] \} \, .$$

The solution for a damped harmonic oscillator initially in a coherent state with $P(\alpha, 0) = \delta^2(\alpha - \alpha_0)$ is

$$P(\alpha, t) = \frac{1}{\pi N(1 - e^{-n})} \exp\left(\frac{-|\alpha - \alpha_0 e^{-n/2}|^2}{N(1 - e^{-n})}\right) \, . \tag{6.79}$$

This represents an initial coherent state undergoing relaxation with a heat bath. Its coherent amplitude decays away and fluctuations from the heat bath cause its P function to assume a Gaussian form characteristic of thermal noise. The width of the distribution grows with time until the oscillator reaches equilibrium with the heat bath.

From the above solution we may construct solutions for all initial conditions which have a non-singular P representation. It is not, however, possible to construct the solution for the oscillator initially in a squeezed state since no non-singular P function exists for such states. Nor can we find the solution for an oscillator damped into a squeezed bath.

We now consider alternative methods of converting the operator master equation to a c-number equation, which can be used for initial squeezed states or a squeezed bath.

6.2.6 Q Representation

A c-number representation for the Q function is obtained by first normally ordering all operator products. We shall make use of the following theorem [3].

If $f(a, a^\dagger)$ is a function which may be expanded in a power series in a and $a^\dagger$, then

$$[a, f(a,a^\dagger)] = \frac{\partial f}{\partial a^\dagger} , \tag{6.80a}$$

$$[a^\dagger, f(a,a^\dagger)] = -\frac{\partial f}{\partial a} . \tag{6.80b}$$

The proof of these relations is as follows: We assume that we may expand f in antinormal order $f^{(a)}$

$$[a^\dagger, f] = \sum_{r,s} f_{r,s}^{(a)} [a^\dagger, a^r (a^\dagger)^s] . \tag{6.81}$$

Using the following result for the commutators

$$[A, BC] = [A,B]C + B[A, C] \tag{6.82}$$

where A, B and C are noncommuting operators, we may write

$$[a^\dagger, f] = \sum_{r,s} f_{r,s}^{(a)} \{[a^\dagger, a^r]a^{\dagger s} + a^r[a^\dagger, a^{\dagger s}]\}$$

$$= -\sum_{r,s} f_{r,s}^{(a)} r a^{r-1} a^{\dagger s} \tag{6.83}$$

$$= -\frac{\partial f}{\partial a} . \tag{6.84}$$

The proof of (6.80a) follows in a similar way.

We consider as an example the term

$$\rho a^\dagger a = a^\dagger \rho a - [a^\dagger, \rho]a . \tag{6.85}$$

Using the result above

$$\rho a^\dagger a = a^\dagger \rho a + \frac{\partial \rho}{\partial a} a . \tag{6.86}$$

Taking matrix elements in coherent states yields

$$\langle\alpha|\rho a^{\dagger}a|\alpha\rangle = \alpha\left\langle\alpha\left|a^{\dagger}\rho + \frac{\partial\rho}{\partial a}\right|\alpha\right\rangle$$

$$= \left(|\alpha|^2 + \alpha\frac{\partial}{\partial\alpha}\right)Q(\alpha). \tag{6.87}$$

Following this procedure we may convert the master equation for the damped harmonic oscillator into a c-number equation for the Q function

$$\frac{\partial Q}{\partial t} = \frac{\gamma}{2}\left(\frac{\partial}{\partial\alpha}\alpha + \frac{\partial}{\partial\alpha^*}\alpha^*\right)Q + \frac{\gamma}{2}\left[M^*\frac{\partial^2}{\partial\alpha^{*2}} + M\frac{\partial^2}{\partial\alpha^2} + 2(N+1)\frac{\partial^2}{\partial\alpha\partial\alpha^*}\right]Q. \tag{6.88}$$

This differs from the corresponding equation of motion for the P function only through the phase independent diffusion coefficient which is $N+1$ rather than N. This is sufficient to give a positive definite diffusion matrix when the bath is in an ideal squeezed state.

To illustrate the use of the Q function consider a damped oscillator which is initially in the squeezed state $|\alpha_0, r\rangle$. Using the *Wang* and *Uhlenbeck* solution for an initial δ-function and convoluting this with the Gaussian Q function for an initial squeezed state we arrive at the result

$$Q(\alpha, t) = \frac{1}{2\pi\sqrt{\det\sigma(t)}}\exp[-\frac{1}{2}\boldsymbol{u}(t)^{\mathrm{T}}\sigma^{-1}(t)\boldsymbol{u}(t)] \tag{6.89}$$

where

$$\boldsymbol{u}(t) = \begin{pmatrix} \alpha - \alpha_0 e^{-\gamma t/2} \\ \alpha^* - \alpha_0^* e^{-\gamma t/2} \end{pmatrix}$$

$$\sigma(t) = \begin{pmatrix} -\sinh 2r & \cosh 2r + 1 \\ \cosh 2r + 1 & -\sinh 2r \end{pmatrix}\frac{e^{-\gamma t}}{2} + \begin{pmatrix} M & N+1 \\ N+1 & M^* \end{pmatrix}(1 - e^{-\gamma t}).$$

The variances for the quadrature phase operators X_i for the oscillator are then easily found to be

$$V(X_1) = \frac{1}{4}[(e^{-2r} - 1)e^{-\gamma t} + 2(N + \mathrm{Re}\{M\})(1 - e^{-\gamma t}) + 1], \tag{6.90a}$$

$$V(X_2) = \frac{1}{4}[(e^{2r} + 1)e^{-\gamma t} + 2(N + \mathrm{Re}\{M\})(1 - e^{-\gamma t}) + 1]. \tag{6.90b}$$

In Fig. 6.1 we depict the evolution of an initial squeezed state coupled to a zero temperature reservoir $(N = M = 0)$. The amplitude of the squeezed state damps to zero and the variances in X_1 and X_2 become equal at the value one corresponding to the vacuum.

Fig. 6.1 Evolution of the error contour of an initial squeezed state of a simple harmonic oscillator damped into a zero temperature heat bath

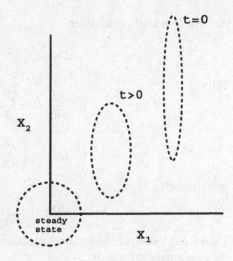

6.2.7 Wigner Function

Alternatively one may convert the operator master equation into a c-number equation via the Wigner function. This is best accomplished by deriving an equation for the characteristic function

$$\chi(\beta) = \mathrm{Tr}\{D\rho\} \tag{6.91}$$

where

$$D = e^{\beta a^\dagger - \beta^* a} \, .$$

An equation of motion for $\chi(\beta)$ may be derived as follows

$$\frac{\partial \chi(\beta)}{\partial t} = \mathrm{Tr}\left\{ D \frac{\partial \rho}{\partial t} \right\} \, . \tag{6.92}$$

To illustrate the technique we shall derive the equation of motion for the Wigner function of a damped harmonic oscillator.

We require some operator rules to convert to differential operators. Writing D in normal order

$$D = e^{-\beta\beta^*/2} e^{\beta a^\dagger} e^{-\beta^* a} \, , \tag{6.93}$$

$$\frac{\partial}{\partial \beta} D = -\frac{\beta^*}{2} D + a^\dagger D \, , \tag{6.94}$$

or

$$a^\dagger D = \left(\frac{\partial}{\partial \beta} + \frac{\beta^*}{2} \right) D \, . \tag{6.95}$$

Similarly we may show

$$Da = \left(\frac{-\beta}{2} - \frac{\partial}{\partial \beta^*} \right) D \, . \tag{6.96}$$

Writing D in antinormal order

$$D = e^{\beta\beta^*/2}e^{-\beta^*a}e^{\beta a^\dagger}.$$ (6.97)

Thus

$$\frac{\partial}{\partial\beta}D = \frac{\beta^*}{2}D + Da^\dagger$$ (6.98)

or

$$Da^\dagger = \left(\frac{\partial}{\partial\beta} - \frac{\beta^*}{2}\right)D$$ (6.99)

and similarly

$$aD = \left(\frac{\beta}{2} - \frac{\partial}{\partial\beta^*}\right)D.$$ (6.100)

Then using these rules the master equation (6.37) yields the following equation for the characteristic function

$$\frac{\partial\chi(\beta)}{\partial t} = \frac{\gamma}{2}\left(-|\beta|^2 - \beta^*\frac{\partial}{\partial\beta^*} - \beta\frac{\partial}{\partial\beta}\right)\chi(\beta) - \gamma N|\beta|^2\chi(\beta)$$

$$- \frac{\gamma M}{2}(\beta^*)^2\chi(\beta) - \frac{\gamma M^*}{2}\beta^2\chi(\beta).$$ (6.101)

The equation for the Wigner function is obtained by taking the Fourier transform of this equation since

$$W(\alpha) = \int e^{\beta^*\alpha - \beta\alpha^*}\chi(\beta)d^2\beta.$$ (6.102)

Thus

$$\int e^{\beta^*\alpha - \alpha^*\beta}\beta^*\beta\chi(\beta)d^2\beta = -\int \frac{\partial}{\partial\alpha}\frac{\partial}{\partial\alpha^*}(e^{\beta^*\alpha - \alpha^*\beta})\chi(\beta)d^2\beta$$

$$= -\frac{\partial^2 W(\alpha)}{\partial\alpha\,\partial\alpha^*},$$ (6.103)

and

$$\int e^{\beta^*\alpha - \alpha^*\beta}\beta^*\frac{\partial}{\partial\beta^*}\chi(\beta)d^2\beta = \frac{\partial}{\partial\alpha}\int (e^{\beta^*\alpha - \alpha^*\beta})\frac{\partial}{\partial\beta^*}\chi(\beta)d^2\beta$$

$$= -\frac{\partial}{\partial\alpha}\int \chi(\beta)\frac{\partial}{\partial\beta^*}(e^{\beta^*\alpha - \alpha^*\beta})d^2\beta$$ (6.104)

$$= -\frac{\partial}{\partial\alpha}[\alpha W(\alpha)].$$ (6.105)

Using these results we may write the equation for the Wigner function as

$$\frac{\partial W(\alpha)}{\partial t} = \frac{\gamma}{2}\left(\frac{\partial}{\partial\alpha}\alpha + \frac{\partial}{\partial\alpha^*}\alpha^*\right)W$$

$$+ \frac{\gamma}{2}\left[M^*\frac{\partial^2}{\partial\alpha^{*2}} + M\frac{\partial^2}{\partial\alpha^2} + 2\left(N + \frac{1}{2}\right)\frac{\partial^2}{\partial\alpha\,\partial\alpha^*}\right]W.$$ (6.106)

A comparison of the three equations (6.63, 6.88 and 6.106) for the P, Q and Wigner functions show that they differ only in the coefficient of the diffusion term being γN, $\gamma(N+1)$ and $\gamma(N+1/2)$, respectively. However, the additional $+\gamma$ and $+\gamma/2$ in the equations for the Q and Wigner function are sufficient to ensure that these equations have positive definite diffusion.

6.2.8 Generalized P Representation

In our study of nonlinear problems we shall find systems which either do not give Fokker–Planck equations in the Q and Wigner representations or no steady state solution may readily be found. For some systems a steady state solution in terms of a Glauber–Sudarshan P representation does not exist. For such systems the complex P representation is sometimes useful in deriving a steady state solution to Fokker–Planck equations. The positive P representation is useful when it is desirable to have a Fokker–Planck equation with a positive definite diffusion term, as is necessary in order to deduce the corresponding stochastic differential equations.

Master equations may be converted to a c-number representation using the complex P representation by an analogous set of operator rules used for the diagonal P representation.

The nondiagonal coherent state projection operator is defined as

$$\Lambda(\alpha) = \frac{|\alpha\rangle\langle\beta^*|}{\langle\beta^*|\alpha\rangle} \tag{6.107}$$

where (α) denotes (α, β). The following identities hold

$$a\Lambda(\alpha) = \alpha\Lambda(\alpha), \quad a^\dagger\Lambda(\alpha) = \left(\beta + \frac{\partial}{\partial\alpha}\right)\Lambda(\alpha),$$

$$\Lambda(\alpha)a^\dagger = \Lambda(\alpha)\beta, \quad \Lambda(\alpha)a = \left(\frac{\partial}{\partial\beta} + \alpha\right)\Lambda(\alpha). \tag{6.108}$$

By substituting the above identities into (4.65) defining the generalized P representation, and using partial integration (providing the boundary terms vanish) these identities can be used to generate operations on the P function depending on the representation.

a) Complex P representation

$$a\rho \leftrightarrow \alpha P(\alpha), \quad a^\dagger\rho \leftrightarrow \left(\beta - \frac{\partial}{\partial\alpha}\right)P(\alpha),$$

$$\rho a^\dagger \leftrightarrow \beta P(\alpha), \quad \rho a \leftrightarrow \left(\alpha - \frac{\partial}{\partial\beta}\right)P(\alpha). \tag{6.109}$$

This procedure yields a very similar equation to that for the Glauber–Sudarshan P function. We assume that, by appropriate reordering of the differential operators, we can reduce an operator master equation to the form [where $(\alpha, \beta) = (\alpha) = (\alpha^{(1)}, \alpha^{(2)})$]

$$\frac{\partial \rho}{\partial t} = \int_c \int_{c'} \Lambda(\alpha) \frac{\partial P(\alpha)}{\partial t} \, d\alpha \, d\beta$$

$$= \int_c \int_{c'} d\alpha^{(1)} \, d\alpha^{(2)} \, P(\alpha) \left[A^\mu(\alpha) \frac{\partial}{\partial \alpha^\mu} + \frac{1}{2} D^{\mu\nu}(\alpha) \frac{\partial}{\partial \alpha^\mu} \frac{\partial}{\partial \alpha^\nu} \right] \Lambda(\alpha) \,.$$

$$(6.110)$$

We now integrate by parts and if we can neglect boundary terms, which may be made possible by an appropriate choice of contours, c and c', at least one solution is obtained by equating the coefficients of $\Lambda(\alpha)$

$$\frac{\partial P(\alpha)}{\partial t} = \left[-\frac{\partial}{\partial \alpha^\mu} A^\mu(\alpha) + \frac{1}{2} \frac{\partial}{\partial \alpha^\mu} \frac{\partial}{\partial \alpha^\nu} D^{\mu\nu}(\alpha) \right] P(\alpha) \,. \qquad (6.111)$$

This equation is sufficient to imply (6.110) but is not a unique equation because the $\Lambda(\alpha)$ are not linearly independent. The Fokker–Planck equation has the same form as that derived using the diagonal P representation with α^* replaced by β.

It should be noted that for the complex P representation, $A^\mu(\alpha)$ and $D^{\mu\nu}(\alpha)$ are always analytic in (α), hence if $P(\alpha)$ is initially analytic (6.111) preserves this analyticity as time develops.

b) Positive P Representation

The operator identities for the positive P representation are the same as (6.109) for the complex P representation. In addition, using the analyticity of $\Lambda(\alpha, \beta)$ and noting that if

$$\alpha = \alpha_x + i\alpha_y, \qquad \beta = \beta_x + i\beta_y,$$

then

$$\frac{\partial}{\partial \alpha} \Lambda(\alpha) = \frac{\partial}{\partial \alpha_x} \Lambda(\alpha) = -i \frac{\partial}{\partial \alpha_y} \Lambda(\alpha)$$

and

$$\frac{\partial}{\partial \beta} \Lambda(\alpha) = \frac{\partial}{\partial \beta_x} \Lambda(\alpha) = -i \frac{\partial}{\partial \beta_y} \Lambda(\alpha) \,. \qquad (6.112)$$

Thus in addition to (6.109) we also have

$$a^\dagger \rho \leftrightarrow \left(\beta - \frac{\partial}{\partial \alpha_x} \right) P(\alpha) \leftrightarrow \left(\beta + i \frac{\partial}{\partial \alpha_y} \right) P(\alpha),$$

$$\rho a \leftrightarrow \left(\alpha - \frac{\partial}{\partial \beta_x} \right) P(\alpha) \leftrightarrow \left(\alpha + i \frac{\partial}{\partial \beta_y} \right) P(\alpha) \,. \qquad (6.113)$$

The positive P representation may be used to give a Fokker–Planck equation with a positive definite diffusion matrix. We shall demonstrate this in the following.

We assume that the same equation (6.64) is being considered but with a positive P representation. The symmetric diffusion matrix can always be factorized in the form

$$\boldsymbol{D}(\alpha) = \boldsymbol{B}(\alpha)\boldsymbol{B}^{\mathrm{T}}(\alpha).$$

We now write

$$\boldsymbol{A}(\alpha) = \boldsymbol{A}_x(\alpha) + i\boldsymbol{A}_y(\alpha), \tag{6.114}$$

$$\boldsymbol{B}(\alpha) = \boldsymbol{B}_x(\alpha) + i\boldsymbol{B}_y(\alpha), \tag{6.115}$$

where $\boldsymbol{A}_x$, $\boldsymbol{A}_y$, $\boldsymbol{B}_x$, $\boldsymbol{B}_y$ are real. We then find that the master equation yields

$$\frac{\partial \rho}{\partial t} = \iint d^2\alpha\, d^2\beta\, \Lambda(\alpha)(\partial P(\alpha)/\partial t)$$

$$= \iint P(\alpha)[A_x^{\mu}(\alpha)\partial_\mu^x + A_y^{\mu}(\alpha)\partial_\mu^y + \frac{1}{2}(B_x^{\mu\sigma}B_x^{\nu\sigma}\partial_\mu^x\partial_\nu^x + B_y^{\mu\sigma}B_y^{\nu\sigma}\partial_\mu^y\partial_\nu^y$$

$$+ 2B_x^{\mu\sigma}B_y^{\nu\sigma}\partial_\mu^x\partial_\nu^y)]\Lambda(\alpha)\mathrm{d}^2\alpha\,\mathrm{d}^2\beta. \tag{6.116}$$

Here we have, for notational simplicity, written $\partial/\partial\alpha_x^{\mu} = \partial_\mu^x$ etc., and have used the analyticity of $\Lambda(\alpha)$ to make either of the replacements

$$\partial/\partial\alpha^{\mu} \leftrightarrow \partial_\mu^x \leftrightarrow -i\partial_\mu^y \tag{6.117}$$

in such a way as to yield (6.116). Now, provided partial integration is permissible, we deduce the Fokker–Planck equation:

$$\partial P(\alpha)/\partial t = [-\partial_\mu^x A_x^{\mu}(\alpha) - \partial_\mu^y A_y^{\mu}(\alpha) + \frac{1}{2}[\partial_\mu^x\partial_\nu^x B_x^{\mu\sigma}(\alpha)B_x^{\nu\sigma}(\alpha)$$

$$+ 2\partial_\mu^x\partial_\nu^y B_x^{\mu\sigma}(\alpha)B_y^{\nu\sigma}(\alpha) + \partial_\mu^y\partial_\nu^y B_y^{\mu\sigma}(\alpha)B_y^{\nu\sigma}(\alpha)]\}\, P(\alpha) \tag{6.118}$$

Again, this is not a unique time-development equation but (6.116) is a consequence of (6.118).

However, the Fokker–Planck equation (6.118) now possesses a positive semidefinite diffusion matrix in a four-dimensional space whose vectors are

$$(\alpha_x^{(1)}, \alpha_x^{(2)}, \alpha_y^{(1)}, \alpha_y^{(2)}) \equiv (\alpha_x, \beta_x, \alpha_y, \beta_y. \tag{6.119}$$

We find the drift vector is

$$\mathscr{A}(\alpha) \equiv (A_x^{(1)}(\alpha), A_x^{(2)}(\alpha), A_y^{(1)}(\alpha), A_y^{(2)}(\alpha)), \tag{6.120}$$

and the diffusion matrix is

$$\mathscr{D}(\alpha) = \begin{pmatrix} \boldsymbol{B}_x\boldsymbol{B}_x^{\mathrm{T}} & \boldsymbol{B}_x\boldsymbol{B}_y^{\mathrm{T}} \\ \boldsymbol{B}_y\boldsymbol{B}_x^{\mathrm{T}} & \boldsymbol{B}_y\boldsymbol{B}_y^{\mathrm{T}} \end{pmatrix}(\alpha) \equiv \mathscr{B}(\alpha)\mathscr{B}^{\mathrm{T}}(\alpha) \tag{6.121}$$

where

$$\mathscr{B}(\alpha) = \begin{pmatrix} B_x & 0 \\ B_y & 0 \end{pmatrix}(\alpha) \tag{6.122}$$

and $\mathscr{D}$ is thus explicitly positive semidefinite.

6.3 Stochastic Differential Equations

A Fokker–Planck equation of the form

$$\frac{\partial P}{\partial t} = -\sum_i \frac{\partial}{\partial x_i} A_i(\boldsymbol{x}, t)P + \frac{1}{2}\sum_{ij}\frac{\partial}{\partial x_i}\frac{\partial}{\partial x_j}[\boldsymbol{B}(\boldsymbol{x}, t)\boldsymbol{B}^{\mathrm{T}}(\boldsymbol{x}, t)]_{ij}P \tag{6.123}$$

may be written in a completely equivalent form as

$$\frac{\mathrm{d}\boldsymbol{x}}{\mathrm{d}t} = A(\boldsymbol{x}, t) + B(\boldsymbol{x})\boldsymbol{E}(t) \tag{6.124}$$

where $\boldsymbol{E}(t)$ are fluctuating forces with zero mean and δ correlated in time

$$\langle E_i(t)E_j(t')\rangle = \delta_{ij}\delta(t - t') . \tag{6.125}$$

We have written (6.124) in the form of a Langevin equation. The relationship between (6.123 and 6.124) may be derived more rigorously in terms of stochastic differential equations where Ito's rules are used. However, the relation quoted in (6.123 and 6.124) is sufficient for our use. The reader is referred to the texts C.W. Gardiner for a complete discourse on stochastic differential equations and their applications to quantum noise problems.

We shall illustrate the use of the stochastic differential equation for a particle undergoing damping and diffusion in one dimension. This motion is described by the Fokker–Planck equation

$$\frac{\partial P(x)}{\partial t} = \kappa\frac{\partial}{\partial x}[xP(x)] + \frac{D}{2}\frac{\partial^2}{\partial x^2}P(x) \tag{6.126}$$

where κ is the damping coefficient and D is the diffusion coefficient. This equation describes an Ornstein–Uhlenbeck process. It may, for example, describe the Brownian motion of a particle under the random influence of collisions from many particles in thermal motion where the variable x represents the particle's velocity.

The Langevin equation equivalent to the Fokker–Planck Equation (6.126) is

$$\dot{x} = -\kappa x + \sqrt{D}E(t) \tag{6.127}$$

where

$$\langle E(t)E(t')\rangle = \delta(t - t') .$$

The solution to this equation is

$$x(t) = x(0)e^{-\kappa t} + \sqrt{D} \int_0^t e^{-\kappa(t-t')} E(t') dt' \,. \tag{6.128}$$

If the initial condition is deterministic or Gaussian distributed, then $x(t)$ is clearly Gaussian, with mean and variance

$$\langle x(t) \rangle = \langle x(0) \rangle e^{-\kappa t} \,, \tag{6.129}$$

$$\mathrm{Var}[x(t)] = \left\langle \left\{ [x(0) - \langle x(0) \rangle] e^{-\kappa t} + \sqrt{D} \int_0^t e^{-\kappa(t-t')} E(t') dt' \right\}^2 \right\rangle \,. \tag{6.130}$$

Assuming the initial condition is independent of $E(t)$, we may write

$$\mathrm{Var}\{x(t)\} = \mathrm{Var}\{x(0)\} e^{-2\kappa t} + D \int_0^t e^{-2\kappa(t-t')} dt'$$

$$= \left[\mathrm{Var}\{x(0)\} - \frac{D}{2\kappa} \right] e^{-2\kappa t} + \frac{D}{2\kappa} \,. \tag{6.131}$$

In the steady state

$$\mathrm{Var}\{x(t)\} = \frac{D}{2\kappa} \,. \tag{6.132}$$

The two time correlation function may be calculated directly, as follows

$$\langle x(t), x(s) \rangle = \langle x(t) x(s) \rangle - \langle x(t) \rangle \langle x(s) \rangle$$

$$= \mathrm{Var}\{x(0)\} e^{-\kappa(t+s)} + D \left\langle \int_0^t e^{-\kappa(t-t')} E(t') dt' \int_0^s e^{-\kappa(s-s')} E(s') ds' \right\rangle$$

$$= \mathrm{Var}\{x(0)\} e^{-\kappa(t+s)} + D \int_0^{\min(t,s)} e^{-\kappa(t+s-2t')} dt'$$

$$= \left[\mathrm{Var}\{x(0)\} - \frac{D}{2\kappa} \right] e^{-\kappa(t+s)} + \frac{D}{2\kappa} e^{-\kappa|t-s|} \,. \tag{6.133}$$

In the stationary state

$$\langle x(t), x(s) \rangle = \frac{D}{2\kappa} e^{-\kappa|t-s|} \,. \tag{6.134}$$

We shall now consider the equivalent Langevin equation for the Fokker–Planck equation for the damped harmonic oscillator. The Fokker–Planck equation for the P representation is

$$\frac{\partial P}{\partial t} = \frac{\gamma}{2} \left(\frac{\partial}{\partial \alpha} \alpha + \frac{\partial}{\partial \alpha^*} \alpha^* \right) P + \gamma N \frac{\partial^2}{\partial \alpha \, \partial \alpha^*} P \,. \tag{6.135}$$

Note that we have set $M = 0$, as the P representation cannot be used with squeezed baths since the diffusion matrix is nonpositive definite. In such cases the Q-function could be used.

Equation (6.135) is an example of an Ornstein–Uhlenbeck process. The diffusion matrix is

$$D = \gamma N \begin{pmatrix} 0 & 1 \\ 1 & 0 \end{pmatrix} \tag{6.136}$$

which may be factored as

$$D = BB^{\mathrm{T}}$$

where

$$B = \left(\frac{\gamma N}{2} \right)^{1/2} \begin{pmatrix} i & 1 \\ -i & 1 \end{pmatrix} . \tag{6.137}$$

Thus the stochastic differential equations become

$$\frac{\mathrm{d}}{\mathrm{d}t} \begin{pmatrix} \alpha \\ \alpha^* \end{pmatrix} = \begin{pmatrix} \frac{-\gamma}{2} & 0 \\ 0 & \frac{-\gamma}{2} \end{pmatrix} \begin{pmatrix} \alpha \\ \alpha^* \end{pmatrix} + \sqrt{\frac{\gamma N}{2}} \begin{pmatrix} i & 1 \\ -i & 1 \end{pmatrix} \begin{pmatrix} \eta_1(t) \\ \eta_2(t) \end{pmatrix} \tag{6.138}$$

where $\eta_1(t)$ and $\eta_2(t)$ are independent stochastic forces which satisfy

$$\langle \eta_i(t) \eta_j(t') \rangle = \delta_{ij} \delta(t - t') . \tag{6.139}$$

Equation (6.138) may be written

$$\frac{\mathrm{d}\alpha}{\mathrm{d}t} = -\frac{\gamma}{2}\alpha + \sqrt{\gamma N}\eta(t),$$

$$\frac{\mathrm{d}\alpha^*}{\mathrm{d}t} = -\frac{\gamma}{2}\alpha + \sqrt{\gamma N}\eta^*(t) \tag{6.140}$$

where

$$\eta(t) \equiv \frac{1}{\sqrt{2}}[\eta_2(t) + i\eta_1(t)]$$

is a complex stochastic force term which satisfies

$$\langle \eta(t)\eta^*(t') \rangle = \delta(t - t') .$$

An alternative factorisation is

$$B = \left(\frac{\gamma N}{2} \right)^{1/2} \begin{pmatrix} e^{i\pi/4} & e^{-i\pi/4} \\ e^{-i\pi/4} & e^{i\pi/4} \end{pmatrix} . \tag{6.141}$$

In this case

$$\frac{\mathrm{d}\alpha}{\mathrm{d}t} = -\frac{\gamma}{2}\alpha + \sqrt{\gamma N}\tilde{\eta} \tag{6.142}$$

where

$$\tilde{\eta} = \frac{1}{\sqrt{2}}(\eta_2 e^{i\pi/4} + \eta_1 e^{-i\pi/4} . \tag{6.143}$$

One easily verifies that $\langle \tilde{\eta}\tilde{\eta}^* \rangle = \delta(t - t')$.

The solutions derived from (6.140) are

$$\langle \alpha(t) \rangle = \langle \alpha(0) \rangle \exp\left(-\frac{\gamma}{2}t\right) ,$$ (6.144)

$$\langle \alpha^*(t)\alpha(t) \rangle = \langle \alpha^*(0)\alpha(0) \rangle e^{-\gamma t} + N(1 - e^{-\gamma t}),$$

$$\langle \alpha^2 \rangle_{ss} = \langle \alpha^{*2} \rangle_{ss} = 0,$$

$$\langle \alpha\alpha^* \rangle_{ss} = \langle \alpha^*\alpha \rangle_{ss} = N ,$$ (6.145)

where ss denotes steady state.

6.3.1 Use of the Positive P Representation

The relationship (6.123 and 6.124) between the Fokker–Planck equation and the stochastic differential equation only holds if the diffusion matrix

$$\mathscr{D}(\boldsymbol{x}, t) = \mathscr{B}(\boldsymbol{x}, t)\mathscr{B}^{\mathrm{T}}(\boldsymbol{x}, t)$$

is positive semidefinite. In some cases use of the Glauber–Sudarshan P representation will result in Fokker–Planck equations with a non-positive semi-definite diffusion matrix, for example, if the bath is squeezed. In such cases use of the positive P representation will give a Fokker–Planck equation with a positive semi-definite diffusion matrix

$$\mathscr{D}(\alpha) = \mathscr{B}(\alpha)\mathscr{B}^{\mathrm{T}}(\alpha)$$ (6.146)

where

$$\mathscr{B}(\alpha) = \begin{pmatrix} B_x & 0 \\ B_y & 0 \end{pmatrix}$$

and $\alpha = (\alpha, \beta)$.

The corresponding stochastic differential equations may be written

$$\frac{\mathrm{d}}{\mathrm{d}t}\begin{pmatrix} \alpha_x \\ \alpha_y \end{pmatrix} = \begin{pmatrix} A_x(\alpha) \\ A_y(\alpha) \end{pmatrix} + \begin{pmatrix} B_x(\alpha)\boldsymbol{E}(t) \\ B_y(\alpha)\boldsymbol{E}(t) \end{pmatrix}$$ (6.147)

on recombining real and imaginary parts

$$\frac{\mathrm{d}\alpha}{\mathrm{d}t} = A(\alpha) + B(\alpha)\boldsymbol{E}(t) .$$ (6.148)

Apart from the substitution $\alpha^* \to \beta$, (6.148) is just the stochastic differential equation which would be obtained by using the Glauber–Sudarshan P representation, and naively converting the Fokker–Planck equation with a non-positive definite diffusion matrix into a stochastic differential equation. In the above derivation the two formal variables (α, α^*) have been replaced by variables in the complex plane (α, β) that are allowed to fluctuate independently. The use of the positive P representation justifies this procedure.

6.4 Linear Processes with Constant Diffusion

For linear processes with constant diffusion coefficients a number of useful results may be proven. These may be derived from the Fokker–Planck equation using the solution (6.78) or the equivalent Langevin equation. We shall quote the results here. The Langevin equations are a useful starting point since for nonlinear processes approximate results may be obtained by a linearization procedure. We consider a process described by the Langevin equation

$$\frac{\mathrm{d}\boldsymbol{x}(t)}{\mathrm{d}t} = -A\boldsymbol{x}(t) + BE(t) \tag{6.149}$$

where A and B are constant matrices. This describes a multivariate Ornstein–Uhlenbeck process. Suppose $AA^{\mathrm{T}} = A^{\mathrm{T}}A$, then we can find an orthogonal matrix S such that

$$SS^{\mathrm{T}} = 1$$

$$SAS^{\mathrm{T}} = SA^{\mathrm{T}}S^{\mathrm{T}} = \mathrm{Diag}\{\lambda_1, \lambda_2 \ldots \lambda_n\} . \tag{6.150}$$

The two time correlation function is given by

$$\langle \boldsymbol{x}(t), \boldsymbol{x}^{\mathrm{T}}(s) \rangle = S^{\mathrm{T}} G(t,s) S$$

where

$$[G(t,s)]_{ij} = \frac{(SBB^{\mathrm{T}}S^{\mathrm{T}})_{ij}}{\lambda_i + \lambda_j} (\mathrm{e}^{-\lambda_i|t-s|} - \mathrm{e}^{-\lambda_i t - \lambda_j s}) . \tag{6.151}$$

In the stationary state the second term in the parentheses is zero and the correlation is only a function of the difference, $\tau \equiv t - s$.

Let us define the stationary covariance matrix σ by

$$\sigma = \langle \boldsymbol{x}_{\mathrm{ss}}(t), \, \boldsymbol{x}_{\mathrm{ss}}^{\mathrm{T}}(t) \rangle . \tag{6.152}$$

Then by setting $\mathrm{d}\boldsymbol{x}(t)/\mathrm{d}t = 0$ in (6.149) we find that σ obeys the equation

$$A\sigma + \sigma A^{\mathrm{T}} = BB^{\mathrm{T}} . \tag{6.153}$$

In the case of a two dimensional problem it may be shown that

$$\sigma = \frac{(\mathrm{Det}\, A)BB^{\mathrm{T}} + [A - (\mathrm{Tr}\, A)I]BB^{\mathrm{T}}[A - (\mathrm{Tr}\, A)I]^{\mathrm{T}}}{2(\mathrm{Tr}\, A)(\mathrm{Det}\, A)} . \tag{6.154}$$

The two time correlation function in the steady state may be shown to obey the same time development equation as the mean. That is

$$\frac{\mathrm{d}}{\mathrm{d}\tau}[G_{\mathrm{ss}}(\tau)] = -AG_{\mathrm{ss}}(\tau) . \tag{6.155}$$

The computation of $G_{ss}(\tau)$, therefore requires the knowledge of $G_{ss}(0) = \sigma$ and the time development equation of the mean.

It is often of more interest to view the noise in the frequency domain. We are thus lead to define the noise spectrum by,

$$S(\omega) = \frac{1}{2\pi} \int_{-\infty}^{\infty} e^{-i\omega\tau} G_{ss}(\tau) d\tau . \tag{6.156}$$

Using (6.155 and 6.153) we find

$$S(\omega) = \frac{1}{2\pi} (A + i\omega)^{-1} BB^{T} (A^{T} - i\omega)^{-1} . \tag{6.157}$$

6.5 Two Time Correlation Functions in Quantum Markov Processes

We shall now demonstrate how two time correlation functions for operators may be derived from the master equation or the equivalent Fokker–Planck equation.

Consider a system coupled to a reservoir. $W(t)$ is the total density operator in the Schrödinger picture and $\mathscr{H}$ is the Hamiltonian, A and B are operators for variables to be measured, then

$$\langle A(t) \rangle = \text{Tr}\{AW(t)\} \tag{6.158}$$

and

$$\langle A(t+\tau)B(t) \rangle = \text{Tr}\left[e^{i\mathscr{H}\tau/\hbar} A e^{-i\mathscr{H}\tau/\hbar} BW(t) \right] \tag{6.159}$$

while this is exact it is not particularly useful. For a system interacting with a heat bath in the Markov approximation we wish to express everything in terms of the Liouvillian for the reduced system in which heat bath variables have been traced out.

Supposing A and B are operators in the system space, then

$$\langle A(t+\tau)B(t) \rangle = \text{Tr}_s\{A \, \text{Tr}_R[e^{-i\mathscr{H}\tau/\hbar} BW(t) e^{i\mathscr{H}\tau/\hbar}]\} . \tag{6.160}$$

The equation of motion for the term

$$X(\tau,t) = e^{-i\mathscr{H}\tau/\hbar} BW(t) e^{i\mathscr{H}\tau/\hbar} \tag{6.161}$$

in terms of τ is

$$i\hbar \frac{\partial}{\partial \tau} X(\tau,t) = [\mathscr{H}, X(\tau,t)] . \tag{6.162}$$

Proceeding in exactly the same way as for the derivation of the Markovian master equation (6.37) which may be written in the form

$$\frac{\partial}{\partial t} \rho(t) = L\rho(t) \tag{6.163}$$

where L is a Liouvillian operator, where $\rho = \mathrm{Tr_R}\{W\}$ is the reduced density operator for the system, we may derive the equation

$$\frac{\partial}{\partial \tau}[\mathrm{Tr_R}\{X(\tau,t)\}] = L\{\mathrm{Tr_R}\, X(\tau,t)\} \tag{6.164}$$

so that the two time correlation function may be expressed as

$$\langle A(t+\tau)B(t)\rangle = \mathrm{Tr}_s\{Ae^{L\tau}B\rho(t)\}\,. \tag{6.165}$$

6.5.1 Quantum Regression Theorem

In cases where the master equation gives linear equations for the mean, we can develop a quantum regression theorem, similar to that for ordinary Markov processes. This result was first derived by *Lax* [4].

Suppose for a certain set of operators Y_i, the master equation can be shown to yield, for any initial ρ

$$\frac{\partial}{\partial t}\langle Y_i(t)\rangle = \sum G_{ij}(t)\langle Y_j(t)\rangle\,. \tag{6.166}$$

Then we assert that

$$\frac{\partial}{\partial t}\langle Y_i(t+\tau)Y_l(t)\rangle = \sum G_{ij}(\tau)\langle Y_j(t+\tau)Y_l(t)\rangle\,. \tag{6.167}$$

For

$$\langle Y_i(t+\tau)Y_l(t)\rangle = \mathrm{Tr}_s\{Y_i e^{L\tau}Y_l\rho(t)\} \tag{6.168}$$

the right-hand side is an average of Y_i at time $t+\tau$, with the choice of initial density matrix

$$\rho_{\mathrm{init}} = Y_l\rho(t)\,. \tag{6.169}$$

Since by hypothesis, any initial ρ is permitted and the equation is linear, we may generate any initial condition whatsoever. Hence, choosing ρ_{init} as defined in (6.169) the hypothesis (6.166) yields the result (6.167) which is the quantum regression theorem.

6.6 Application to Systems with a P Representation

For systems where a P representation exists the following results for normally ordered time correlation functions may be proved

$$G^{(1)}(t,\tau) = \langle a^\dagger(t+\tau)a(t)\rangle = \langle \alpha^*(t+\tau)\alpha(t)\rangle\,, \tag{6.170}$$

$$G^{(2)}(t,\tau) = \langle a^\dagger(t)a^\dagger(t+\tau)a(t+\tau)a(t)\rangle,$$

$$= \langle |\alpha(t+\tau)|^2|\alpha(t)|^2\rangle\,. \tag{6.171}$$

In these cases the measured correlation functions correspond to the same correlation function for the variables in the P representation. For non-normally ordered correlation functions the result is not as simple.

6.7 Stochastic Unravellings

The master equation describes the dynamics of a subsystem by averaging over (tracing out) the properties of the larger "bath" to which it is coupled. Solving the master equation typically results in a mixed state. Any mixed state admits infinitely many decompositions into convex combinations of (non-orthogonal) pure states. In a stochastic unravelling of a master equation we represent the solution at any time as a convex combination of pure states each evolving under a stochastic Schrödinger equation such that if we average over the noise we obtain the solution to the original master equation. This approach leads to a powerful numerical simulation tool as much less memory is required to store a pure quantum state at each time step. In Chap. 15, we give an alternative interpretation of an unravelling in terms of conditional states conditioned on a continuously recorded sequence of measurement results.

Consider a simple harmonic oscillator coupled to a zero temperature heat bath. The dynamics, in the interaction picture, is given by the master equation, (6.37) with $N = M = 0$. Solving this equation over a small time interval dt we can write

$$\rho(t + dt) = \left[\rho(t) - \frac{\gamma}{2}(a^\dagger a\rho(t) + \rho(t)a^\dagger a)dt\right] + \gamma a\rho(t)a^\dagger dt \qquad (6.172)$$

We can think of this as describing photons leaking from a single mode cavity at Poisson distributed times. Suppose there were exactly n photons in the cavity at time t so that $\rho(t) = |n\rangle\langle n|$. Then (6.172) would become

$$\rho(t + dt) = (1 - \gamma n dt)|n\rangle\langle n| + \gamma n dt|n-1\rangle\langle n-1| \qquad (6.173)$$

We can think of this as follows. In a small increment of time dt, two events are possible: either a single photon is lost or no photon is lost. If a photon is lost, the state of the field has one less photon so that it changes from $|n\rangle$ to $|n-1\rangle$ and this event will occur with probability $\gamma n dt$. This form results from the last term of (6.172). On the other hand, if no photon is lost the state is unchanged, and this will occur with probability $1 - \gamma n dt$, which arises from the first term in (6.172). Thus (6.172) describes a statistical mixture of the two events that can occur in a small time step dt: the first term in square brackets describes the change in the state of the cavity field given that no photon is lost in time interval dt, while the second term describes what happens to the state of the field if one photon is lost in a time interval dt.

If this interpretation is correct it suggests an answer to conditional questions such as: if no photon is lost from time t to $t + dt$, what is the conditional state of the field?

In this case we have no contribution from the last term in (6.172) so the conditional state is the solution to

$$\rho(t+dt)_c = \frac{[\rho(t)_c - \frac{\gamma}{2}(a^\dagger a\rho(t)_c + \rho(t)_c a^\dagger a)dt]}{\text{Tr}[\rho(t)_c - \frac{\gamma}{2}(a^\dagger a\rho(t)_c + \rho(t)_c a^\dagger a)dt]}$$

$$\approx \rho_c(t) - \gamma dt\left[\frac{1}{2}\left(a^\dagger a\rho_c(t) + \rho_c(t)a^\dagger a\right) - \langle a^\dagger a\rangle_c(t)\rho_c(t)\right]$$

to linear order in dt, where the subscript c is to remind us that we are dealing with a particular conditional state conditioned on a rather special history of null events and $\langle a^\dagger a\rangle_c(t)$ is the conditional average of the photon number in the state $\rho_c(t)$.

We can now introduce a *classical* stochastic process, a conditional Poisson process, $dN(t)$ which is the number of photons lost in time dt. Clearly

$$dN(t)^2 = dN(t) \tag{6.174}$$

$$\mathscr{E}[dN(t)] = \gamma\langle a^\dagger a\rangle_c(t) \tag{6.175}$$

where $\mathscr{E}$ is an average over the classical stochastic variable. In terms of $dN(t)$ we can now define a *stochastic master equation*

$$d\rho_c(t) = dN(t)\mathscr{G}[a]\rho_c(t) - \gamma dt\,\mathscr{H}[a^\dagger a]\rho_c(t) \tag{6.176}$$

where we have defined two new super-operators (that map density operators to density operators),

$$\mathscr{G}[A]\rho = \frac{A\rho A^\dagger}{\text{Tr}[A\rho c^\dagger]} - \rho \tag{6.177}$$

$$\mathscr{H}[A]\rho = A\rho + \rho A^\dagger - \text{Tr}[A\rho + \rho A^\dagger] \tag{6.178}$$

for any operator A. Note that if we take the classical ensemble average over the noise process $dN(t)$ we recover the original unconditional master equation in (6.172). The solution to (6.176) is the conditional state at time t conditioned on an entire fine-grained history of jump events (that is to say, the total number of jumps and the time of each jump event). Denote such a history as the sequence of jump times on the interval $[0,t)$ as $h[t] = \{t_1, t_2, \ldots, t_m\}$. The unconditional state is a sum over all such histories

$$\rho(t) = \sum_{h[t]} Pr(h[t])\rho_c(h[t]) \tag{6.179}$$

where we have explicitly indicated that the conditional state ρ_c is conditioned on the history of jump events, $h[t]$ in the time interval of interest and $Pr(h[t])$ is the probability for each history. We have *unravelled* the solution to the master equation in terms of conditional stochastic events. For a point process as considered here the sum over histories has an explicit form in terms of time ordered integrals [5]

$$\rho(t) = \sum_{m=0}^{\infty} \int_0^t dt_m \int_0^{t_m} \cdots \int_0^{t_1} \mathscr{S}(t-t_m)\,\mathscr{J}\,\mathscr{S}(t_m - t_{m-1})\cdots\mathscr{J}\,\mathscr{S}(t_1)\rho(0) \quad (6.180)$$

where the super operators are defined by

$$\mathscr{S}(t)\rho = e^{-\frac{\gamma}{2}ta^\dagger a}\rho e^{-\frac{\gamma}{2}ta^\dagger a} \quad (6.181)$$

$$\mathscr{J} = \gamma a \rho a^\dagger \quad (6.182)$$

Clearly the probability of a specific jump history is given by

$$Pr(h[t]) = \mathrm{Tr}[\mathscr{S}(t-t_m)\,\mathscr{J}\,\mathscr{S}(t_m - t_{m-1})\cdots\mathscr{J}\,\mathscr{S}(t_1)\rho(0)] \quad (6.183)$$

The form of (6.180) indicates that if we start in a pure state, and have access to the entire history of photon loss events $h[t]$, the conditional state $\rho_c(h[t])$ must still be a pure state. This implies that we can write a stochastic Schrödinger for the damped harmonic oscillator:

$$d|\psi_c(t)\rangle = \left[dN_c(t)\left(\frac{a}{\sqrt{\langle a^\dagger a\rangle_c(t)}} - 1 \right) + \gamma dt\left(\frac{\langle a^\dagger a\rangle_c(t)}{2} - \frac{a^\dagger a}{2} \right) - iH dt \right] |\psi_c(t)\rangle \quad (6.184)$$

where we have now included the possibility of a hamiltonian part to the dynamics. We can show the equivalence between this equation and the stochastic master equation by considering the Ito-like expansion

$$d(|\psi_c(t)\rangle\langle\psi_c(t)|) = (d|\psi_c(t)\rangle)\langle\psi_c(t)| + |\psi_c(t)\rangle(d\langle\psi_c(t)|) + (d(|\psi_c(t)\rangle))(d\langle\psi_c(t)|) \quad (6.185)$$

and retaining all terms to first order in dt, noting that $dN^2 = dN$.

A point process with a large rate parameter γ can be well approximated on a time scale long compared to γ^{-1} by a white noise process. This suggests that it must be possible to unravel the master equation in terms of white noise processes as well as the point process $dN(t)$. In Chap. 15 we will see that such master equations give the conditional dynamics conditioned on homodyne and heterodyne measurements on the field leaving the cavity. Here we simply quote the result and show that averaging over the classical noise returns us to the unconditioned master equation.

In the case of the real valued Weiner process $dW(t)$ we can write the *homodyne* stochastic master equation for a dampled simple as

$$d\rho_c(t) = -i[H, \rho_c(t)]dt + \mathscr{D}[a]\rho_c(t)dt + dW(t)\mathscr{H}[a]\rho_c(t) \quad (6.186)$$

where

$$\mathscr{D}[A]\rho = A\rho A^\dagger + \frac{1}{2}(A^\dagger A\rho + \rho A^\dagger A) \quad (6.187)$$

and $\mathscr{H}[a]$ is given in (6.178)

In terms of a complex valued white noise process, $dW(t) = dW_1(t) + idW_2(t)$ where $dW_i(t)$ are independent Winer processes, we can write the *heterodyne* stochastic master equation

$$d\rho_c(t) = -i[H,\rho_c(t)]dt + \mathscr{D}[a]\rho_c(t)dt + \frac{1}{\sqrt{2}}(dW_1(t)\mathscr{H}[a]\rho_c(t)$$

$$+dW_2(t)\mathscr{H}[-ia])\rho_c(t) \tag{6.188}$$

There is a connection between the quantum jump process and the two-time correlation function discussed in Sect. 6.5. We first define a new stochastic process, the rate or the current, as

$$i(t) = \frac{dN}{dt} \tag{6.189}$$

This is a rather singular stochastic process, comprised of a series of delta functions concentrated at the actual jump times. In physical terms this is intended to model the output of an ideal photon counting detector, with infinite response bandwidth, that detects every photon that is lost from the cavity. Define the classical current two-time correlation function

$$G(\tau,t) = \mathscr{E}(i(t+\tau)i(t)) \tag{6.190}$$

Given the nature of a Poisson jump process, $dN(t)$ can only take the values 0 or 1, so it is easy to see that we can write the two-time correlation function in terms of the conditional probability to get $dN(t+\tau) = 1$ given a jump at time t,

$$G(\tau,t)dt^2 = \Pr(dN(t+\tau) = 1|dN(t) = 1) \tag{6.191}$$

This conditional probability is given by

$$\Pr(dN(t+\tau) = 1|dN(t) = 1) = \gamma^2 \mathrm{Tr}[a^\dagger a e^{\mathscr{L}\tau} a\rho(t)a^\dagger]dt^2 \tag{6.192}$$

where $e^{\mathscr{L}t}$ is the formal solution to the unconditional master equation evolution written in terms of the abstract generator $\mathscr{L}$ as $\dot{\rho} = \mathscr{L}\rho$. The two time correlation function is then given by

$$G(\tau,t) = \gamma^2 \mathrm{Tr}[a^\dagger a e^{\mathscr{L}\tau} a\rho(t)a^\dagger] \tag{6.193}$$

Note that the so-called regression theorem follows directly from the definition of G,

$$\frac{dG(\tau,t)}{d\tau} = \mathscr{L}G(\tau,t) \tag{6.194}$$

We usually deal with driven damped harmonic oscillators for which the system settles into a steady state, emitting photons according to the conditional Poisson process derived from the steady state solution $\rho_\infty = \lim_{t\to\infty}\rho(t)$, so we define the *stationary* two-time correlation function for the current as

$$G(\tau) = \gamma^2 \mathrm{Tr}\left[a^\dagger a e^{\mathscr{L}\tau} a\rho_\infty a^\dagger\right] \tag{6.195}$$

6.7.1 Simulating Quantum Trajectories

A mixed state for system with a Hilbert space of dimension N requires that we specify N^2 complex matrix elements. On the other hand a pure state requires that we specify only N complex numbers. For this reason numerically solving the master equation is more computationally difficult than solving the Schödinger equation. We can use the unravelling of a master equation in terms of a stochastic Schödinger equation to make the numerical solution of master equations more tractable. In this numerical setting, the method of quantum trajectories was independently developed as the *Monte-Carlo wavefunction* method [6]. We will illustrate the method using the jump process.

Suppose the state at time t is $|\psi(t)\rangle$. Then in a time interval δt, sufficiently short compared to γ^{-1}, the system will evolve to the (unnormalised) state conditioned on no-jump having occurred,

$$|\tilde{\psi}(t+\delta t)\rangle = e^{-iH\delta t - \gamma a^\dagger a \delta t/2}|\psi(t)\rangle \tag{6.196}$$

To compute this we implement a routine to solve the Schrödinger equation with the effective non-hermitian Hamiltonian

$$K = H - i\frac{\gamma}{2}a^\dagger a \tag{6.197}$$

The norm of this state is the probability that no-jump has occurred in the time interval δt,

$$p_0 = \langle \tilde{\psi}(t+\delta t)|\tilde{\psi}(t+\delta t)\rangle \tag{6.198}$$
$$= 1 - p \tag{6.199}$$

where it is easy to see that

$$p = \gamma\delta t \langle \psi(t)|a^\dagger a|\psi(t)\rangle \tag{6.200}$$

which we understand to be the probability that a jump takes place in this time interval. We need to ensure that $p \ll 1$.

Let us now chose a random number r uniformly distributed on the unit interval. At the end of the time interval, we compare p and r. If $p < r$ (usually the case) we normalise the state

$$|\psi(t+\delta t)\rangle = \frac{|\tilde{\psi}(t+\delta t)\rangle}{\sqrt{p_0}} \tag{6.201}$$

and continue the non-hermitian evolution for a further time step. If however $p > r$, we implement a quantum jump via

$$|\tilde{\psi}(t+\delta t)\rangle \rightarrow |\psi(t+\delta t)\rangle = \frac{\sqrt{\gamma}a|\tilde{\psi}(t+\delta t)\rangle}{p/\delta t} \tag{6.202}$$

Based on our previous discussions we see that $p/\delta t = \gamma \langle \psi(t) | a^\dagger a | \psi(t) \rangle$, and the jump operation is as described by the first term in (6.184). As the simulation proceeds we accumulate the record of times at which particular jump events occur. That is to say, we have access to a sample fine grained history of the jump process, $h[t]$. However we are primarily interested in solving the master equation. We thus run K trials up to time t, starting from an identical initial state each time, and then form the mixed state

$$\bar{\rho}(t) = K^{-1} \sum_{k=1}^{k} |\psi_k(t)\rangle \qquad (6.203)$$

as a uniform average over the K trials. Strictly speaking the probability of each of the K trials is not uniform, however one can show that for K sufficiently large $\bar{\rho}(t) \approx \rho(t)$ with a error that scales as $K^{-1/2}$. In Mølmer et al. [6] more general cases are discussed including how to simulate non zero temperature master equations or master equations with multiple jump processes.

Exercises

6.1 The photon number distribution for a laser may be shown to obey the master equation

$$\frac{d}{dt}P(n) = \frac{An}{1+n/n_s}P(n-1) - \frac{A(n+1)}{1+(n+1)/n_s}P(n) - \gamma n P(n) + \gamma(n+1)P(n+1),$$

where A is related to the gain, n_s is the saturation photon number and γ is the cavity loss rate.

Use detailed balance to show that the steady state solution is

$$P_{ss}(n) = N \frac{\left(\frac{An_s}{\gamma}\right)^n}{(n+n_s)!}$$

where N is a normalisation constant.

6.2 The interaction picture master equation for a damped harmonic oscillator driven by a resonant linear force is

$$\frac{d\rho}{dt} = i\varepsilon[a+a^\dagger, \rho] + \frac{\gamma}{2}(2a\rho a^\dagger - a^\dagger a \rho - \rho a^\dagger a).$$

Show that the steady state solution is the coherent state $|2i\varepsilon/\gamma\rangle$.

6.3 A model for phase diffusion of a simple harmonic oscillator is provided by the master equation

$$\frac{d\rho}{dt} = -\Gamma[a^\dagger a, [a^\dagger a, \rho]].$$

Show that the Q function obeys the Fokker–Planck equation.

$$\frac{\partial Q}{\partial t} = \frac{\Gamma}{2}\left(\frac{\partial}{\partial\alpha}\alpha Q + \frac{\partial}{\partial\alpha^*}\alpha^* Q + 2\frac{\partial^2}{\partial\alpha\partial\alpha^*}|\alpha|^2 Q - \frac{\partial^2}{\partial\alpha^2}\alpha^2 Q - \frac{\partial^2}{\partial\alpha^{*2}}\alpha^{*2}Q\right).$$

Thus show that while the mean amplitude decays the energy remains constant. Using intensity and phase variables $\alpha = I^{1/2}e^{i\theta}$ show that the model is simply a diffusion process for the phase.

6.4 Show that in terms of the quadrature operators $X_1 = a + a^\dagger$, $X_2 = -i(a - a^\dagger)$, the master equation (6.37) may be written

$$\frac{d\rho}{dt} = i\frac{\gamma}{8}[X_2, \{X_1, \rho\}] - i\frac{\gamma}{8}[X_1, \{X_2, \rho\}]$$

$$- \frac{\gamma}{8}e^{2r}[X_1, [X_1, \rho]] - \frac{\gamma}{8}e^{-2r}[X_2, [X_2, \rho]]$$

where $\{,\}$ is an anticommutator and we have taken $N = \sinh^2 r$, $M = \sinh r \cosh r$ for an ideal squeezed bath. Show that the first and second terms describe damping in X_1 and X_2 respectively, while the third and fourth terms describe diffusion in X_2 and X_1, respectively.

6.5 Show that the homodyne conditional master equation for a driven simple harmonic oscillator, damped into a zero temperature heat bath, has the same pure steady state as the unconditional case (Exercise 6.2).

6.6 If a cavity mode starts in a state for which the P-representation is Gaussian, show that under the conditional dynamics of the homodyne master in (6.15), the state continues to have a Gaussian P representation

References

1. F. Haake: (private communication)
2. W.H. Louisell: *Quantum Statistical Properties of Radiation* (Wiley, New York 1973)
3. M.C. Wang, G.E. Uhlenbeck: Rev. Mod. Phys. **17**, 323 (1945)
4. M. Lax: In *Brandeis University Summer Institute Lectures*, ed. by M. Chretien, E.P. Gross, S. Deser (Gordon and Breach, New York 1966) Vol. 2
5. M.D. Srinivas and E.B. Davies: J. Mod. Opt. **28**, 981 (1981)
6. K. Mølmer, Y. Castin and J. Dalibard, J. Opt. Soc. Am. B, **10**, 524 (1993)

Further Reading

Agarwal, G.S.: *Quantum Statistical Theories of Spontaneous Emission and Their Relation to Other Approaches*, Springer Tracts Mod. Phy., Vol. 70 (Springer, 1974)

Carmichael, H.J.: *Statistical Methods in Quantum Optics 1, Master Equations and Fokker Planck Equations* (Springer, 1999)

Gardiner, C.W.: *Handbook of Stochastic Processes* (Springer, 1985)
Gardiner, C.W. and Zoller, P.: *Quantum Noise: A Handbook of Markovian and Non-Markovian Quantum Stochastic Methods with Applications to Quantum Optics*, Springer Series in Synergetics, (Springer, 2004)
Risken, H.: *The Fokker Planck Equation* (Springer 1984)

Chapter 7
Input–Output Formulation of Optical Cavities

Abstract In preceding chapters we have used a master-equation treatment to calculate the photon statistics inside an optical cavity when the internal field is damped. This approach is based on treating the field external to the cavity, to which the system is coupled, as a heat bath. The heat bath is simply a passive system with which the system gradually comes into equilibrium. In this chapter we will explicitly treat the heat bath as the external cavity field, our object being to determine the effect of the intracavity dynamics on the quantum statistics of the output field. Within this perspective we will also treat the field input to the cavity explicitly. This approach is necessary in the case of squeezed state generation due to interference effects at the interface between the intracavity field and the output field.

An input–output formulation is also required if the input field state is specified as other than simply a vacuum or thermal state. In particular, we will want to discuss the case of an input squeezed state.

7.1 Cavity Modes

We will consider a single cavity mode interacting with an external multi-mode field. To being with we will assume the cavity has only one partially transmitting mirror that couples the intracavity mode to the external field. The geometry of the cavity and the nature of the dielectric interface at the mirror determines which output modes couple to the intracavity mode. It is usually the case that the emission is strongly direction. We will assume that the only modes that are excited have the same plane polarisation and are all propagating in the same direction, which we take to be the positive x-direction. The positive frequency components of the quantum electric field for these modes are then

$$E^{(+)}(x,t) = i \sum_{n=0}^{\infty} \left(\frac{\hbar \omega_n}{2\varepsilon_0 V} \right)^{1/2} b_n e^{-i\omega_n(t - x/c)} \tag{7.1}$$

In ignoring all the other modes, we are implicitly assuming that they remain in the vacuum state.

127

Let us further assume that all excited modes of this form have frequencies centered on the cavity resonance frequency and we call this the *carrier frequency* of $\Omega \gg 1$. Then we can approximate the positive frequency components by

$$E^{(+)}(x,t) = i\left(\frac{\hbar\Omega_n}{2\varepsilon_0 Ac}\right)^{1/2}\sqrt{\frac{c}{L}}\sum_{n=0}^{\infty} b_n e^{-i\omega_n(t-x/c)} \tag{7.2}$$

where A is a characteristic transverse area. This operator has dimensions of electric field. In order to simplify the dimensions we now define a field operator that has dimensions of $s^{-1/2}$. Taking the continuum limit we thus define the positive frequency operator for modes propagating in the *positive x–direction*,

$$b(x,t) = e^{-i\Omega(t-x/c)}\frac{1}{\sqrt{2\pi}}\int_{-\infty}^{\infty} d\omega\, b(\omega)e^{-i\omega(t-x/c)} \tag{7.3}$$

where we have made a change of variable $\omega \mapsto \Omega + \omega'$ and used the fact that $\Omega \gg 1$ to set the lower limit of integration to minus infinity, and

$$[b(\omega_1), b^\dagger(\omega_2)] = \delta(\omega_1 - \omega_2) \tag{7.4}$$

In this form the moment $n(x,t) = \langle b^\dagger(x,t)b(x,t)\rangle$ has units of s^{-1}. This moment determines the probability per unit time (the count rate) to count a photon at space-time point (x,t).

Consider now the single side cavity geometry depicted in Fig. 7.1. The field operators at some external position, $b(t) = b(x > 0,t)e^{i\Omega t}$ and $b^\dagger(t) = b^\dagger(x > 0,t)e^{-i\Omega t}$ can be taken to describe the field, in the interaction picture with frequency Ω. As the cavity is confined to some region of space, we need to determine how the field outside the cavity responds to the presence of the cavity and any matter it may contain. The interaction Hamiltonian between the cavity field, represented by the harmonic oscillator annihilation and creation operators a, $a^\dagger$, and the external field in the rotating wave approximation is given by (6.14). Restricting the sum to only the modes of interest and taking the continuum limit, we can write this as

$$V(t) = i\hbar\int_{-\infty}^{\infty} d\omega\, g(\omega)[b(\omega)a^\dagger - ab^\dagger(\omega)] \tag{7.5}$$

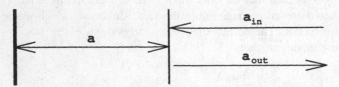

Fig. 7.1 A schematic representation of the cavity field and the input and output fields for a single-sided cavity

with $[a, a^\dagger] = 1$ and $g(\omega)$ is the coupling strength as a function of frequency which is typically peaked around $\omega = 0$ (which corresponds to $\omega = \Omega$ in the original non-rotating frame). In fact $g(\omega)$ is the Fourier transform of a spatially varying coupling constant that describes the local nature of the cavity/field interaction (see [1]). If the cavity contains matter, the field inside the cavity may acquire some non trivial dynamics which then forces the external fields to have a time dependence different from the free field dynamics. This leads to an explicit time dependence in the frequency space operators, $b(t, \omega)$, in the Heisenberg picture.

We now follow the approach of *Collett and Gardiner* [1]. The Heisenberg equation of motion for $b(t, \omega)$, in the interaction picture, is

$$\dot{b}(t, \omega) = -i\omega b(\omega) + g(\omega)a \tag{7.6}$$

The solution to this equation can be written in two ways depending on weather we choose to solve in terms of the initial conditions at time $t_0 < t$ (the *input*) or in terms of the final conditions at times $t_1 > t$, (the *output*). The two solutions are respectively

$$b(t, \omega) = e^{-i\omega(t-t_0)}b_0(\omega) + g(\omega) \int_0^t e^{-i\omega(t-t')}a(t')dt' \tag{7.7}$$

where $t_0 < t$ and $b_0(\omega) = b(t = t_0, \omega)$, and

$$b(t, \omega) = e^{-i\omega(t-t_1)}b_1(\omega) - g(\omega) \int_t^{t_1} e^{-i\omega(t-t')}a(t')dt' \tag{7.8}$$

where $t < t_1$ and $b_1(\omega) = b(t = t_1, \omega)$. In physical terms $b_0(\omega)$ and $b_1(\omega)$ are usually specified at $-\infty$ and $+\infty$ respectively, that is, for times such that the field is simply a free field, however here we only require $t_0 < t < t_1$.

The cavity field operator obeys the equation

$$\dot{a} = -\frac{i}{\hbar}[\mathcal{H}_S, a] - \int_{-\infty}^{\infty} d\omega \, g(\omega)b(t, \omega) \tag{7.9}$$

where $\mathcal{H}_S$ is the Hamiltonian for the cavity field alone. In terms of the solution with initial conditions, (7.7), this equation becomes

$$\dot{a} = -\frac{i}{\hbar}[\mathcal{H}_S, a] - \int_{-\infty}^{\infty} d\omega \, g(\omega)e^{-i\omega(t-t_0)}b_0(\omega)$$

$$- \int_{-\infty}^{\infty} d\omega \, g(\omega)^2 \int_{t_0}^{t} e^{-i\omega(t-t')}a(t') \tag{7.10}$$

We now assume that $g(\omega)$ is independent of frequency over a wide range of frequencies around $\omega = 0$ (that is around $\omega = \Omega$ in non rotating frame). This is the first approximation we need to get a Markov quantum stochastic process. Thus we set

$$g(\omega)^2 = \gamma/2\pi \qquad (7.11)$$

We also define an *input field* operator by

$$a_{\text{IN}}(t) = -\frac{1}{2\pi} \int\limits_{-\infty}^{\infty} d\omega e^{-i\omega(t-t_0)} b_0(\omega) \qquad (7.12)$$

(the minus sign is a phase convention: left-going fields are negative, right-going fields are positive). Using the relation

$$\int\limits_{-\infty}^{\infty} d\omega e^{-i\omega(t-t')} = 2\pi\delta(t-t') \qquad (7.13)$$

the input field may be shown to satisfy the commutation relations

$$[a_{\text{IN}}(t), a_{\text{IN}}^{\dagger}(t')] = \delta(t-t') \qquad (7.14)$$

When (7.13) is achieved as the limit of an integral of a function which goes smoothly to zero at $\pm\infty$ (for example, a Gaussian), the following result also holds

$$\int\limits_{t_0}^{t} f(t')\delta(t-t')dt' = \int\limits_{t}^{t_1} f(t')\delta(t-t')dt' = \frac{1}{2}f(t), \quad (t_0 < t < t_1) \qquad (7.15)$$

Interchanging the order of time and frequency integration in the last term in (7.10) and using (7.15) gives

$$\dot{a}(t) = -\frac{i}{\hbar}[a(t), \mathcal{H}_{\text{SYS}}] - \frac{\gamma}{2}a(t) + \sqrt{\gamma}a_{\text{IN}}(t) \qquad (7.16)$$

Equation (7.16) is a *quantum stochastic differential equation* (qsde) for the intra-cavity field, $a(t)$. The quantum noise term appears explicitly as the input field to the cavity.

In a similar manner we may substitute the solution in terms of final conditions, (7.8) into (7.10) to obtain the time-reversed qsde as

$$\dot{a}(t) = -\frac{i}{\hbar}[a(t), \mathcal{H}_{\text{SYS}}] + \frac{\gamma}{2}a(t) - \sqrt{\gamma}a_{\text{IN}}(t) \qquad (7.17)$$

where we define the output field operator as

$$a_{\text{OUT}}(t) = \frac{1}{\sqrt{2\pi}} \int\limits_{-\infty}^{\infty} \mathrm{d}\omega e^{-i\omega(t-t_1)} b_1(\omega) \tag{7.18}$$

(Note that the phase convention between left going and right going external fields required for the boundary condition has been explicitly incorporated in the definitions of a_{IN}, a_{OUT}). The input and output fields are then seen to be related by

$$a_{\text{IN}}(t) + a_{\text{OUT}}(t) = \sqrt{\gamma} a(t) \tag{7.19}$$

This represents a boundary condition relating each of the far field amplitudes outside the cavity to the internal cavity field. Interference terms between the input and the cavity field may contribute to the observed moments when measurements are made on a_{OUT}.

7.2 Linear Systems

For many systems of interest the Heisenberg equations of motion are linear and may be written in the form

$$\frac{\mathrm{d}}{\mathrm{d}t} \boldsymbol{a}(t) = A\boldsymbol{a}(t) - \frac{\gamma}{2}\boldsymbol{a}(t) + \sqrt{\gamma}\boldsymbol{a}_{\text{IN}}(t) \,, \tag{7.20}$$

where

$$\boldsymbol{a}(t) = \begin{pmatrix} a(t) \\ a^\dagger(t) \end{pmatrix} \,, \tag{7.21}$$

$$\boldsymbol{a}_{\text{IN}}(t) = \begin{pmatrix} a_{\text{IN}}(t) \\ a_{\text{IN}}^\dagger(t) \end{pmatrix} \,, \tag{7.22}$$

Define the Fourier components of the intracavity field by

$$a(t) = \frac{1}{\sqrt{2\pi}} \int\limits_{-\infty}^{\infty} e^{-i\omega(t-t_0)} a(\omega) \mathrm{d}\omega \tag{7.23}$$

and a frequency component vector

$$\boldsymbol{a}(\omega) = \begin{pmatrix} a(\omega) \\ a^\dagger(\omega) \end{pmatrix} \tag{7.24}$$

where $a^\dagger(\omega)$ is the Fourier transform of $a^\dagger(t)$.

The equations of motion become

$$\left[A + \left(i\omega - \frac{\gamma}{2}\right) I \right] \boldsymbol{a}(\omega) = -\sqrt{\gamma} \boldsymbol{a}_{\text{IN}}(\omega) \,. \tag{7.25}$$

However, we may use (7.18) to eliminate the internal modes to obtain

$$a_{\mathrm{OUT}}(\omega) = -\left[A + \left(i\omega + \frac{\gamma}{2}\right)I\right]\left[A + \left(i\omega - \frac{\gamma}{2}\right)I\right]^{-1}a_{\mathrm{IN}}(\omega) . \qquad (7.26)$$

To illustrate the use of this result we shall apply it to the case of an empty one-sided cavity. In this case the only source of loss in the cavity is through the mirror which couples the input and output fields. The system Hamiltonian is

$$\mathscr{H}_{\mathrm{SYS}} = \hbar\omega_0 a^\dagger a .$$

Thus

$$A = \begin{pmatrix} -i\omega_0 & 0 \\ 0 & i\omega_0 \end{pmatrix} . \qquad (7.27)$$

Equation (7.26) then gives

$$a_{\mathrm{OUT}}(\omega) = \frac{\frac{\gamma}{2} + i(\omega - \omega_0)}{\frac{\gamma}{2} - i(\omega - \omega_0)}a_{\mathrm{IN}}(\omega) . \qquad (7.28)$$

Thus there is a frequency dependent phase shift between the output and input. The relationship between the input and the internal field is

$$a(\omega) = \frac{\sqrt{\gamma}}{\frac{\gamma}{2} - i(\omega - \omega_0)}a_{\mathrm{IN}}(\omega) , \qquad (7.29)$$

which leads to a Lorentzian of width $\gamma/2$ for the intensity transmission function.

7.3 Two-Sided Cavity

A two-sided cavity has two partially transparent mirrors with associated loss coefficients γ_1 and γ_2, as shown in Fig. 7.2. In this case there are two input ports and two output ports. The equation of motion for the internal field is then given by an obvious generalisation as

$$\frac{da(t)}{dt} = -i\omega_0 a(t) - \frac{1}{2}(\gamma_1 + \gamma_2)a(t) + \sqrt{\gamma_1}a_{\mathrm{IN}}(t) + \sqrt{\gamma_2}b_{\mathrm{IN}}(t) . \qquad (7.30)$$

Fig. 7.2 A schematic representation of the cavity field and the input and output fields for a double-sided cavity

The relationship between the internal and input field frequency components for an empty cavity is then

$$a(\omega) = \frac{\sqrt{\gamma_1}a_{\text{IN}}(\omega) + \sqrt{\gamma_2}b_{\text{IN}}(\omega)}{\left(\frac{\gamma_1+\gamma_2}{2}\right) - i(\omega - \omega_0)}. \tag{7.31}$$

The relationship between the input and output modes may be found using the boundary conditions at each mirror, see (7.19),

$$a_{\text{OUT}}(t) + a_{\text{IN}}(t) = \sqrt{\gamma_1}a(t), \tag{7.32a}$$

$$b_{\text{OUT}}(t) + b_{\text{IN}}(t) = \sqrt{\gamma_2}a(t). \tag{7.32b}$$

We find

$$a_{\text{OUT}}(\omega) = \frac{\left[\frac{\gamma_1-\gamma_2}{2} + i(\omega - \omega_0)\right]a_{\text{IN}}(\omega) + \sqrt{\gamma_1\gamma_2}b_{\text{IN}}(\omega)}{\frac{\gamma_1+\gamma_2}{2} - i(\omega - \omega_0)} \tag{7.33}$$

For equally reflecting mirrors $\gamma_1 = \gamma_2 = \gamma$ this expression simplifies to

$$a_{\text{OUT}}(\omega) = \frac{i(\omega - \omega_0)a_{\text{IN}}(\omega) + \gamma b_{\text{IN}}(\omega)}{\gamma - i(\omega - \omega_0)}. \tag{7.34}$$

Near to resonance this is approximately a through pass Lorentzian filter

$$a_{\text{OUT}}(\omega) \approx \frac{\gamma b_{\text{IN}}(\omega)}{\gamma - i(\omega - \omega_0)}, \tag{7.35}$$

This is only an approximate result, the neglected terms are needed to preserve the commutation relations. Away from resonance there is an increasing amount of backscatter. In the limit $|\omega - \omega_0| \gg \gamma$ the field is completely reflected

$$a_{\text{OUT}}(\omega) = -a_{\text{IN}}(\omega). \tag{7.36}$$

Before going on to consider interactions within the cavity we shall derive some general relations connecting the two time correlation functions inside and outside the cavity.

7.4 Two Time Correlation Functions

Integrating (7.7) over ω, and using (7.13) gives

$$a_{\text{IN}}(t) = \frac{\sqrt{\gamma}}{2}a(t) - \frac{1}{\sqrt{2\pi}}\int_{-\infty}^{\infty} d\omega b(\omega,t). \tag{7.37}$$

Let $c(t)$ be any system operator. Then

$$[c(t), \sqrt{\gamma} a_{\mathrm{IN}}(t)] = \frac{\gamma}{2}[c(t), a(t)] \ . \tag{7.38}$$

Now since $c(t)$ can only be a function of $a_{\mathrm{IN}}(t')$ for earlier times $t' < t$ and the input field operators must commute at different times we have

$$[c(t), \sqrt{\gamma} a_{\mathrm{IN}}(t')] = 0, \quad t' > t \ . \tag{7.39}$$

Similarly

$$[c(t), \sqrt{\gamma} a_{\mathrm{OUT}}(t')] = 0, \quad t' < t \ . \tag{7.40}$$

From (7.40 and 7.18) we may show that

$$[c(t), \sqrt{\gamma} a_{\mathrm{IN}}(t')] = \gamma[c(t), a(t)], \quad t' < t \ . \tag{7.41}$$

Combining (7.38–7.41) we then have

$$[c(t), \sqrt{\gamma} a_{\mathrm{IN}}(t')] = \gamma \theta(t - t')[c(t), a(t')] \ , \tag{7.42}$$

where $\theta(t)$ is the step function

$$\theta(t) = \begin{cases} 1 & t > 0, \\ \frac{1}{2} & t = 0, \\ 0 & t < 0. \end{cases} \tag{7.43}$$

The commutator for the output field may now be calculated to be

$$[a_{\mathrm{OUT}}(t), a_{\mathrm{OUT}}^{\dagger}(t')] = [a_{\mathrm{IN}}(t), a_{\mathrm{IN}}^{\dagger}(t')] \tag{7.44}$$

as required.

For the case of a coherent or vacuum input it is now possible to express variances of the output field entirely in terms of those of the internal system. For an input field of this type all moments of the form $\langle a_{\mathrm{IN}}^{\dagger}(t) a_{\mathrm{IN}}(t') \rangle$, $\langle a(t) a_{\mathrm{IN}}(t') \rangle$, $\langle a^{\dagger}(t) a_{\mathrm{IN}}(t') \rangle$, $\langle a_{\mathrm{IN}}^{\dagger}(t) a(t') \rangle$, and $\langle a_{\mathrm{IN}}^{\dagger}(t) a^{\dagger}(t') \rangle$ will factorise. Using (7.18) we find

$$\langle a_{\mathrm{OUT}}^{\dagger}(t), a_{\mathrm{OUT}}(t') \rangle = \gamma \langle a^{\dagger}(t), a(t') \rangle \ , \tag{7.45}$$

where

$$\langle U, V \rangle \equiv \langle U V \rangle - \langle U \rangle \langle V \rangle \ . \tag{7.46}$$

In this case there is a direct relationship between the two time correlation of the output field and the internal field. Consider now the phase dependent two time correlation function

$$\langle a_{\text{OUT}}(t), a_{\text{OUT}}(t')\rangle = \langle a_{\text{IN}}(t) - \sqrt{\gamma}a(t), a_{\text{IN}}(t') - \sqrt{\gamma}a(t')\rangle$$

$$= \gamma\langle a(t), a(t')\rangle - \sqrt{\gamma}\langle[a_{\text{IN}}(t'), a(t)]\rangle$$

$$= \gamma\langle a(t), a(t')\rangle + \gamma\theta(t' - t)\langle[a(t'), a(t)]\rangle$$

$$= \gamma\langle a(\max(t,t')), a(\min(t,t'))\rangle . \qquad (7.47)$$

In this case the two time correlation functions of the output field are related to the time ordered two time correlation functions of the cavity field.

These results mean that the usual spectrum of the output field, as given by the Fourier transform of (7.45), will be identical to the spectrum of the cavity field. The photon statistics of the output field will also be the same as the intracavity field. Where a difference will arise, is in phase-sensitive spectrum such as in squeezing experiments.

7.5 Spectrum of Squeezing

The output field from the cavity is a multi mode field. Phase-dependent properties of this field are measured by mixing the field, on a beam splitter, with a known coherent field – the local oscillator, as discussed in Sect. 3.8. The resulting field may then be directed to a photodetector and the measured photocurrent directed to various devices such as a noise-power spectrum analyser to produce a spectrum, $S(\omega)$. If we write the signal field as $a_{\text{out}}(t)$ and the local oscillator is $a_{\text{LO}}(t)$, the average photo current is proportional to

$$\overline{i(t)} = (1 - \eta)\langle a_{\text{LO}}^{\dagger}(t)a_{\text{LO}}(t)\rangle + \sqrt{\eta(1-\eta)}\langle a_{\text{OUT}}(t)a_{\text{LO}}^{\dagger}(t) + a_{\text{OUT}}^{\dagger}(t)a_{\text{LO}}(t)\rangle$$
$$+ \eta\langle a_{\text{OUT}}^{\dagger}(t)a_{\text{OUT}}(t)\rangle \qquad (7.48)$$

If $(1 - \eta)\langle a_{\text{LO}}^{\dagger}(t)a_{\text{LO}}(t)\rangle >> \eta\langle a_{\text{OUT}}^{\dagger}(t)a_{\text{OUT}}(t)\rangle$, we can neglect the last term in (7.48). If the local oscillator is in a coherent state $\langle a_{\text{LO}}(t)\rangle = |\beta|e^{i\theta} e^{-i\Omega t}$, then not only the average current, but all its moments are determined by the quantum statistics of the quadrature phase operator

$$X_{\theta}^{\text{OUT}} = a_{\text{OUT}}e^{-i(\theta - \Omega t)} + a_{\text{OUT}}^{\dagger}e^{-i(\theta - \Omega t)} \qquad (7.49)$$

In particular, the noise power spectrum of the photocurrent is given by

$$S(\omega, \theta) = \int_{-\infty}^{\infty} dt\, e^{-i\omega t} \langle : X_{\theta}^{\text{OUT}}(t), X_{\theta}^{\text{OUT}}(0) :\rangle \qquad (7.50)$$

where : indicates normal ordering. The combination $a_{\text{OUT}}e^{i\Omega t}$ is simply the definition of the output field in the interaction picture defined at frequency Ω. Using (7.47) and (7.49) this may be written in terms of the intracavity field as

$$S(\omega,\theta) = \gamma \int_{-\infty}^{\infty} dt\ e^{-i\omega t} T \langle : X_\theta(t), X_\theta(0) : \rangle \tag{7.51}$$

where T denotes time-ordering and $X_\theta(t)$ intracavity quadrature phase operator in an interaction picture at frequency, Ω, defined by the local oscillator frequency,

$$X_\theta(t) = a(t)e^{-i\theta} + a^\dagger(t)e^{i\theta} \tag{7.52}$$

Conventionally we define the in-phase and quadrature-phase operators as $X_1 = X_{\theta=0}$, $X_2 = X_{\theta=\pi/2}$.

7.6 Parametric Oscillator

We shall now proceed to calculate the squeezing spectrum from the output of a parametric oscillator. Below threshold the equations for the parametric oscillator are linear and hence we can directly apply the linear operator techniques. When the equations are nonlinear such as for the parametric oscillator above threshold, then linearization procedures must be used. One procedure using the Fokker–Planck equation is described in Chap. 8.

Below threshold the pump mode of the parametric oscillator may be treated classically. It can then be described by the Hamiltonian

$$\mathscr{H} = \hbar\omega a^\dagger a + \frac{i\hbar}{2}(\varepsilon a^{\dagger 2} - \varepsilon^* a^2) + a\Gamma^\dagger + a^\dagger \Gamma, \tag{7.53}$$

where $\varepsilon = \varepsilon_p \chi$ and ε_p is the amplitude of the pump, and χ is proportional to the non-linear susceptibility of the medium. Γ is the reservoir operator representing cavity losses. We consider here the case of a single ended cavity with loss rate γ_1.

The Heisenberg equations of motion for $a(t)$ are linear and given by (7.20) where

$$A = \begin{pmatrix} \frac{\gamma_1}{2} & -\varepsilon \\ -\varepsilon^* & \frac{\gamma_1}{2} \end{pmatrix}. \tag{7.54}$$

We can obtain an expression for the Fourier components of the output field from (7.26)

$$a_{\text{OUT}}(\omega) = \frac{1}{\left[\left(\frac{\gamma_1}{2} - i\omega\right)^2 - |\varepsilon|^2\right]} \left\{ \left[\left(\frac{\gamma_1}{2}\right)^2 + \omega^2 + |\varepsilon|^2 \right] \right.$$
$$\left. \times a_{\text{IN}}(\omega) + \varepsilon\gamma_1 a_{\text{IN}}^\dagger(-\omega) \right\}. \tag{7.55}$$

Defining the quadrature phase operators by

$$2a_{\text{OUT}} = e^{i\theta/2}(X_1^{\text{OUT}} + iX_2^{\text{OUT}}), \tag{7.56}$$

where θ is the phase of the pump, we find the following correlations:

$$\langle : X_1^{OUT}(\omega), X_1^{OUT}(\omega') : \rangle = \frac{2\gamma_1|\varepsilon|}{\left(\frac{\gamma_1}{2} - |\varepsilon|\right)^2 + \omega^2} \delta(\omega + \omega'), \qquad (7.57)$$

$$\langle : X_2^{OUT}(\omega), X_2^{OUT}(\omega') : \rangle = \frac{-2\gamma_1|\varepsilon|}{\left(\frac{\gamma_1}{2} + |\varepsilon|\right)^2 + \omega^2} \delta(\omega + \omega'), \qquad (7.58)$$

where the input field a_{IN} has been taken to be in the vacuum.

The δ function in (7.57 and 7.58) may be removed by integrating over ω' to give the normally ordered spectrum: $S^{OUT}(\omega)$:. The final result for the squeezing spectra of the quadrature is

$$S_1^{OUT}(\omega) = 1 + : S_1^{OUT}(\omega) := 1 + \frac{2\gamma_1|\varepsilon|}{\left(\frac{\gamma_1}{2} - |\varepsilon|\right)^2 + \omega^2}, \qquad (7.59)$$

$$S_2^{OUT}(\omega) = 1 + : S_2^{OUT}(\omega) := 1 - \frac{2\gamma_1|\varepsilon|}{\left(\frac{\gamma_1}{2} + |\varepsilon|\right)^2 + \omega^2}, \qquad (7.60)$$

These spectra are defined in a frame of frequency Ω so that $\omega = 0$ is on cavity resonance.

The maximum squeezing occurs at the threshold for parametric oscillation $|\varepsilon| = \gamma_1/2$ where

$$S_1^{OUT}(\omega) = 1 + \left(\frac{\gamma_1}{\omega}\right)^2, \qquad (7.61)$$

$$S_2^{OUT}(\omega) = 1 - \frac{\gamma_1^2}{\gamma_1^2 + \omega^2}, \qquad (7.62)$$

Thus the squeezing occurs in the X_2 quadrature which is $\pi/2$ out of phase with the pump. The light generated in parametric oscillation is therefore said to be phase squeezed.

In Fig. 7.3 we plot $S_2^{OUT}(\omega)$ at threshold. We see that at $\omega = 0$, that is the cavity resonance, the fluctuations in the X_2 quadrature tend to zero. The fluctuations in the X_1 quadrature on the other hand diverge at $\omega = 0$. This is characteristic of critical fluctuations which diverge at a critical point. In this case however the critical

Fig. 7.3 A plot of the spectrum of the squeezed quadrature for a cavity containing a parametric amplifier with a classical pump. *Solid*: single-sided cavity with $\gamma_1 = \gamma_2$, *dashed*: double-sided cavity

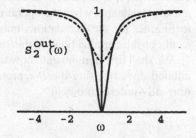

flucuations are phase dependent. As the fluctuations in one phase are reduced to zero the fluctuations in the other phase necessarily diverge. This characteristic of good squeezing near critical points is found in other phase dependent nonlinear optical systems [2]. This behaviour is in contrast to the threshold for laser oscillation where the critical fluctuations are random in phase.

7.7 Squeezing in the Total Field

The squeezing in the total field may be found by integrating (7.62) over ω. At threshold we find

$$S_2^{\text{TOT}} = \int \left(1 - \frac{\gamma_1^2}{\gamma_1^2 + \omega^2}\right) d\omega = \frac{\gamma_1}{2} \,. \tag{7.63}$$

The squeezing in the total field is given by the equal time correlation functions

$$\langle a, a \rangle_{\text{OUT}} = \gamma_1 \langle a, a \rangle,$$
$$\langle a, a^\dagger \rangle_{\text{OUT}} = \gamma_1 \langle a, a^\dagger \rangle \,. \tag{7.64}$$

Hence, the squeezing in the internal field is

$$V(X_2) = \frac{1}{2} \,. \tag{7.65}$$

Thus the internal field mode is 50% squeezed, in agreement with the calculations of *Milburn* and *Walls* [3]. This can be surpassed in the individual frequency components of the output field which have 100% squeezing for $\omega = 0$. It is the squeezing in the individual frequency components of the output field which may be measured by a spectrum analyser following a homodyne detection scheme.

7.8 Fokker–Planck Equation

We shall now give an alternative method for evaluating the squeezing spectrum. This converts the operator master equation to a c-number Fokker–Planck equation. This is a useful technique when the operator equations are nonlinear. Standard linearization techniques for the fluctuations may be made in the Fokker–Planck equation. We shall consider applications of this technique to nonlinear systems in Chap. 8.

We shall first demonstrate how time and normally-ordered moments may be calculated directly using the P representation. We consider the following time- and normally-ordered moment

$$T \langle : X_1(t)X_1(0) : \rangle = e^{-2i\theta} \langle a(t)a(0) \rangle + e^{2i\theta} \langle a^\dagger(0)a^\dagger(t) \rangle$$
$$+ \langle a^\dagger(t)a(0) \rangle + \langle a^\dagger(0)a(t) \rangle . \tag{7.66}$$

The two-time correlation functions may be evaluated using the P representation which determines normally-ordered moments. Thus equal time moments of the c-number variables give the required normally-ordered operator moments. The two time moments imply precisely the time ordering of the internal operators that are required to compute the output moments. This can be seen by noting that the evolution of the system will in general mix $a^\dagger$ and a. Hence $a(t+\tau)$ contains both $a(t)$ and $a^\dagger(t)$, $\tau > 0$. In a normally-ordered two time product $a(t+\tau)$ must therefore stand to the left of $a(t)$, similarly $a^\dagger(t+\tau)$ must stand to the right of $a^\dagger(t)$. Thus

$$\langle \alpha(t+\tau)\alpha(t) \rangle_p = \langle a(t+\tau)a(t) \rangle , \tag{7.67}$$

$$\langle \alpha^*(t+\tau)\alpha^*(t) \rangle_p = \langle a^\dagger(t)a^\dagger(t+\tau) \rangle , \tag{7.68}$$

where the left-hand side of these equations represent averages of c-number variables over the P representation. The normally-ordered output correlation matrix defined by

$$: C^{\text{OUT}}(\tau) := \begin{pmatrix} \langle a_{\text{OUT}}(t+\tau), a_{\text{OUT}}(t) \rangle & \langle a_{\text{OUT}}^\dagger(t), a_{\text{OUT}}(t+\tau) \rangle \\ \langle a_{\text{OUT}}^\dagger(t+\tau), a_{\text{OUT}}(t) \rangle & \langle a_{\text{OUT}}^\dagger(t+\tau), a_{\text{OUT}}^\dagger(t) \rangle \end{pmatrix} \tag{7.69}$$

is given by

$$: C^{\text{OUT}}(\tau) := \gamma \begin{pmatrix} \langle \alpha(t+\tau), \alpha(t) \rangle & \langle \alpha(t+\tau), \alpha^*(t) \rangle \\ \langle \alpha^*(t+\tau), \alpha(t) \rangle & \langle \alpha^*(t+\tau), \alpha^*(t) \rangle \end{pmatrix}$$
$$\equiv \gamma C_p(\tau) . \tag{7.70}$$

The two time correlation functions for the output field may be calculated directly from the correlation functions of the stochastic variables describing the internal field using the P representation.

For nonlinear optical processes the Fokker–Planck equation for the P function may have nonlinear drift terms and nonconstant diffusion. In such circumstances we first linearise the equation about the deterministic steady states, to obtain a linear Fokker–Planck equation of the form

$$\frac{\partial P}{\partial t}(\alpha) = \left(\frac{\partial}{\partial \alpha_i} A_i \alpha_i + \frac{1}{2} \frac{\partial^2}{\partial \alpha_i \partial \alpha_j} D_{ij} \right) P(\alpha) , \tag{7.71}$$

where A is the drift matrix, and D is the diffusion matrix. The linearised description is expected to give the correct descriptions away from instabilities in the deterministic equations of motion. For fields exhibiting quantum behaviour, such as squeezing, D is non-positive definite and a Fokker–Planck equation is not defined for the Glauber–Sudarshan P function. In these cases a Fokker–Planck equation is defined for the positive P representation, where α^* is replaced by $\alpha^\dagger$ an independent complex variable as described in Chap. 6.

The spectral matrix $S(\omega)$ is defined as the Fourier transform of $C_p(\tau)$. In a linearised analysis it is given by

$$S(\omega) = \gamma(A + i\omega I)^{-1} D (A^T - i\omega I)^{-1} . \tag{7.72}$$

The squeezing spectrum for each quadrature phase is then given by

$$: S_1^{OUT}(\omega) := \gamma[e^{-2i\theta} S_{11}(\omega) + e^{2i\theta} S_{22}(\omega) + S_{12}(\omega) + S_{21}(\omega)] \tag{7.73}$$

$$: S_2^{OUT}(\omega) := \gamma[-e^{-2i\theta} S_{11}(\omega) - e^{2i\theta} S_{22}(\omega) + S_{12}(\omega) + S_{21}(\omega)] \tag{7.74}$$

These spectra are defined in a frame of frequency Ω (the cavity-resonance frequency) so that $\omega = 0$ corresponds to the cavity resonance.

It should be noted that in the above derivation there is only one input field and one output field, that is, there is only one source of cavity loss. Thus the above results only apply to a single-ended cavity; one in which losses accrue only at one mirror.

If there are other significant losses from the cavity the γ appearing in (7.60 and 7.61) is not the total loss but only the loss from the mirror through which the output field of interest is transmitted.

The above procedure enables one to calculate the squeezing in the output field from an optical cavity, provided the internal field may be described by the linear Fokker–Planck equation (7.71).

Alternatively the squeezing spectrum for the parametric oscillator may be calculated using the Fokker–Planck equation. The Fokker–Planck equation for the distribution $P(\alpha)$ for the system described by the Hamiltonian (7.53) may be derived using the techniques of Chap. 6.

$$\frac{\partial P(\alpha)}{\partial t} = -\left\{ \left(\varepsilon^* \frac{\partial}{\partial \alpha^*} \alpha + \varepsilon \frac{\partial}{\partial \alpha} \alpha^* \right) + \frac{\gamma_1}{2} \left(\frac{\partial}{\partial \alpha^*} \alpha^* + \frac{\partial}{\partial \alpha} \alpha \right) \right.$$
$$\left. + \frac{1}{2} \left[\varepsilon^* \frac{\partial^2}{\partial \alpha^{*2}} + \varepsilon \frac{\partial^2}{\partial \alpha^2} \right] \right\} P(\alpha) \tag{7.75}$$

The drift and diffusion matrices are

$$A = \begin{pmatrix} \frac{\gamma_1}{2} & -\varepsilon \\ -\varepsilon^* & \frac{\gamma_1}{2} \end{pmatrix}, \quad D = \begin{pmatrix} \varepsilon & 0 \\ 0 & \varepsilon^* \end{pmatrix} . \tag{7.76}$$

Direct application of (7.72–7.74) yields the squeezing spectra given by (7.59 and 7.60).

Exercises

7.1 Calculate the squeezing spectrum for a degenerate parametric oscillator with losses γ_1 and γ_2 at the end mirrors.

7.2 Calculate the squeezing spectrum for a non-degenerate parametric oscillator. [Hint: Use the quadratures for a two mode system described in (5.49)].

References

1. W. Gardiner, IBM J. Res. Dev. **32**, 127 (1988); M.J. Collett and C.W. Gardiner, Phys. Rev. **30**, 1386 (1984)
2. M.J. Collett, D.F. Walls: Phys. Rev A **32** 2887 (1985)
3. G.J. Milburn, D.F. Walls: Optics Commun. **39**, 401 (1981)

Further Reading

Gardiner, C.W.: *Quantum Noise* (Springer, Berlin, Heidelberg 1991)
Reynaud, S.: A. Heidman: Optics Comm. **71**, 209 (1989)
Yurke, B.: Phys. Rev. A **32**, 300 (1985)

Chapter 8
Generation and Applications of Squeezed Light

Abstract In this chapter we shall describe how the squeezing spectrum may be calculated for intracavity nonlinear optical processes. We shall confine the examples to processes described by an effective Hamiltonian where the medium is treated classically. We are able to extend the treatment o squeezing in the parametric oscillator to the above threshold regime. In addition, we calculate the squeezing spectrum for second harmonic generation and dispersive optical bistability. We also consider the non degenerate parametric oscillator where it is possible to achieve intensity fluctuations below the shot-noise level for the difference in the signal and idle intensities. Two applications of squeezed light will be discussed: interferometric detection of gravitational radiation and sub-shot-noise phase measurements.

8.1 Parametric Oscillation and Second Harmonic Generation

We consider the interaction of a light mode at frequency ω_1 with its second harmonic at frequency $2\omega_1$. The nonlinear medium is placed within a Fabry–Perot cavity driven coherently either at frequency $2\omega_1$ (parametric oscillation or frequency ω_1 (second harmonic generation)). We shall begin by including driving fields both at frequency ω_1 and $2\omega_1$ so that both situations may be described within the one formalism. We write the Hamiltonian as [1]

$$\mathcal{H} = \mathcal{H}_1 + \mathcal{H}_2,$$

$$\mathcal{H}_1 = \hbar\omega_1 a_1^\dagger a_1 + 2\hbar\omega_1 a_2^\dagger a_2 + i\frac{\hbar\kappa}{2}(a_1^{\dagger 2}a_2 - a_1^2 a_2^\dagger) + i\hbar(E_1 a_1^\dagger e^{-i\omega_1 t}$$
$$- E_1^* a_1 e^{i\omega_1 t}) + i\hbar(E_2 a_2^\dagger e^{-2i\omega_1 t} - E_2^* a_2 e^{2i\omega_1 t}),$$

$$\mathcal{H}_2 = a_1 \Gamma_1^\dagger + a_1^\dagger \Gamma_1 + a_2 \Gamma_2^\dagger + a_2^\dagger \Gamma_2,$$

where a_1 and a_2 are the Boson operators for modes of frequency ω_1 and $2\omega_1$, respectively, κ is the coupling constant for the interaction between the two modes and the spatial mode functions are chosen so that κ is real, Γ_1, Γ_2 are heat bath operators which represent cavity losses for the two modes and E_1 and E_2 are proportional to the coherent driving field amplitudes.

The master equation for the density operator of the two cavity modes after tracing out over the reservoirs is

$$\frac{\partial \rho}{\partial t} = \frac{1}{i\hbar}[\mathcal{H}_1, \rho] + (L_1 + L_2)\rho , \tag{8.2}$$

where

$$L_i \rho = \gamma_i (2a_i \rho a_i^\dagger - a_i^\dagger a_i \rho - \rho a_i^\dagger a_i) ,$$

and γ_i are the cavity damping rates of the modes.

This master equation may be converted to a c-number Fokker–Planck equation in the generalised P representation. The generalised P representation must be used since the c-number equation would have a non-positive definite diffusion matrix if the Glauber–Sudarshan P representation were used. The result is

$$\frac{\partial}{\partial t} P(a) = \left\{ \frac{\partial}{\partial \alpha_1}(\gamma_1 \alpha_1 - E_1 - \kappa \alpha_1^\dagger \alpha_2) + \frac{\partial}{\partial \alpha_1^\dagger}(\gamma_1 \alpha_1^\dagger - E_1^* - \kappa \alpha_1 \alpha_2^\dagger) \right.$$
$$+ \frac{\partial}{\partial \alpha_2}\left(\gamma_2 \alpha_2 - E_2 + \frac{\kappa}{2}\alpha_1^2\right) + \frac{\partial}{\partial \alpha_2^\dagger}\left(\gamma_2 \alpha_2^\dagger - E_2^* + \frac{\kappa}{2}\alpha_1^{\dagger 2}\right)$$
$$\left. + \frac{1}{2}\left[\frac{\partial^2}{\partial \alpha_1^2}(\kappa \alpha_2) + \frac{\partial^2}{\partial \alpha_1^{\dagger 2}}(\kappa \alpha_2^\dagger)\right] \right\} P(a) , \tag{8.3}$$

where $a = [\alpha_1, \alpha_1^\dagger, \alpha_2, \alpha_2^\dagger]$, and we have made the following transformation to the rotating frames of the driving fields

$$\alpha_1 \to \alpha_1 \exp(-i\omega_1 t), \qquad \alpha_2 \to \alpha_2 \exp(-2i\omega_1 t) .$$

In the generalized P representation α and $\alpha^\dagger$ are independent complex variables and the Fokker–Planck equation has a positive semi-definite diffusion matrix in an eight-dimensional space. This allows us to define equivalent stochastic differential equations using the Ito rules

$$\frac{\partial}{\partial t}\begin{pmatrix} \alpha_1 \\ \alpha_1^\dagger \end{pmatrix} = \begin{pmatrix} E_1 + \kappa \alpha_1^\dagger \alpha_2 - \gamma_1 \alpha_1 \\ E_1^* + \kappa \alpha_1 \alpha_2^\dagger - \gamma_1 \alpha_1^\dagger \end{pmatrix} + \begin{pmatrix} \kappa \alpha_2 & 0 \\ 0 & \kappa \alpha_2^\dagger \end{pmatrix}^{1/2} \begin{pmatrix} \eta_1(t) \\ \eta_1^\dagger(t) \end{pmatrix} , \tag{8.4}$$

$$\frac{\partial}{\partial t}\begin{pmatrix} \alpha_2 \\ \alpha_2^\dagger \end{pmatrix} = \begin{pmatrix} E_2 - \frac{\kappa}{2}\alpha_1^2 - \gamma_2 \alpha_2 \\ E_2^* - \frac{\kappa}{2}\alpha_1^{\dagger 2} - \gamma_2 \alpha_2^\dagger \end{pmatrix} , \tag{8.5}$$

where $\eta_1(t)$, $\eta_1^\dagger(t)$ are delta correlated stochastic forces with zero mean, namely

$$\langle \eta_1(t) \rangle = 0$$
$$\langle \eta_1(t)\eta_1(t') \rangle = \delta(t-t') \qquad (8.6)$$
$$\langle \eta_1(t)\eta^\dagger(t') \rangle = 0 .$$

8.1.1 Semi-Classical Steady States and Stability Analysis

The semi-classical or mean value equations follow directly from (8.4 and 8.5) with the replacement of $\alpha_i^\dagger$ by α_i^*.

$$\frac{\partial}{\partial t}\alpha_1 = E_1 + \kappa\alpha_1^*\alpha_2 - \gamma_1\alpha_1 , \qquad (8.7)$$

$$\frac{\partial \alpha_2}{\partial t} = E_2 + \frac{\kappa}{2}\alpha_1^2 - \gamma_2\alpha_2 . \qquad (8.8)$$

We shall investigate the steady states of these equations and their stability. The stability of the steady states may be determined by a linearized analysis for small perturbations around the steady state

$$\alpha_1 = \alpha_1^0 + \delta\alpha_1 , \qquad \alpha_2 = \alpha_2^0 + \delta\alpha_2 , \qquad (8.9)$$

where α_1^0, α_2^0 are the steady-state solutions of (8.7 and 8.8). The linearized equations for the fluctuations are

$$\frac{\partial}{\partial t}\begin{pmatrix} \delta\alpha_1 \\ \delta\alpha_1^* \\ \delta\alpha_2 \\ \delta\alpha_2^* \end{pmatrix} = \begin{pmatrix} -\gamma_1 & \kappa\alpha_2^0 & \kappa\alpha_1^0 & 0 \\ \kappa\alpha_2^{0*} & -\gamma_1 & 0 & \kappa\alpha_1^0 \\ -\kappa\alpha_1^{0*} & 0 & -\gamma_2 & 0 \\ 0 & -\kappa\alpha_1^{0*} & 0 & -\gamma_2 \end{pmatrix}\begin{pmatrix} \delta\alpha_1 \\ \delta\alpha_1^* \\ \delta\alpha_2 \\ \delta\alpha_2^* \end{pmatrix} . \qquad (8.10)$$

The four eigenvalues of these equations are

$$\lambda_1, \lambda_2 = -\frac{1}{2}(-|\kappa\alpha_2^0| + \gamma_1 + \gamma_2) \pm \frac{1}{2}[(-|\kappa\alpha_2^0| + \gamma_1 - \gamma_2)^2 - 4|\kappa\alpha_1^0|^2]^{1/2} ,$$

$$\lambda_3, \lambda_4 = -\frac{1}{2}(|\kappa\alpha_2^0| + \gamma_1 + \gamma_2) \pm \frac{1}{2}[(|\kappa\alpha_2^0| + \gamma_1 - \gamma_2)^2 - 4|\kappa\alpha_1^0|^2]^{1/2} . \qquad (8.11)$$

The fixed points become unstable when one or more of these eigenvalues has a positive real part. If a fixed point changes its stability as one of the parameters is varied we call this a bifurcation. In this problem the nature of bifurcations exhibited come in many forms including a fixed point to limit cycle transition.

We shall consider the cases of parametric oscillation and second harmonic generation separately.

Fig. 8.1 Steady state amplitude of the fundamental mode
versus pump field amplitude
for parametric oscillation,
$\kappa = 1.0$, $E_2^c = 4.0$

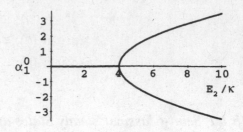

8.1.2 Parametric Oscillation

For parametric oscillation only the mode at frequency $2\omega_1$ is pumped so we set $E_1 = 0$. The stable steady state solutions for the mode amplitudes are below threshold $E_2 < E_2^c$

$$\alpha_1^0 = 0, \qquad \alpha_2^0 = \frac{E_2}{\gamma_2}, \tag{8.12}$$

above threshold $E_2 > E_2^c$

$$\alpha_1^0 = \pm \left[\frac{2}{\kappa} (E_2 - E_2^c) \right]^{1/2}, \qquad \alpha_2^0 = \frac{\gamma_1}{\kappa}, \tag{8.13}$$

where $E_2^c = \gamma_1 \gamma_2 / \kappa$, and we have taken E_2 to be positive. Thus the system exhibits behaviour analogous to a second-order phase transition at $E_2 = E_2^c$ where the below-threshold solution $\alpha_1^0 = 0$ becomes unstable and the system moves onto a new stable branch. Above threshold there exist two solutions with equal amplitude and opposite phase. In Fig. 8.1 we plot the amplitude α_1^0 versus E_2.

8.1.3 Second Harmonic Generation

For second harmonic generation only the cavity mode at frequency ω is pumped so we set $E_2 = 0$. Equations (8.7 and 8.8) then yield the following equation for the steady state amplitude of the second harmonic

$$-2\gamma_2(\kappa\alpha_2^0)^3 + 4\gamma_1\gamma_2(\kappa\alpha_2^0) - 2\gamma_1^2\gamma_2(\kappa\alpha_2^0) = |\kappa E_1|^2. \tag{8.14}$$

This gives a solution for α_2^0 which is negative and the intensity $|\alpha_2^0|^2$ is a monotonically increasing, single valued function of $|E_1|^2$.

However from the stability analysis we find that the eigenvalues

$$\lambda_1, \lambda_2 \to 0 \pm i\omega, \tag{8.15}$$

where $\omega = [\gamma_2(2\gamma_1 + \gamma_2)]^{1/2}$ when the driving field E_1 reaches the critical value

$$E_1^c = \frac{1}{\kappa}(2\gamma_1 + \gamma_2)[2\gamma_2(\gamma_1 + \gamma_2)]^{1/2} . \tag{8.16}$$

Thus the light modes in the cavity undergo a hard mode transition, where the steady state given by (8.14) becomes unstable and is replaced by periodic limit cycle behaviour. This behaviour is illustrated in Fig. 8.2 which shows the time development of the mode intensities above the instability point.

8.1.4 Squeezing Spectrum

We shall calculate the squeezing spectrum using a linearized fluctuation analysis about the steady state solutions [2, 3]. The linearized drift and diffusion matrices for the Fokker–Planck equation (8.3) are

$$A = \begin{pmatrix} \gamma_1 & -\varepsilon_2 & -\varepsilon_1^* & 0 \\ -\varepsilon_2^* & \gamma_1 & 0 & -\varepsilon_1 \\ \varepsilon_1 & 0 & \gamma_2 & 0 \\ 0 & \varepsilon_1^* & 0 & \gamma_2 \end{pmatrix}, \tag{8.17}$$

$$D = \begin{pmatrix} \varepsilon_2 & 0 & 0 & 0 \\ 0 & \varepsilon_2^* & 0 & 0 \\ 0 & 0 & 0 & 0 \\ 0 & 0 & 0 & 0 \end{pmatrix}. \tag{8.18}$$

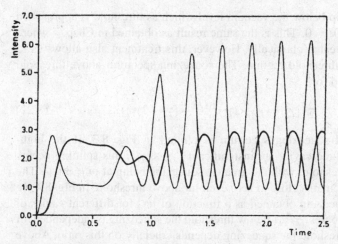

Fig. 8.2 Self pulsing in second harmonic generation: $|\alpha_1|^2$ (light), $|\alpha_2|^2$ (heavy) as functions of time. Numerical solutions of (8.7 and 8.8) with $\kappa = 10.0$, $\gamma_1 = \gamma_2 = 3.4$, $\varepsilon_1 = 20.0$, $\varepsilon_2 = 0.0$ and initial conditions $\alpha_1 = 0.1 + 0.1i$, $\alpha_2 = 0.0$

where $\varepsilon_2 = \kappa \alpha_2^0$, $\varepsilon_1 = \kappa \alpha_1^0$, and we have replaced α_2 in the diffusion matrix by its steady state value. We may then use (7.72) to calculate the spectral matrix $S(\omega)$.

The results for the squeezing in the amplitude and phase quadratures follow from (8.17 and 8.18). The squeezing in the low frequency mode (ω_1) is

$$S_{1\pm}^{\text{out}}(\omega) = 1 \pm \frac{4\gamma_1 |\varepsilon_2|(\gamma_2^2 + \omega^2)}{[\gamma_2(\gamma_1 \mp |\varepsilon_2|) + |\varepsilon_1|^2 - \omega^2]^2 + \omega^2(\gamma_1 \mp |\varepsilon_2| + \gamma_2)^2}, \tag{8.19}$$

where the $+$ and $-$ refer to the unsqueezed and squeezed quadratures, respectively. The squeezing in the high frequency mode ($2\omega_1$) is

$$S_{2\pm}^{\text{out}}(\omega) = 1 \pm \frac{4\gamma_2 |\varepsilon_2||\varepsilon_1|^2}{[\gamma_2(\gamma_1 \mp |\varepsilon_2|) + |\varepsilon_1|^2 - \omega^2]^2 + \omega^2(\gamma_1 \mp |\varepsilon_2| + \gamma_2)^2}, \tag{8.20}$$

The above results are general for two driving field ε_1 and ε_2. We now consider the special cases of parametric oscillation and second harmonic generation.

8.1.5 Parametric Oscillation

For parametric oscillation $\varepsilon_1 = 0$ and below threshold the expression for the squeezing spectrum simplifies considerably. The phase quadrature is squeezed with

$$S_{1-}^{\text{out}}(\omega) = 1 - \frac{4\gamma_1 |\varepsilon_2|}{(\gamma_1 + |\varepsilon_2|)^2 + \omega^2}. \tag{8.21}$$

This is a Lorentzian dip below the vacuum level which as threshold is approached $|\varepsilon_2| = \gamma$ gives $S_1^{\text{out}} - (0) = 0$. This is the same result as obtained in Chap. 7 where the pump mode was treated classically. However, this treatment also allows us to investigate the above-threshold regime. The squeezing spectrum above threshold becomes double peaked for

$$|\varepsilon_1|^2 > \gamma_2^2 \{[\gamma_2^2 + (\gamma_2 + 2\gamma_1)^2]^{1/2} - (\gamma_2 + 2\gamma_1)\}. \tag{8.22}$$

The double-peaked squeezing spectrum is plotted in Fig. 8.3. If the high-frequency losses from the cavity are insignificant ($\gamma_2 \ll \gamma_1$), this splitting occurs immediately above threshold, with the greatest squeezing being at $\omega = \pm|\varepsilon_1|$. The value of $S_1^{\text{out}} - (|\varepsilon_1|)$ remains close to zero even far above threshold. In Fig. 8.4 we plot the maximum squeezing obtained as a function of $|\varepsilon_1|$ for different values of the ratio of the cavity losses γ_2/γ_1. Below threshold the squeezing is independent of this ratio but above threshold the squeezing depends crucially on this ratio. Above threshold the pump is depleted, and noise from the pump enters the signal field. If the cavity losses at the pump frequency are significant, then uncorrelated vacuum fluctuations will feed through into the signal and degrade the squeezing. Thus a low

Fig. 8.3 The squeezing spectrum for parametric oscillation with $\gamma_1 = 2\gamma_2$. *Solid line*: on threshold with $\varepsilon_2 = \gamma_1$. *Dashed line*: above threshold with $\varepsilon_1 = \gamma_2$

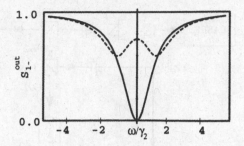

cavity loss at the pump frequency by comparison with the signal loss is needed to obtain good squeezing above threshold in the parametric oscillator.

We may also consider the squeezing in the pump mode. Below threshold this mode is not squeezed. Above threshold the peak squeezing (at $\omega = 0$) increases to a maximum value of 50% at $|\varepsilon_1| = 2\gamma_1 \gamma_2$. When $|\varepsilon_1| = 2\gamma_1^2 + \frac{1}{2}\gamma_2^2$, we again find a splitting into a double peak.

8.1.6 Experiments

The first experiment to demonstrate the generation of squeezed light in an optical parametric oscillator below threshold was been performed by *Wu* et al. [4]. They demonstrated reductions in photocurrent noise greater than 60% (4 dB) below the limit set by the vacuum fluctuations of the field are observed in a balanced homodyne detector. *Lam* et al. [5] reported 7 dB of measured vacuum squeezing. A schematic of their experiment is shown in Fig. 8.5. The experiment used a monolithic MgO:LiNbO$_3$ nonlinear crystal as the nonlinear medium. This was pumped at a wavelength of 532 nm from a second harmonic source (a hemilithic crystal of MgO:LiNbO$_3$). Squeezed light is generated at 1064 nm. The squeezing cavity output coupler is 4% reflective to 532 nm and 95.6% reflective to 1064 nm. The other end is a high reflector with 99.96% for both wavelengths. The cavity finesse was $F = 136$, and a free spectral range $FSR = 9$ GHz. The cavity linewidth was 67 MHz. The output of the OPO is directed to a pure TEM$_{00}$ mode cleaning cavity with a

Fig. 8.4 Maximum squeezing above threshold as a function of the amplitude of the fundamental mode $|\varepsilon_1|$, for different values of the cavity losses γ_2/γ_1; (a) 0.02, (b) 0.1, (c) 1.0

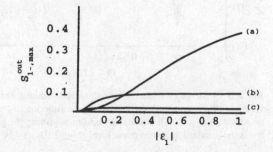

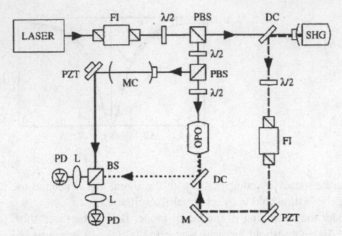

Fig. 8.5 The experimental scheme of Lam et al. for producing vacuum squeezing in an optical parametric oscillator below threshold. Solid, dashed and dotted lines are the 1,064 nm laser, second harmonic and vacuum squeezed light beams, respectively. M: mirror, FI: Faraday isolator, PZT: piezo-electric actuator, DC: dichroic beamsplitter, L: lens, PD: photodetector, (P)BS: (polarizing) beamsplitter, '/2: half-wave plate, SHG: second-harmonic generator and MC: mode cleaner cavity

finesse of 5,000 and a line width of 176 kHZ. This allows the squeezing generated by the OPO to be optimized by tuning the mode cleaner length. The final homodyne detection used a pair of ETX-500 InGaAs photodiodes with a quantum efficiency of 0.94 ± 0.02 and a 6 mW optical local oscillator. The dark noise of the photodetectors

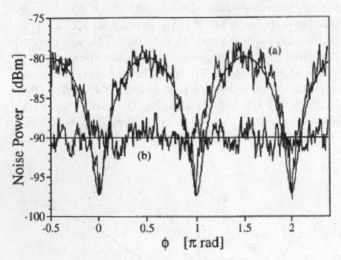

Fig. 8.6 Quadrature variance of the squeezed vacuum. Trace (a) shows experimental results of the variance of the squeezed vacuum state as a function of local oscillator phase. The smooth line is fitted values of a 7.1 dB squeezed vacuum assuming the given experimental efficiencies. In curve (b) is the standard quantum noise level at −90 dB

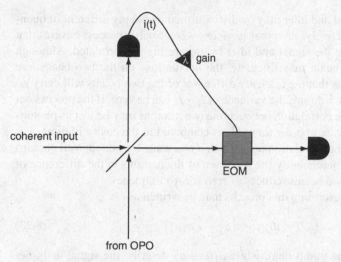

coherent input

from OPO

Fig. 8.7 An electro-optic feed-forward scheme used by Lam et al. to produce a bright squeezed state. The electro-optic modulator (EOM) is controlled be the photo-current $i(t)$

was 10 dB below the measured vacuum quantum noise level and hence the squeezed vacuum measurement does not require any electronic noise floor correction.

In Fig. 8.6 are the results of Lam et al. At a pump power of around $60 \pm 10\%$, they found an optimal vacuum squeezing of more than $7.0 \pm 0.2\,\text{dB}$.

In some applications it is desirable to have a squeezed stater with a non zero coherent amplitude, a *bright* squeezed state. The conventional way to do this would be to simply mix the squeezed vacuum state produced by the OPO with a coherent state on a beam splitter (see Exercise 8.1). However the transmitivity of the squeezed vacuum state must be very close to unity in order not to loose the squeezing. This means that very little of the coherent light is reflected and the scheme is rather wasteful of power. Lam et al. also showed two alternative methods to produce a squeezed state with a significant coherent amplitude. In the first method a small seed coherent beam was injected into the back face of the OPO. This gave an amplitude squeezed state with 4 dB of squeezing. The second method was based on electro-optical feed-forward to transfer the squeezing onto a coherent beam, see Fig. 8.7. By carefully adjusting the gain on the controlling photon current to the EOM a bright squeezed beam with squeezing in the amplitude quadrature corresponding to a reduction of intensity noise of 4db below shot noise.

8.2 Twin Beam Generation and Intensity Correlations

Another second-order process which can produce non-classical states is non-degenerate down conversion. A pump photon with frequency 2ω creates a signal and an idler photon each with frequency ω but different polarisations.

Alternatively the signal and idler may be distinguished by having different frequencies ω_1 and ω_2, respectively, such that $\omega_1 + \omega_2 = 2\omega$. Such a process ensures that the photon numbers in the signal and idler beams are highly correlated. Although the intensity of each beam may fluctuate, the fluctuations on the two beams are identical. This suggests that the intensity difference of the two beams will carry no fluctuations at all. That is to say, the variance of $I_1 - I_2$ can be zero. If the process occurs inside a cavity the correlation between the two photons may be lost as photons escape the cavity. This is true for times short compared to the cavity lifetime. For long times, however, the correlation is restored; if one waits long enough all photons will exit the cavity. Consequently the spectrum of fluctuations in the difference of the intensities in the two beams reduces to zero at zero frequency.

The Hamiltonian describing this process may be written as

$$\mathscr{H}_I = i\hbar\chi(a_0 a_1^\dagger a_2^\dagger - a_0^\dagger a_1 a_2) ,\tag{8.23}$$

where a_0 describes the pump field, while a_1 and a_2 describe the signal and idler fields. The pump field is driven by a coherent field external to the cavity with amplitude ε. The damping rates for the three cavity modes a_0, a_1 and a_2 are κ_0, κ_1 and κ_2, respectively.

Following from the Fokker–Planck equation for the positive P representation we establish the c-number stochastic differential equations [6]

$$\dot{\alpha}_0 = -\kappa_0\alpha_0 + \varepsilon - \chi\alpha_1\alpha_2 ,$$
$$\dot{\alpha}_1 = -\kappa_1\alpha_1 + \chi\alpha_0\alpha_2^\dagger + R_1(t) ,$$
$$\dot{\alpha}_2 = -\kappa_2\alpha_2 + \chi\alpha_0\alpha_1^\dagger + R_2(t) .\tag{8.24}$$

In our treatment we will assume for simplicity that $\kappa_1 = \kappa_2 = \kappa$, where the only non-zero noise correlation functions are

$$\langle R_1(t)R_2(t')\rangle = \chi\langle\alpha_0\rangle\delta(t-t'),$$
$$\langle R_1^\dagger(t)R_2^\dagger(t')\rangle = \chi\langle\alpha_0^\dagger\rangle\delta(t-t') .\tag{8.25}$$

The semi-classical steady state solutions depend on whether the driving field ε is above or below a critical "threshold" value given by

$$\varepsilon_{\text{thr}} = \frac{\kappa_0\kappa}{\chi} .\tag{8.26}$$

Above threshold one of the eigenvalues of the drift matrix is zero. This is associated with a phase instability. To see this we use an amplitude and phase representation:

$$\alpha_j(t) = r_j[1 + \mu_j(t)]e^{-i(\phi_j + \psi_j(t))}\tag{8.27}$$

where r_j, ϕ_j are the steady state solutions, and $\mu_j(t)$ and $\psi_j(t)$ represent small fluctuations around the steady state. Solving for the steady state below threshold we have

$$r_1 = r_2 = 0, \qquad r_0 = \frac{|\varepsilon|}{\kappa_0}, \qquad \phi_0 = \phi_p, \qquad (8.28)$$

where we have used $\varepsilon = |\varepsilon|e^{i\phi_p}$, with ϕ_p denoting the phase of the coherent pump. Above threshold

$$r_0 = \frac{\kappa}{\chi}, \quad \phi_0 = \phi_p,$$

$$r_1 = r_2 = \frac{\sqrt{\kappa_0}\kappa}{\chi}(E-1)^{1/2}, \qquad \phi_1 + \phi_2 = \phi_p, \qquad (8.29)$$

with

$$E = \frac{|\varepsilon|}{|\varepsilon_{\text{thr}}|}. \qquad (8.30)$$

Note that in the above threshold solution only the sum of the signal and idler phases is defined. No steady state exists for the phase difference. It is the phase difference variable which is associated with the zero eigenvalue.

We now turn to an analysis of the intensity fluctuations above threshold. The nonlinear dynamics of (8.24) is approximated by a linear dynamics for intensity fluctuations about the steady states above threshold. Define the new variables by

$$\Delta I_j = \alpha_j^\dagger \alpha_j - I_j^{ss}, \qquad (8.31)$$

where I_j^{ss} is the steady state intensity above threshold for each of the three modes. It is more convenient to work with scaled "intensity-sum" and 'intensity-difference' variables defined by

$$\Delta I_s = \kappa(\Delta I_1 + \Delta I_2), \qquad \Delta I_D = \kappa(\Delta I_1 - \Delta I_2), \qquad (8.32)$$

The linear stochastic differential equations are then given by

$$\Delta \dot{I}_0 = -\kappa_0 \Delta I_0 - \Delta I_s, \qquad (8.33)$$

$$\Delta \dot{I}_s = 2\kappa\kappa_0(E-1)\Delta I_0 + F_s(t), \qquad (8.34)$$

$$\Delta \dot{I}_D = -2\kappa \Delta I_D + F_D(t), \qquad (8.35)$$

where the non zero noise correlations are

$$\langle F_s(t)F_s(t')\rangle = -\langle F_D(t)F_D(t')\rangle = 4\frac{\kappa_0\kappa^4}{\chi^2}(E-1)\delta(t-t'). \qquad (8.36)$$

We are now in a position to calculate the spectrum of fluctuations in the intensity difference in the signal and idler modes outside the cavity. The equation for the intensity difference fluctuations may be solved immediately to give

$$\Delta I_0(t) = \Delta I_D(0)e^{-2\kappa t} + \int_0^t dt' e^{-2\kappa(t-t')} F_D(t'). \qquad (8.37)$$

Thus the steady state two-time correlation function is found to be

$$\langle I_D(\tau), I_D(0)\rangle = \langle \Delta I_D(\tau)\Delta I_D(0)\rangle = \frac{\kappa_0\kappa^3}{\chi^2}(E-1)e^{-2\kappa\tau} \tag{8.38}$$

with $\langle A, B\rangle = \langle AB\rangle - \langle A\rangle\langle B\rangle$.

The spectrum of fluctuations in the intensity difference field outside the cavity is defined by

$$S_D(\omega) = \int d\tau e^{-i\omega\tau}\langle \hat{I}_1(\tau) - \hat{I}_2(\tau), \hat{I}_1(0) - \hat{I}_2(0)\rangle_{ss}, \tag{8.39}$$

where $\hat{I}_j(t)$ are the external intensity operators. However, from Chap. 7, we can relate this operator average to the c-number averages for $\alpha_j(t)$ inside the cavity. The result is

$$\langle \hat{I}_j(\tau), \hat{I}_k(0)\rangle = 2\delta_{jk}\kappa\delta(\tau)\langle I_j(0)\rangle + 4\kappa^2\langle I_j(\tau), I_k(0)\rangle, \tag{8.40}$$

where $I_j \equiv \alpha_j^\dagger(t)\alpha_j(t)$. Finally, we write the result directly in terms of the valuables $\Delta I_D(t)$,

$$S_D(\omega) = S_0 + 4\int d\tau e^{-i\omega\tau}\langle \Delta I_D(\tau)\Delta I_D(0)\rangle, \tag{8.41}$$

where

$$S_0 = 2\kappa(\langle I_1\rangle_{ss} + \langle I_2\rangle_{ss}) = \frac{4\kappa_0\kappa^3}{\chi^2}(E-1). \tag{8.42}$$

The frequency independent term S_0 represents the contribution of the shot-noise from each beam. Thus to quantify the degree of reduction below the shot-noise level we define the 'normalised' intensity difference spectrum

$$\bar{S}_D(\omega) \equiv \frac{S_D(\omega)}{S_0}. \tag{8.43}$$

Substituting (8.38 and 8.42) into (8.41), and integrating we obtain

$$\bar{S}_D(\omega) \equiv \frac{\omega^2}{\omega^2 + 4\kappa^2}. \tag{8.44}$$

This is a simple inverted Lorentzian with a width 2κ. As expected, at zero frequency there is perfect noise suppression in the intensity difference between the signal and idler. This result was first obtained by *Reynaud* et al. [7].

The above results assume no additional cavity losses beyond those corresponding to the (equal) transmitivities at the output mirror. When additional losses are included the correlation between the signal and idler is no longer perfect as one of the pair of photons may be lost otherwise than through the output mirror. In that case there is no longer perfect suppression of quantum noise at zero frequency [8]. The result is shown in Fig. 8.8a. Furthermore, the spectrum of intensity difference fluctuations is very sensitive to any asymmetry in the loss for each mode [8]. In Fig. 8.8b we depict the effect of introducing different intracavity absorption rates

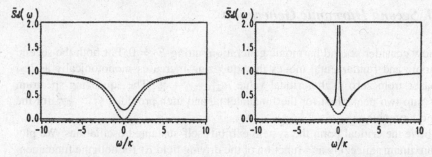

Fig. 8.8 The effect of intracavity absorption on the intensity difference spectrum. (**a**) A plot of the normalized spectrum when the total cavity losses for each mode are equal to the total loss from the pump mode, and greater than the output loss rate κ (*solid line*). The dashed line shows the perfect case when the only losses are through the output mirrors. (**b**) The effect of asymmetrical intracavity absorption. The total loss for the idler is equal to the idler damping rate but the damping rate for the signal is only 80% of the total loss for that mode. $E = 1.05$ solid line, $E = 2.0$, dashed line. [8]

for each mode. This phenomenon could form the basis of a sub-shot-noise absorption spectrometer.

The prediction of noise suppression in the differenced intensity has been confirmed in the experiment of *Heidmann* et al. [9]. They used a type II phase-matched potassium triphosphate, known as KTP, crystal placed inside an optical cavity, thus forming an Optical Parametric Oscillator (OPO).

The damping constant at the pump frequency was much greater than for the signal and idler. Above threshold the OPO emits two cross polarised twin beams at approximately the same frequency. The twin beams are separated by polarising beam splitters and then focussed on two photodiodes which have quantum efficiencies of 90%. The two photo-currents are amplified and then subtracted with a 180° power combiner. The noise on the resulting difference current is then monitored by a spectrum analyser.

A maximum noise reduction of 30% ± 5% is observed at a frequency of 8 MHz. The noise reduction is better than 15% from 3 to 13 MHz. The noise reduction is limited due to other losses inside the OPO and various detector inefficiencies.

Using an α-cut KTP crystal, *Gao* et al. [10] achieved a noise reduction in the intensity difference of 88% below the shot noise level, corresponding to a 9.2 dB reduction. They also showed how such highly correlated intensities could be used to enhance the signal-to-noise ratio for relative absorption measurements when one beam passes through an absorber. The improvement was about 7 dB.

The bandwidth of squeezing in cavity experiments is restricted to the cavity bandwidth. Single-pass experiments are feasible using the higher intensity possible with pulsed light. Pulsed twin beams of light have been generated by means of an optical parametric amplifier that is pumped by the second harmonic of a mode-locked and Q-switched Nd : YAG laser [11]. While the noise levels of the individual signal and idler beams exceed their coherent state limits by about 11 dB the correlation is so strong that the noise in the difference current falls below the quantum limit by more than 6 dB (75%).

8.2.1 Second Harmonic Generation

We now consider second harmonic generation setting $E_2 = 0$. For both the second harmonic and fundamental modes the squeezing increases monotonically as $|\varepsilon_2|$ increases from zero to the critical value $|\varepsilon_2| = \gamma_1 + \gamma_2$. The squeezing spectrum splits into two peaks first for the fundamental and then provided $\gamma_2^2 > \frac{1}{2}\gamma_1^2$ for the second harmonic.

Above the critical point the system exhibits self-sustained oscillations. We plot the maximum squeezing as a function of the driving field E_1 for both the fundamental and second harmonic in Fig. 8.9 [3].

In both the cases considered the maximum squeezing occurs as an instability point is approached. This is an example of critical quantum fluctuations which are asymmetric in the two quadrature phases. It is clear that in order to approach zero fluctuations in one quadrature the fluctuations in the other must diverge. At the critical point itself, with the critical frequency being $\omega_c^2 = \gamma_2(\gamma_2 + 2\gamma_1)$ (which is, in fact, the initial frequency of the hard mode oscillations) we have for the amplitude quadrature of the fundamental.

$$S_{1+}^{OUT}(\omega_c) = 1 - \frac{\gamma_1}{\gamma_1 + \gamma_2} , \tag{8.45}$$

which gives perfect squeezing for $\gamma_1 \gg \gamma_2$ at $\omega = \pm\sqrt{2\gamma_1\gamma_2}$, and for the amplitude quadrature of the second harmonic

$$S_{2+}^{OUT}(\omega_c) = 1 - \frac{\gamma_2}{\gamma_1 + \gamma_2} , \tag{8.46}$$

this gives perfect squeezing for $\gamma_2 \gg \gamma_1$ at $\omega = \gamma_2$. The squeezing spectra for the two modes at the critical point is shown in Fig. 8.10. The fluctuations in the phase quadrature must tend to infinity, a characteristic of critical fluctuations. We note that the linearization procedure we have used will break down in the vicinity of the critical point and in practice the systems will operate some distance from the critical point.

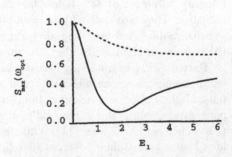

Fig. 8.9 A plot of the maximum squeezing versus driving field amplitude for second harmonic generation. The solid line is the fundamental, the dashed line is the second-harmonic. $\gamma_1 = \gamma_2 = 1.0$

Fig. 8.10 The squeezing
spectrum for the fundamental
and the second harmonic at
the critical point for oscilla-
tion. $\gamma_1 = 1.0$, **(a)** $\gamma_2 = 0.1$,
(b) $= \gamma_2 = 10.0$

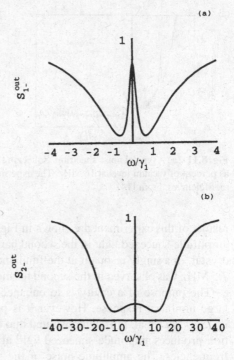

8.2.2 Experiments

The earliest experiments to demonstrate amplitude squeezing in second harmonic
generation were done by *Pereira* et al. [12] and by *Sizman* et al. [13]. Both experiments used a crystal of MgO:LiNbO$_3$ and driven by a frequency stabilised Nd:Yag
laser. In second harmonic generation the squeezing appears in the amplitude quadrature so a direct detection scheme can be employed. In the case of Pereira et al. the
nonlinear crystal was inside an optical cavity. They looked for squeezing at the
fundamental frequency. Sensitivity to phase noise in both the pump laser and from
scattering processes in the crystal limited the observed squeezing to 13% reduction
relative to vacuum. Sizman et al. used a monolithic crystal cavity. The end faces of
the crystal have dielectric mirror coatings with a high reflectivity for both the fundamental and second harmonic modes. They reported a 40% reduction in the intensity
fluctuations of the second harmonic light. These two schemes used a doubly resonant cavity, i.e. both the fundamental and the second harmonic are resonant with the
cavity. This poses a number of technical difficulties not least of which is maintaining
the double resonance condition for extended periods. *Paschotta* et al. [14] demonstrated a singly resonant cavity at the fundamental frequency for generating amplitude squeezed light in the second harmonic mode. Using a MgO:LiNbO3 monolithic
standing-wave cavity they measured 30% noise reduction (1.5 dB) at 532 nm. The

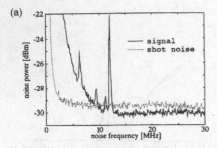 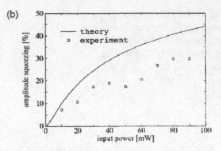

Fig. 8.11 (a) A typical noise spectrum. Squeezing is apparent above about 12 MHz. (b) Squeezing as percent of vacuum level at 16 MHz. The experimental results (*squares*) are corrected for known inefficiencies. From [14]

results of this experiment are shown in Fig. 8.11. *Tsuchida* et al. also demonstrated amplitude-squeezed light in the second harmonic mode at 431 nm in with a $KNbO_3$ crystal, in a singly resonant at the fundamental mode. Noise reduction of 2.4 dB at 7.5 MHz was observed in the second harmonic mode.

The purpose of a cavity is to enhance the pump power and get a sufficiently large nonlinear response. However it is possible to get squeezing without a cavity. *Serkland* et al. [15] demonstrated that traveling-wave second-harmonic generation produces amplitude-squeezed light at both the fundamental and the harmonic frequencies. The amplitude noise of the transmitted fundamental field was measured to be 0.8 dB below the shot-noise level, and the generated 0.765-mm harmonic light was measured to be amplitude squeezed by 0.35 dB. The conversion-efficiency dependence of the observed squeezing at both wavelengths agrees with theoretical predictions.

8.3 Applications of Squeezed Light

8.3.1 Interferometric Detection of Gravitational Radiation

Interest in the practical generation of squeezed states of light became significant when *Caves* [16] suggested in 1981 that such light might be used to achieve better sensitivity in the interferometric detection of gravitational radiation. The result of *Caves* indicated that while squeezed light would not increase the maximum sensitivity of the device, it would enable maximum sensitivity to be achieved at lower laser power. Later analyses [17, 18, 19, 20] demonstrated that by an optimum choice of the phase of the squeezing it is possible to increase the maximum sensitivity of the interferometer. This result was established by a full nonlinear quantum theory of the entire interferometer, including the action of the light pressure on the end mirrors. We shall demonstrate this following the treatment of *Pace* et al. [20].

A schematic illustration of a laser interferometer for the detection of gravitational radiation is shown in Fig. 8.12. To understand how the device works we need to

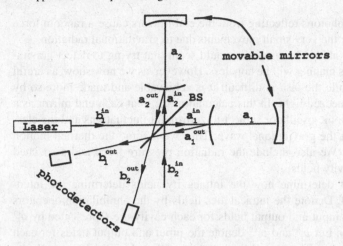

Fig. 8.12 Schematic representation of a laser interferometer for the detection of gravitational radiation

recall some properties of gravitational radiation. A gravitational wave induces weak tidal forces, in a plane perpendicular to the direction of propagation. A gravitational wave passing normal to a circular arrangement of masses would periodically force the circle into an ellipse [21]. In the case of the interferometer depicted in Fig. 8.12, the end mirrors of the two cavities are constrained by a weak harmonic potential, and lie on a circular arc separated by 90°. Thus, when a gravitational wave passes orthogonal to the plane of the interferometer, one cavity will be shortened as the other cavity is lengthened. If the intensity difference of the light leaving each arm of the interferometer is monitored, the asymmetric detuning of each cavity caused by the moving end mirrors causes this intensity to be modulated at the frequency of the gravitational wave.

While this scheme sounds very promising it suffers from a big problem. Even though gravitational radiation reaching terrestrial detectors is highly classical (many quanta of excitation) it interacts very weakly with the end mirrors. The relative change in the length of each cavity is then so small that it is easily lost amid a multitude of noise sources, which must somehow be reduced if any systematic effect is to be observed. To begin with, it is necessary to isolate the end mirrors from external vibrations and seismic forces. Then one must ensure that the random thermal motion of the end mirrors is negligible. Ultimately as each end mirror is essentially an oscillator, there is the zero-point motion to take account of. Quite apart from the intrinsic noise in the motion of the end mirrors, noise due to the light also limits the sensitivity of the device. The light noise can be separated into two contributions. Firstly the measurement we ultimately perform is an intensity measurement which is limited by shot-noise. In the case of shot-noise, however, the signal-to-noise ratio scales as the square root of the input power, thus one might attempt to avoid this noise source by simply raising the input power. Unfortunately, increasing the input power increases the contribution from another source – radiation pressure. Individual photons reflecting from the end mirrors cause a random force large enough to mask the very small movements due to gravitational radiation.

In the light of the above discussion it would seem that trying to detect gravitational radiation in this manner will be hopeless. However, as we now show, a careful study reveals that while the task is difficult it is achievable and made more so by the careful use of squeezed light. In this calculation we treat each end mirror as a damped simple harmonic oscillator subject to zero-point fluctuations and the classical driving force of the gravitational wave. Thus we assume the thermal motion has been eliminated. We also include the radiation pressure force and associated fluctuations in the cavity fields.

To begin we first determine how the intracavity fields determine the intensity difference signal. Denote the intracavities fields by the annihilation operators a_i ($i = 1, 2$) and the input and output fields for each cavity are represented by a_i^{in} and a_i^{out}, respectively. Let b_i^{in} and b_i^{out} denote the input and output fields for each arm of the interferometer. The central beam-splitter (BS in Fig. 8.12) connects the cavity inputs and outputs to the interferometer inputs and outputs by

$$a_1^{\text{in}} = \frac{1}{\sqrt{2}}(b_1^{\text{in}} + \mathrm{i}b_2^{\text{in}}) \,, \tag{8.47}$$

$$a_2^{\text{in}} = \frac{1}{\sqrt{2}}(b_1^{\text{in}} + \mathrm{i}b_2^{\text{in}}) \,, \tag{8.48}$$

$$b_1^{\text{out}} = \frac{1}{\sqrt{2}}(a_1^{\text{out}} + \mathrm{i}a_2^{\text{out}}\mathrm{e}^{\mathrm{i}\phi}) \,, \tag{8.49}$$

$$b_2^{\text{out}} = \frac{1}{\sqrt{2}}(a_1^{\text{out}} + \mathrm{i}a_2^{\text{out}}\mathrm{e}^{\mathrm{i}\phi}) \,, \tag{8.50}$$

where ϕ is a controlled phase shift inserted in arm 2 of the interferometer to enable the dc contribution to the output intensity difference to be eliminated.

The measured signal is then represented by the operator

$$\begin{aligned} I_-(t) &= (b_1^{\text{out}})^{\dagger}b_1^{\text{out}} - (b_2^{\text{out}})^{\dagger}b_2^{\text{out}} \\ &= -\mathrm{i}[(a_2^{\text{out}})^{\dagger}a_1^{\text{out}}\mathrm{e}^{-\mathrm{i}\phi} - \text{h.c.}] \,. \end{aligned} \tag{8.51}$$

Now the relationship between the cavity fields and the respective input and output fields is given by

$$a_i^{\text{out}} = \sqrt{\gamma}a_i - a_i^{\text{in}} \quad (i = 1, 2) \,, \tag{8.52}$$

where we assume the damping rate for each cavity, γ, is the same.

We now assume that arm one of the interferometer is driven by a classical coherent source with amplitude $E/\sqrt{\gamma}$ in units such that the intensity of the input is measured in photons/second. The scaling $\gamma^{-1/2}$ is introduced, for convenience. Then from (8.47 and 8.48), each cavity is driven with the same amplitude $\varepsilon/\sqrt{\gamma}$, where $\varepsilon = E/\sqrt{2}$. That is

$$\langle a_1^{\text{in}} \rangle = \langle a_2^{\text{in}} \rangle = \frac{\varepsilon}{\sqrt{\gamma}} \,. \tag{8.53}$$

As we show below, it is possible to operate the device in such a way that in the absence of gravitational radiation, a stable deterministic steady state amplitude α_0 is established in each cavity. This steady state is then randomly modulated by fluctuations in the cavity fields and deterministically modulated by the moving end mirrors of each cavity. Both these effects are of similar magnitude. It thus becomes possible to linearise the output fields around the stationary states. With this in mind we now define the fluctuation operators δa_i and δa_i^{in} for each cavity ($i = 1, 2$)

$$\delta a_i = a_i - \alpha_0 , \tag{8.54}$$

$$\delta a_i^{in} = a_i^{in} - \frac{\varepsilon}{\sqrt{\gamma}} . \tag{8.55}$$

Using these definitions, together with (8.47–8.50), in (8.51), the output signal is then described by the operator

$$I_-(t) = \frac{\gamma\alpha_0}{2}[\delta y_1(t) - \delta y_2(t)] - \frac{\sqrt{\gamma}\alpha_0}{2}[\delta y_1^{in}(t) - \delta y_2^{in}(t)] , \tag{8.56}$$

where

$$\delta y_i(t) \equiv i(\delta a_i - \delta a_i^\dagger) , \tag{8.57}$$

$$\delta y_i^{in}(t) \equiv -i[\delta a_i^{in} - (\delta a_i^{in})^\dagger] . \tag{8.58}$$

We have chosen the arbitrary phase reference so that the input amplitude, and thus the steady state amplitude α_0, is real.

Equation (8.56) indicates that the signal is carried by the phase quadrature not the amplitude quadrature. Thus we must determine $y_i(t)$.

We turn now to a description of the intracavity dynamics. The end mirror is treated as a quantised simple harmonic oscillator with position and momentum operators (Q, P). The radiation pressure force is proportional to the intracavity photon number. The total Hamiltonian for the system may then be written [20]

$$\mathcal{H} = \hbar\Delta a^\dagger a + \frac{P^2}{2M} + \frac{M\Omega^2}{2}Q^2 - \hbar\frac{\omega_0}{L}a^\dagger a Q + F(t)Q , \tag{8.59}$$

where M is the mass of the end mirror, Ω is the oscillator frequency of the end mirror, L is the cavity length, Δ is the cavity detuning, and $F(t)$ is the driving force on the end mirror due to the gravitational wave. If we assume the acceleration produced by the gravitational wave is

$$g(t) = g \cos(\omega_g t) , \tag{8.60}$$

the force $F(t)$ may be written as

$$F(t) = -MhL\omega_g^2 S(t) \tag{8.61}$$

where h is defined to be the maximum fractional change in the cavity length, L, produced by the gravitational wave in the absence of all other forces, and $s(t) = co(\omega_g t)$.

It is convenient to define the dimensionless position q and the momentum variables p for the mirror, which are the analogue of the quadrature phase operators for the field,

$$q = \left(\frac{2\hbar}{M\Omega}\right)^{1/2} Q,\tag{8.62}$$

$$p = (2\hbar M\Omega)^{-1/2}P.\tag{8.63}$$

The commutation relations for these new variables is $[q, p] = i/2$. Thus in the ground state, the variance in q and p are both equal to 1/4.

The quantum stochastic differential equations for this system may now be written

$$\frac{da}{dt} = \varepsilon - i(\Delta + 2\kappa q)a - \frac{\gamma}{2}a + \sqrt{\gamma}a^{in},\tag{8.64}$$

$$\frac{dq}{dt} = \Omega p - \frac{\Gamma}{2}q + \sqrt{\Gamma}q^{in},\tag{8.65}$$

$$\frac{dp}{dt} = -\Omega q - \kappa a^{\dagger}a - \kappa s(t) - \frac{\Gamma}{2}p + \sqrt{\Gamma}p^{in},\tag{8.66}$$

where

$$\kappa \equiv \frac{-\omega_0}{L}\left(\frac{\hbar}{2M\Omega}\right)^{1/2},\tag{8.67}$$

$$k = -hL\omega_g^2\left(\frac{M}{2\hbar\Omega}\right)^{1/2},\tag{8.68}$$

and $\gamma/2$ is the damping rate for the intracavity field, while $\Gamma/2$ is the damping rate for the end mirrors. Note that the form of the stochastic equation for the mirror is that for a zero-temperature, under-damped oscillator and will thus only be valid provided $\Gamma \ll \Omega$.

Let us first consider the corresponding deterministic semi-classical equations

$$\dot{\alpha} = \varepsilon - i(\Delta + 2\kappa q)\alpha - \frac{\gamma}{2}\alpha,\tag{8.69}$$

$$\dot{q} = \Omega p - \frac{\Gamma}{2}q,\tag{8.70}$$

$$\dot{p} = -\Omega q - \kappa|\alpha|^2 - ks(t) - \frac{\Gamma}{2}p.\tag{8.71}$$

These equations represent a pair of nonlinearly coupled harmonically driven oscillators, and as such are candidates for unstable, chaotic behaviour. However, the amplitude of the driving, k, is so small that one expects the system to remain very close to the steady state in the absence of driving. The first step is thus to determine the steady state values, α_0, q_0 and p_0. If we choose Δ such that $\Delta = -2\kappa q_0$ (so the cavity is always on resonance), then

$$\alpha_0 = \frac{2\varepsilon}{\gamma}. \tag{8.72}$$

Of course, this steady state itself may be unstable. To check this we linearise the undriven dynamics around the steady state. Define the variables

$$\delta x(t) = \text{Re}\{\alpha(t) - \alpha_0\}, \tag{8.73}$$
$$\delta y(t) = \text{Im}\{\alpha(t) - \alpha_0\}, \tag{8.74}$$
$$\delta q(t) = q(t) - q_0, \tag{8.75}$$
$$\delta p(t) = p(t) - p_0. \tag{8.76}$$

Then

$$\frac{\mathrm{d}}{\mathrm{d}t} \begin{pmatrix} \delta x \\ \delta y \\ \delta q \\ \delta p \end{pmatrix} = \begin{pmatrix} -\frac{\gamma}{2} & 0 & 0 & 0 \\ 0 & -\frac{\gamma}{2} & -\mu & 0 \\ 0 & 0 & -\frac{\Gamma}{2} & \Omega \\ -\mu & 0 & -\Omega & -\frac{\Gamma}{2} \end{pmatrix} \begin{pmatrix} \delta x \\ \delta y \\ \delta q \\ \delta p \end{pmatrix}, \tag{8.77}$$

where $\mu = 4\kappa\alpha_0$ and we have assumed ε and thus α_0 are real. The eigenvalues of the linear dynamics are then found to be $(-\gamma/2, -\gamma/2, -\Gamma/2 + i\Omega, -\Gamma/2 - i\Omega)$, so clearly the steady state is stable in the absence of the gravitational wave.

We shall point out the interesting features of (8.77). First we note that the quadrature carrying the coherent excitation (δx) is totally isolated from all other variables. Thus $\delta x(t) = \delta x(0)e^{-\gamma t/2}$. However, as fluctuations evolve from the steady state $\delta x(0) = 0$, one can completely neglect the variables $\delta x(t)$ for the deterministic part of the motion. Secondly we note the mirror position fluctuations δq feed directly into the field variable $\delta y(t)$ and thus directly determine the output intensity difference signal by (8.56). Finally, we note the fluctuations of the in-phase field variable δx drive the fluctuating momentum of the mirror. This is, of course, the radiation pressure contribution. However, for the deterministic part of the dynamics $\delta x(t) = 0$, as discussed above, so the mirror dynamics is especially simple – a damped harmonic oscillator. In the presence of the gravitational wave the deterministic dynamics for the end mirrors is then

$$\begin{pmatrix} \delta\dot{q} \\ \delta\dot{p} \end{pmatrix} = \begin{pmatrix} -\frac{\Gamma}{2} & \Omega \\ -\Omega & -\frac{\Gamma}{2} \end{pmatrix} \begin{pmatrix} \delta q \\ \delta p \end{pmatrix} - \begin{pmatrix} 0 \\ ks(t) \end{pmatrix}, \tag{8.78}$$

with the initial conditions $\delta q(0) = \delta p(0)$ the solution for $\delta q(t)$ is

$$\delta q(t) = R \cos(\omega_g t + \phi), \tag{8.79}$$

with

$$R = \frac{k\Omega}{\left|\frac{\Gamma}{2} + i(\omega_g - \Omega)\right| \left|\frac{\Gamma}{2} + i(\omega_g + \Omega)\right|}, \tag{8.80}$$

$$\phi = \arctan\left(\frac{-\Gamma\omega_g}{\frac{\Gamma^2}{4} + \Omega^2 - \omega_g^2}\right). \tag{8.81}$$

Substituting this solution into the equation for $\delta y(t)$ and solving, again with $\delta y(0) = 0$, we find

$$\delta y(t) = \frac{-4\kappa\alpha_0 R}{\left|\frac{\gamma}{2} + i\omega_g\right|} \cos(\omega_g t + \theta + \phi),\tag{8.82}$$

where

$$\theta = \arctan\left(\frac{\alpha\omega_g}{\gamma}\right).\tag{8.83}$$

We have neglected an initial decaying transient. Apart from the phase shifts θ and ϕ, the out-of-phase field quadrature follows the displacements of the end mirror induced by the gravitational wave.

Due to the tidal nature of the gravitational wave if one cavity end mirror experiences a force $F(t)$, the other experiences $-F(t)$. Thus $\delta y_1(t) = -\delta y_2(t)$ and the mean signal is

$$\langle I_-(t)\rangle = -\frac{16\kappa IR\cos(\omega_g t + \phi + \theta)}{\left|\frac{\gamma}{2} + i\omega_g\right|},\tag{8.84}$$

where the output intensity I is defined by

$$I = |\langle a_i^{\text{out}}\rangle|^2 = \frac{\gamma\alpha_0^2}{4}.\tag{8.85}$$

Using the definitions in (8.80, 8.67 and 8.68) we find

$$\langle I_-(t)\rangle = \frac{-8hI\omega_0\omega_g^2\cos(\omega_g t + \theta + \phi)}{\left|\frac{\gamma}{2} + i\omega_g\right|\left|\frac{\Gamma}{2} + i(\omega_g - \Omega)\right|\left|\frac{\Gamma}{2} + i(\omega_g + \Omega)\right|}\tag{8.86}$$

and the signal is directly proportional to the mirror displacement h.

Before we consider a noise analysis of the interferometer it is instructive to look at the frequency components of variable $\delta y(t)$ by

$$\delta y(\omega) = \int_{-\infty}^{\infty} dt\, e^{i\omega t}\delta y(t).\tag{8.87}$$

As $\delta y(t)$ is real we have that $\delta y(t) = \delta y^*(-\omega)$. This relationship enables us to write

$$\delta y(t) = \int_0^{\infty} d\omega[\delta y(\omega)e^{-i\omega t} + \delta y(\omega)^* e^{i\omega t}],\tag{8.88}$$

thus distinguishing positive and negative frequency components. Inspection of (8.84) immediately gives that

$$\delta y(\omega) = \frac{-2\kappa\alpha_0 Re^{-i(\theta+\phi)}}{\left|\frac{\gamma}{2} + i\omega_g\right|}\delta(\omega - \omega_g).\tag{8.89}$$

Thus

$$|\langle I_-(\omega)\rangle| = hS(\omega_g)\delta(\omega - \omega_g),\tag{8.90}$$

where

$$S(\omega_g) = \frac{8hI\omega_0\omega_g^2}{\left|\frac{\gamma}{2} + i\omega_g\right|\left|\frac{\Gamma}{2} + i(\omega_g - \omega)\right|\left|\frac{\Gamma}{2} + i(\omega_g + \Omega)\right|} \ . \tag{8.91}$$

We now analyse the noise response of the interferometer. As the gravitational wave provides an entirely classical driving of the mirrors it can only effect the deterministic part of the dynamics, which we have already described above. To analyse the noise component we must consider the fluctuation *operators* δx, δy, δq and δp defined by $\delta x = x(t) - x_s(t)$, where x_s is the semi-classical solution. In this way the deterministic contribution is removed.

The quantum stochastic differential equations are then

$$\frac{d}{dt}\delta x(t) = -\frac{\gamma}{2}\delta x(t) + \sqrt{\gamma}\delta x^{in}(t) \ , \tag{8.92}$$

$$\frac{d}{dt}\delta y(t) = -\frac{\gamma}{2}\delta y(t) - \mu\delta q(t) + \sqrt{\gamma}\delta y^{in}(t) \ , \tag{8.93}$$

$$\frac{d}{dt}q(t) = -\frac{\Gamma}{2}q(t) + \Omega p(t) + \sqrt{\Gamma}q^{in}(t) \ , \tag{8.94}$$

$$\frac{d}{dt}p(t) = -\frac{\Gamma}{2}p(t) - \Omega q(t) - \mu x(t) + \sqrt{\Gamma}p^{in}(t) \ , \tag{8.95}$$

with the only non-zero noise correlations being

$$\langle\delta x^{in}(t)\delta x^{in}(t')\rangle = \langle\delta y^{in}(t)\delta y^{in}(t')\rangle = \delta(t - t') \ , \tag{8.96}$$

$$\langle\delta x^{in}(t)\delta y^{in}(t')\rangle = \langle\delta y^{in}(t)\delta x^{in}(t')\rangle^* = i\delta(t - t') \ , \tag{8.97}$$

$$\langle q^{in}(t)q^{in}(t')\rangle = \langle p^{in}(t)p^{in}(t')\rangle = \delta(t - t') \ , \tag{8.98}$$

$$\langle q^{in}(t)p^{in}(t')\rangle = \langle p^{in}(t)q^{in}(t')\rangle^* = i\delta(t - t') \ , \tag{8.99}$$

From an experimental perspective the noise response in the frequency domain is more useful. Thus we define

$$\delta y(\omega) = \int_{-\infty}^{\infty} dt \ e^{i\omega t}\delta y(t) \tag{8.100}$$

and similar expressions for the other variables. As $\delta y(t)$ is Hermitian we have $\delta y(\omega) = \delta y(-\omega)^\dagger$. The two time correlation functions for the variables are then determined by

$$\langle\delta y(t)\delta y(0)\rangle = \int_{-\infty}^{\infty} d\omega e^{-i\omega t}\langle\delta y(\omega)\delta y^\dagger(\omega)\rangle \tag{8.101}$$

and similar expressions for the other quantities. Thus our objective is to calculate the signal variance

$$V_{I_-}(\omega) = \langle I_-(\omega)I_-(\omega)^\dagger\rangle \ . \tag{8.102}$$

In order to reproduce the δ-correlated noise terms of (8.96–8.99), the correlation function in the frequency domain must be

$$\langle \delta x^{\mathrm{in}}(\omega) \delta x^{\mathrm{in}}(\omega')^\dagger \rangle = \langle \delta y^{\mathrm{in}}(\omega) \delta y^{\mathrm{in}}(\omega') \rangle = \delta(\omega - \omega') \,, \tag{8.103}$$

$$\langle \delta x^{\mathrm{in}}(\omega) \delta y^{\mathrm{in}}(\omega')^\dagger \rangle = \langle \delta y^{\mathrm{in}}(\omega) \delta x^{\mathrm{in}}(\omega')^\dagger \rangle^* = \mathrm{i}\delta(\omega - \omega') \,, \tag{8.104}$$

$$\langle q^{\mathrm{in}}(\omega) q^{\mathrm{in}}(\omega')^\dagger \rangle = \langle p^{\mathrm{in}}(\omega) p^{\mathrm{in}}(\omega')^\dagger \rangle = \delta(\omega - \omega') \,, \tag{8.105}$$

$$\langle q^{\mathrm{in}}(\omega) p^{\mathrm{in}}(\omega')^\dagger \rangle = \langle p^{\mathrm{in}}(\omega) q^{\mathrm{in}}(\omega')^\dagger \rangle^* = \mathrm{i}\delta(\omega - \omega') \,, \tag{8.106}$$

We now directy transform the equations of motion and solve the resulting algebraic equations for the frequency components. The result for the crucial field variable is

$$\delta y(\omega) = A \delta x^{\mathrm{in}}(\omega) + B \delta y^{\mathrm{in}}(\omega) + C_q^{\mathrm{in}}(\omega) + D_p^{\mathrm{in}}(\omega) \,, \tag{8.107}$$

where

$$A = \frac{\mu^2 \Omega \sqrt{\gamma}}{\Lambda(\omega) \left(\frac{\gamma}{2} - \mathrm{i}\omega \right)} \,,$$

$$B = \frac{\sqrt{\gamma}}{\frac{\gamma}{2} - \mathrm{i}\omega} \,,$$

$$C = \frac{-\mu \sqrt{\Gamma} \left(\frac{\Gamma}{2} - \mathrm{i}\omega \right)}{\Lambda(\omega) \left(\frac{\gamma}{2} - \mathrm{i}\omega \right)} \,,$$

$$D = \frac{-\mu \sqrt{\Gamma} \Omega}{\Lambda(\omega) \left(\frac{\gamma}{2} - \mathrm{i}\omega \right)} \,. \tag{8.108}$$

$$\Lambda(\omega) = \left(\frac{\Gamma}{2} - \mathrm{i}\omega \right)^2 + \Omega^2 \,. \tag{8.109}$$

Thus

$$\langle y(\omega) y^\dagger(\omega) \rangle = |A|^2 \langle \delta x^{\mathrm{in}}(\omega) \delta x^{\mathrm{in}}(\omega)^\dagger \rangle + |B|^2 \langle \delta y^{\mathrm{in}}(\omega) \delta y^{\mathrm{in}}(\omega)^\dagger \rangle$$
$$+ |C|^2 \langle q^{\mathrm{in}}(\omega) q^{\mathrm{in}}(\omega)^\dagger \rangle + |D|^2 \langle p^{\mathrm{in}}(\omega) p^{\mathrm{in}}(\omega') \rangle$$
$$+ (AB^* \langle \delta x^{\mathrm{in}}(\omega) \delta y^{\mathrm{in}}(\omega)^\dagger \rangle + \mathrm{c.c.})$$
$$+ (CD^* \langle q^{\mathrm{in}}(\omega) p^{\mathrm{in}}(\omega)^\dagger \rangle + \mathrm{c.c.}) \tag{8.110}$$

It is now constructive to consider the physical interpretation of each term. The first term proportional to the in-phase field amplitude is the error in the output intensity due to radiation pressure fluctuations. The second term is the error due to the out-of-phase amplitude of the field, i.e. the intrinsic phase fluctuations. The second and third terms are the fluctuations in mirror position and momentum due to intrinsic mirror fluctuations and radiation pressure. The fourth term represents correlations between the amplitude and the phase of the field due to radiation pressure modulating the length of the cavity. In a similar way the final term is the correlation between the position and momentum of the mirror as the radiation pressure changes the momentum which is coupled back to the position under free evolution.

Define the normalised variance by

$$N(\omega) = \frac{V_{I-}(\omega)}{2I} , \qquad (8.111)$$

where I is the output intensity from each cavity. This quantity is given by

$$N(\omega) = 1 + \frac{16\kappa^2 I \Gamma \left(\frac{\Gamma^2}{4} + \Omega^2 + \omega^2 \right)}{|\Lambda(\omega)|^2 \left| \frac{\gamma}{2} - i\omega \right|^2} + \frac{(16\kappa^2 I)^2 \Omega^2}{|\Lambda(\omega)|^2 \left| \frac{\gamma}{2} - i\omega \right|^4} . \qquad (8.112)$$

The first term in (8.112) is the shot-noise of the incident light on the detector, the second term arises from the intrinsic (zero-point) fluctuations in the positions of the end mirrors, while the last term represents the radiation pressure noise.

In Fig. 8.13 we display the total noise $N(\omega)$ as a function of frequency (a) (solid line) together with the contributions to the noise from: (b) photon-counting noise (dashed line); (c) mirror noise (dash-dot line); (d) radiation-pressure noise (dotted line). Typical interferometer parameters, summarised in Table 8.1 were used.

From signal processing theory, a measurement at frequency ω_g of duration τ entails an error Δh in the displacement h given by

$$\Delta h^2 = \frac{2S(\omega_g)}{\tau V_{I-}(\omega_g)} . \qquad (8.113)$$

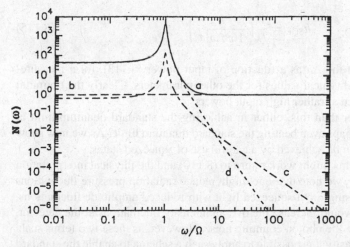

Fig. 8.13 The normalized variance for the fluctuations in the intensity difference versus frequency. The solid line (a) represents the total noise, (b) represents the photon counting noise, (c) represents the mirror noise and (d) represents the radiation pressure noise. The interferometer parameters used are given in Table 8.1

Table 8.1 The values of the experimental parameters used in the graphs

Quantity	Symbol	Value
Mass of mirror	M	10 kg
Mirror characteristic angular frequency	Ω	20π rad s^{-1}
Mirror damping	γ_b	2π rad s^{-1}
Length of cavity	L	4 m
Reflectivity	R	0.98
Laser power	P	10 W
Laser angular frequency	ω_0	3.66×10^5 rad s^{-1}
Gravity-wave-angular frequency	ω_g	2000π rad s^{-1}

We may now substitute the expressions for the signal frequency components $S(\omega_g)$ and the noise at this frequency to obtain an error which depends on the input intensity I (or equivalently the input power $P = 2\hbar\omega_0 I$). The error may then be minimised with respect to I to give minimum detectable displacement $h_{\min}$. In the limit $\omega_g^2 \gg \Gamma^2 + \Omega^2$, the appropriate limit for practical interferometers we find

$$h_{\min}^2 = \frac{\hbar}{32 M \omega_g^2 L^2 \tau \Omega}(2\Omega + \Gamma). \tag{8.114}$$

The first term in this expression is due to the light fluctuations whereas the second term is due to the intrinsic quantum noise in the end mirrors. If we neglect the mirror-noise contribution we find the 'standard quantum limit'

$$h_{SQL} = \frac{1}{L}\left(\frac{\hbar}{16 M \omega_g^2 \tau}\right)^{1/2}. \tag{8.115}$$

In Fig. 8.14 we plot the Δh as a function of input power (8.113), for a measurement time of 1 s, and typical values for the other parameters. Clearly the optimum sensitivity is achieved at rather high input powers.

Can one do better than this, either in achieving the standard quantum limit at lower powers or perhaps even beating the standard quantum limit? As we now show both these results can be achieved by a careful use of squeezed states.

To see now how this might work return to (8.110) and the physical interpretation of each term. Firstly, we note that one might reduce radiation pressure fluctuations (the first term) by using input squeezed light with reduced amplitude fluctuations. Unfortunately, this would increase the overall intensity fluctuations at the detector, i.e. it would increase the photon counting noise. However, as these two terms scale differently with intensity it is possible to apply such a scheme to enable the standard quantum limit to be achieved at lower input power. This is indeed the conclusion of *Caves* [16] in a calculation which focussed entirely on these terms. However, one can actually do better by using squeezed states to induce correlations between the

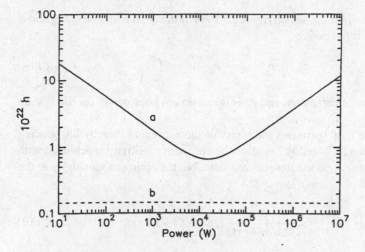

Fig. 8.14 The error in the fractional length change versus input power for a measurement time of one second. Parameters are as in Table 8.1

in-phase and out-of-phase quadratures of the field. In fact, if one chooses the phase of the squeezing (with respect to the input laser) carefully the fifth term in (8.110) can be made negative with a consequent improvement in the overall sensitivity of the device.

We will not present the details of this calculation [20], but summarise the results with reference to Fig. 8.15. Firstly, if we simply squeeze the fluctuations in $\hat{x}^{in}$ without changing the vacuum correlations between $\hat{x}^{in}$ and $\hat{y}^{in}$, the standard quantum

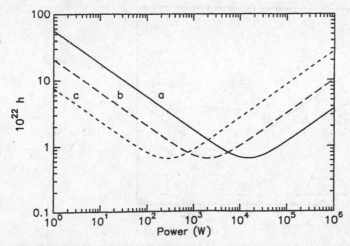

Fig. 8.15 The minimum possible detectable gravitational wave amplitude h as a function of power using amplitude squeezed light at the input and for three different squeeze parameters; (a) $r = 0$; (b) $r = 1$; (c) $r = 2$

limit (8.115) is the optimum sensitivity regardless of the degree of squeezing and it is achieved for the input power

$$P_{ss} = e^{-2r} P_0 , \tag{8.116}$$

where r is the squeeze parameter, and P_0 is the optimum laser power for the system with no squeezing.

However, if one now optimises the phase of the squeezing thereby introducing correlations between $\delta \hat{x}^{in}$ and $\delta \hat{y}^{in}$ we find the optimum sensitivity is achieved with the *same* input power P_0 as the unsqueezed state, but the optimum sensitivity in the appropriate limit is

$$h_{min}^2 \approx \frac{\hbar}{32 M \omega_g^2 L^2 \tau \Omega} (2e^{-2|r|} \Omega + \Gamma) . \tag{8.117}$$

clearly this may be made much smaller than the standard quantum limit. For Lightly squeezed input light the sensitivity is ultimately limited by the intrinsic quantum fluctuations in the positions of the end mirrors. The optimum phase of squeezing is $\pi/4$ which is the angle at which maximum correlation between $\hat{x}^{in}$ and $\hat{y}^{in}$ occurs, i.e., the error ellipse has the same projection onto the in-phase and out-of-phase directions. The exact results are shown in Fig. 8.15 for the same parameters, as employed in Fig. 8.15. Shown is the minimum-possible value of h detectable as a function of power at the optimum phase of squeezing, for three different values of the squeeze parameter. Also exhibited is the noise floor due to the intrinsic quantum fluctuations of the mirror positions.

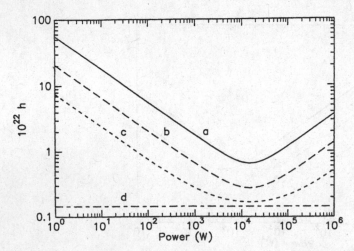

Fig. 8.16 The minimum possible detectable amplitude h as a function of input power when the phase of the input squeezed light is optimized, for three different values of the squeeze parameter (a) $r = 0$; (b) $r = 1$; (c) $r = 2$. Also shown is the mirror noise contribution (d)

In summary, the experimentalist can apply a squeezed input to a gravitational wave interferometer in two ways. Either the maximum sensitivity of the device can be greatly increased but achieved at a rather high input power, or the standard quantum limit can be achieved at input powers less threatening to the life of the optical components of the interferometer.

8.3.2 Sub-Shot-Noise Phase Measurements

The second major application of squeezed light is to the detection of very small phase shifts. A Mach–Zehnder interferometer (Fig. 8.17) can be used to determine a phase shift introduced in one arm.

Assuming 50:50 beam splitters the relationship between the input and output field operators is

$$a_0 = e^{i\theta/2}\left(\cos\frac{\theta}{2}a_i + \sin\frac{\theta}{2}b_i\right), \tag{8.118}$$

$$b_0 = e^{i\theta/2}\left(\cos\frac{\theta}{2}b_i + \sin\frac{\theta}{2}a_i\right), \tag{8.119}$$

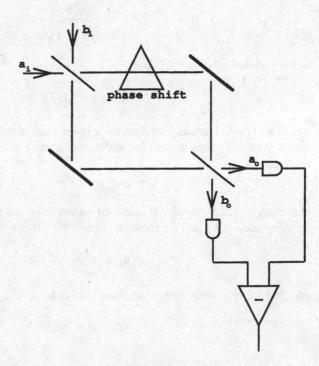

Fig. 8.17 Schematic representation of an experiment designed to measure a phase shift below the shot-noise limit

where θ is the phase difference between the two arms. The two output fields are directed onto two photo-detectors and the resulting currents combined with a $180°$ power combiner. This realises a measurement of the photon number difference

$$c_o^\dagger c_o = a_o^\dagger a_o - b_o^\dagger b_o$$
$$= \cos\theta(a_i^\dagger a_i - b_i^\dagger b_i) - \mathrm{i}\sin\theta(a_i b_i^\dagger - a_i^\dagger b_i) . \tag{8.120}$$

In standard interferometry the input a_i is a stabilised cw laser while b_i is the vacuum state. However, as we shall show, smaller phase changes may be detected if b_i is prepared in a squeezed vacuum state.

Assuming a_i is in the coherent state $|\alpha\rangle$ while b_i is in the squeezed state $|0, r\rangle$, the mean and variance of the photon number difference at the output is

$$\langle c_-^\dagger c_-\rangle = \cos\theta(|\alpha|^2 - \sinh^2 r) \tag{8.121}$$

$$V(c_-^\dagger c_-) = \cos^2\theta(|\alpha|^2 + \sinh^2 r\cosh 2r) + \sin^2\theta[|\alpha|^2(1 - 2\sinh^2 r)]$$
$$- \frac{1}{2}(\alpha^2 + \alpha^{*2})\sinh 2r + \sinh^2 r] . \tag{8.122}$$

If we now set $\theta = \pi/2 + \delta\theta$, then when phase shift $\delta\theta$ is zero, the mean signal is zero. That is we operate on a null fringe. The Signal-to-Noise Ratio (SNR) is defined by

$$\mathrm{SNR} = \frac{\langle c_-^\dagger c_-\rangle}{\sqrt{V(c_-^\dagger c_-)}} . \tag{8.123}$$

In the standard scheme $r = 0$ and

$$\mathrm{SNR} = \bar{n}^{1/2}\sin\delta\theta \tag{8.124}$$

where $\bar{n} = |\alpha|^2$. The smallest detectable phase shift is defined to be that phase for which $\mathrm{SNR} = 1$. Thus the minimum detectable phase shift for coherent state interferometry is

$$\delta\theta_{\min} = \bar{n}^{-1/2} . \tag{8.125}$$

However, if b_i is prepared in a squeezed vacuum state with squeezing in phase with the amplitude α we find for moderate squeezing ($|\alpha|^2 \gg \sinh^2 r$)

$$\mathrm{SNR}_{ss} = \bar{n}^{1/2}e^r\sin\delta\theta \tag{8.126}$$

and thus the minimum detectable phase change is

$$\delta\theta_{\min} = \bar{n}^{1/2}e^r . \tag{8.127}$$

The minimum detectable phase change may thus be much smaller than for coherent state interferometry, provided we choose $r < 0$ i.e. *phase squeezing*.

Such an enhancement has been reported by *Xiao* et al. [22] in an experiment on the measurement of phase modulation in a Mach Zehnder interferometer. They reported on an increase in the signal-to-noise ratio of 3 dB relative to the shot-noise limit when squeezed light from an optical parametric oscillator is injected into a port of the interferometer. A comparison of the fluctuations in the difference current for the cases of squeezed and a vacuum input is shown in Fig. 8.18. A similar experiment was performed by *Grangier* et al. [23] employing a polarization interferometer which is equivalent to a Mach–Zehnder scheme. In their experiment an enhancement factor of 2 dB was achieved.

8.3.3 Quantum Information

Squeezed states are being applied to new protocols in quantum information which we discuss in. In Chap. 16. Quantum information is concerned with communication and computational tasks enabled by quantum states of light, including squeezed states. One such application is quantum teleportation in which an unknown quantum state is transferred from one subsystem to another using the correlations inherent in a two mode squeezed state.

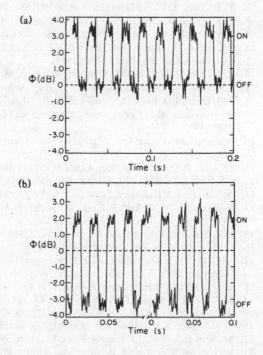

Fig. 8.18 A comparison of the level of fluctuations in the differenced-photocurrent for a Mach–Zehnder interferometer versus time as the phase difference is varied at a frequency of 1.6 MHz. Curve in (a) is for the case of vacuum state input, curve (b) uses squeezed state input. The dashed line gives the vacuum level with no phase modulation [21]

Exercises

8.1 One of the input modes of a beam splitter, with transmitivity T, is prepared in a coherent state, $|\alpha\rangle$ and the other in a squeezed vacuum state $|0, r\rangle$. Show that in the limit $T \to 1$, $|\alpha| \to \infty$ with $\sqrt{(1-T)}|\alpha| = \beta$ fixed, one of the output states is a squeezed state with a coherent amplitude β.

8.2 Calculate the squeezing spectrum for parametric oscillation in a cavity that has different losses at each mirror for the fundamental frequency, ω_1.

8.3 Calculate the spectrum of fluctuations in the difference intensity, $I_1 - I_2$, if an intracavity loss is present at the idler frequency.

8.4 Two photon absorption inside a cavity can be modeled by coupling the cavity mode to a bath via the interaction Hamiltonian

$$\mathcal{H} = a^2 \Gamma^\dagger + (a^\dagger)^2 \Gamma$$

where Γ is a reservoir operator and the reservoir is at zero temperature. Determine the squeezing spectrum.

References

1. P.D. Drummond, K.J. McNeil, D.F. Walls: Optica Acta **28**, 211 (1981)
2. M.J. Collett, C.W. Gardiner: Phys. Rev. **30**, 1386 (1984)
3. M.J. Collett, D.F. Walls: Phys. Rev. A **32**. 2887 (1985)
4. L.A. Wu, M. Xiao, H.J. Kimble: J. Opt. Soc. Am. B **4**, 1465 (1987)
5. P.K. Lam, T.C. Ralph, B.C. Buchler, D. E. McClelland, H.-A. Bachor, J. Gao: J. Opt. B: Quantum Semi-class. Opt. **1**, 469 (1999)
6. A.S. Lane, M.D. Reid, D.F. Walls: Phys. Rev. A **38**, 788 (1988)
7. S. Reynaud, C. Fabre, E. Giacobino: J. OPt. Soc Am. B **4**, 1520 (1987)
8. A.S. Lane, M.D. Reid, D.F. Walls: Phys. Rev. Lett. **60**, 1940 (1988)
9. A. Heidmann, R.J. Horowicz, S. Reynaud, E. Giacobino, C. Fabre: Phys. Rev. Lett. **59**, 2555 (1987)
10. Jiangrui Gao, Fuyun Cui, Chenyang Xue, Changde Xie, Peng Kunchi: Opts. Letts. **23**, 870 (1998)
11. O. Aytür, P. Kumar: Phys. Rev. Lett. **65**, 1551 (1990)
12. S.F. Pereira, M. Xiao, H.J. Kimble, J.L. Hall: Phys. Rev. A, **38**, 4931 (1989)
13. A. Sizman, R.J. Horowicz, G. wagner, G. Leuchs: Opt. Comms. **80**, 138 (1990)
14. R. Paschotta, M. Collett, P. Kurz, K. Fiedler, H.A. Bachor, J. Mlynek: Phys. Rev. Lett. **72**, 3807 (1994)
15. D.K. Serkland, P. Kumar, M.A. Arbore, M.M. Fejer: Opts. Letts., **22**, 1497 (1997)
16. C.M. Caves: Phys. Rev. Lett. **45**, 75 (1980)
17. W.G. Unruh: *In Quantum Optics, Experimental Gravitation and Measurement Thery*, ed. by P. Meystre, M.O. Scully (Plenum, New York 1983) p.647
18. R.S. Bondurant, J.H. Shapiro: Phys. Rev. D **30**, 2548 (1984)
19. M.T. Jaekel, S. Reynaud: Europhys. Lett. **13**, 301 (1990)
20. A.F. Pace, M.J. Collett, D.F. Walls: Phys. Rev. A **47**, 3173 (1993)
21. C.W. Misner, A.S. Thorne, J.A. Wheller: *Gravitation* (Freeman, San Francisco 1973)
22. M. Xiao, L.A. Wu, H.J. Kimble: Phys. Rev. Lett. **59**, 278 (1987)
23. P. Grangier, R.E. Slusher, B. Yurke, A. La Porta: Phys. Rev. Lett. **59**, 2153 (1987)

Further Reading

Bachor, H.A., J.C. Ralph: *A Guide to Experiments in Quantum Optics* (Wiley-VCH, 2004)

Giacobino, E., C. Fabre (guest eds.): *Quantum noise reduction in optical systems-experiments*, Appl. Phys. B **55**(3) (1992)

Kimble, H.J., D.F. Walls (guest eds.): *Special Issue on Squeeze States of the Electromagnetic Field*, J. Opt. Soc. Am. B **4**(10) (1987)

Chapter 9
Nonlinear Quantum Dissipative Systems

Abstract In the preceding chapter we derived linearised solutions to the quantum fluctuations occurring in some nonlinear systems in optical cavities. In these solutions the quantum noise has been treated as a small perturbation to the solutions of the corresponding nonlinear classical problem. It is not possible, in general, to find exact solutions to the nonlinear quantum equations which arise in nonlinear optical interactions. It has, however, been possible to find solutions to some specific systems. These solutions provide a test of the region of validity of the linearised solutions especially in the region of an instability. Furthermore they allow us to consider the situation where the quantum noise is large and may no longer be treated as a perturbation. In this case, manifestly quantum mechanical states may be produced in a nonlinear dissipative system.

We shall give solutions to the nonlinear quantum equations for two of the problems considered in Chap. 8, namely, the parametric oscillator and dispersive optical bistability.

9.1 Optical Parametric Oscillator: Complex P Function

We shall first solve for the steady state of the parametric oscillator using the complex P function. Then, we show, using the positive P function, that the steady state subharmonic field is in a superposition state. We go on to calculate the tunnelling time between the two states in the superposition.

We consider the degenerate parametric oscillator described in Chap. 8, following the treatment of *Drummond* et al. [1]. The Hamiltonian is

$$\mathcal{H} = \sum_{i=0}^{3} \mathcal{H}_i \tag{9.1}$$

where

$$\mathcal{H}_0 = \hbar\omega a_1^\dagger a_1 + 2\hbar\omega a_2^\dagger a_2 , \tag{9.2}$$

$$\mathcal{H}_1 = i\hbar\frac{\kappa}{2}(a_1^{\dagger 2}a_2 - a_1^2 a_2^\dagger)\,, \tag{9.3}$$

$$\mathcal{H}_2 = i\hbar(\varepsilon_2 a_2^\dagger e^{-2i\omega t} - \varepsilon_2^* a_2 e^{2i\omega t})\,, \tag{9.4}$$

$$\mathcal{H}_3 = a_1\Gamma_1^\dagger + a_2\Gamma_2^\dagger + \text{h.c}\,. \tag{9.5}$$

where a_1 and a_2 are the boson operators for two cavity modes of frequency ω and 2ω, respectively. κ is the coupling constant for the nonlinear coupling between the modes. The cavity is driven externally by a coherent driving field with frequency 2ω and amplitude ε_2. Γ_1, Γ_2 are the bath operators describing the cavity damping of the two modes.

We recall from Chap. 8 that there are two stable steady state solutions depending on whether the driving field amplitude is above or below the threshold amplitude $\varepsilon_2^c = \gamma_1\gamma_2/\kappa$. In particular, the steady states for the low frequency mode α_1 are

$$\alpha_1^0 = 0, \qquad \varepsilon_2 < \varepsilon_2^c\,,$$
$$\alpha_1^0 = \pm\left[\frac{2}{\kappa}(\varepsilon_2 - \varepsilon_2^c)\right]^{1/2}, \qquad \varepsilon_2 \geq \varepsilon_2^c\,. \tag{9.6}$$

The master equation for the density operator of the two modes is

$$\frac{\partial}{\partial t}\rho = \frac{1}{i\hbar}[\mathcal{H}_0 + \mathcal{H}_1 + \mathcal{H}_2, \rho] + \gamma_1(2a_1\rho a_1^\dagger - a_1^\dagger a_1\rho - \rho a_1^\dagger a_1)$$
$$+ \gamma_2(2a_2\rho a_2^\dagger - a_2^\dagger a_2\rho - \rho a_2^\dagger a_2) \tag{9.7}$$

where the irreversible part of the master equation follows from (6.44) for a zero-temperature bath. γ_1, γ_2 are the cavity damping rates.

This equation may be converted to a c-number Fokker–Planck equation using the generalized P representation discussed in Chap. 6. Using the operator-algebra rules described in Chap. 6, we arrive at the Fokker–Planck equation

$$\frac{\partial}{\partial t}P(\alpha) = \left\{\frac{\partial}{\partial\alpha_1}(\gamma_1\alpha_1 - \kappa\beta_1\alpha_2) + \frac{\partial}{\partial\beta_1}(\gamma_1\beta_1 - \kappa\alpha_1\beta_2)\right.$$
$$+ \frac{\partial}{\partial\alpha_2}\left(\gamma_2\alpha_2 - \varepsilon_2 + \frac{\kappa}{2}\alpha_1^2\right) + \frac{\partial}{\partial\beta_2}\left(\gamma_2\beta_2 - \varepsilon_2^* + \frac{\kappa}{2}\beta_1^2\right)$$
$$\left. + \frac{1}{2}\left[\frac{\partial^2}{\partial\alpha_1^2}(\kappa\alpha_2) + \frac{\partial^2}{\partial\beta_1^2}(\kappa\beta_2)\right]\right\}P(\alpha) \tag{9.8}$$

where $(\alpha) = [\alpha_1, \beta_1, \alpha_2, \beta_2]$.

An attempt to find the steady state solution of this equation by means of a potential solution fails since the potential conditions (6.73) are not satisfied.

We proceed by adiabatically eliminating the high-frequency mode. This may be accomplished best in the Langevin equations equivalent to (9.8).

$$\frac{\partial}{\partial t}\begin{pmatrix}\alpha_1\\\beta_1\end{pmatrix} = \begin{pmatrix}\kappa\beta_1\alpha_2 - \gamma_1\alpha_1 + \sqrt{\kappa\alpha_2}[\eta_1(t)]\\\kappa\alpha_1\beta_2 - \gamma_1\beta_1 + \sqrt{\kappa\beta_2}[\tilde{\eta}_1(t)]\end{pmatrix}$$

$$\frac{\partial}{\partial t}\begin{pmatrix}\alpha_2\\\beta_2\end{pmatrix} = \begin{pmatrix}\varepsilon_2 - \frac{\kappa}{2}\alpha_1^2 - \gamma_2\alpha_2\\\varepsilon_2^* - \frac{\kappa}{2}\beta_1^2 - \gamma_2\beta_2\end{pmatrix} \qquad (9.9)$$

where $\eta_1(t)$, $\tilde{\eta}_1(t)$ are delta correlated stochastic forces with zero mean

$$\langle\eta_1(t)\rangle = \langle\tilde{\eta}_1(t)\rangle = \langle\eta_1(t)\eta_1(t')\rangle = \langle\tilde{\eta}_1(t)\tilde{\eta}_1(t')\rangle = 0, \qquad (9.10)$$

$$\langle\eta_1(t)\tilde{\eta}_1(t)\rangle = \delta(t - t'). \qquad (9.11)$$

Under the conditions $\gamma_2 \gg \gamma_1$ we can adiabatically eliminate α_2 and β_2 which gives the resultant Langevin equation for α_1 and β_1

$$\frac{\partial}{\partial t}\begin{pmatrix}\alpha_1\\\beta_1\end{pmatrix} = \begin{pmatrix}\frac{\kappa}{\gamma_2}\left(\varepsilon_2 - \frac{\kappa}{2}\alpha_1^2\right)\beta_1 - \gamma_1\alpha_1\\\frac{\kappa}{\gamma_2}\left(\varepsilon_2^* - \frac{\kappa}{2}\beta_1^2\right)\alpha_1 - \gamma_1\beta_1\end{pmatrix} + \begin{pmatrix}\left[\frac{\kappa}{\gamma_2}\left(\varepsilon_2 - \frac{\kappa}{2}\alpha_1^2\right)\right]^{1/2}\eta_1(t)\\\left[\frac{\kappa}{\gamma_2}\left(\varepsilon_2^* - \frac{\kappa}{2}\beta_1^2\right)\right]^{1/2}\tilde{\eta}_1(t)\end{pmatrix}. \qquad (9.12)$$

The Fokker–Planck equation corresponding to these equations is

$$\frac{\partial}{\partial t}P(\alpha_1,\beta_1) = \left\{\frac{\partial}{\partial\alpha_1}\left[\gamma_1\alpha_1 - \frac{\kappa}{\gamma_2}\left(\varepsilon_2 - \frac{\kappa}{2}\alpha_1^2\right)\beta_1\right]\right.$$
$$+ \frac{\partial}{\partial\beta_1}\left[\gamma_1\beta_1 - \frac{\kappa}{\gamma_2}\left(\varepsilon_2^* - \frac{\kappa}{2}\beta_1^2\right)\alpha_1\right]$$
$$\left. + \frac{1}{2}\left[\frac{\partial^2}{\partial\alpha_1^2}\frac{\kappa}{\gamma_2}\left(\varepsilon_2 - \frac{\kappa}{2}\alpha_1^2\right) + \frac{\partial}{\partial\beta_1^2}\frac{\kappa}{\gamma_2}\left(\varepsilon_2^* - \frac{\kappa}{2}\beta_1^2\right)\right]\right\}P(\alpha_1\beta_1).$$
$$(9.13)$$

We set $\frac{\partial}{\partial t}P(\alpha_1,\beta_1) = 0$ and attempt to find a potential solution as given by (6.72). It is found as

$$F_1 = -2\left(\beta_1 - \frac{2\gamma_2\left(\gamma_1 - \frac{\kappa^2}{2\gamma_2}\right)\alpha_1}{2\kappa\varepsilon_2 - \kappa^2\alpha_1^2}\right) \qquad (9.14)$$

$$F_2 = -2\left(\alpha_1 - \frac{2\gamma_2\left(\gamma_1 - \frac{\kappa^2}{2\gamma_2}\right)\beta_1}{2\kappa\varepsilon_2^* - \kappa^2\beta_1^2}\right) \qquad (9.15)$$

It follows that the potential conditions

$$\frac{\partial F_1}{\partial\alpha_1} = \frac{\partial F_2}{\partial\beta_1} \qquad (9.16)$$

are satisfied.

The potential is obtained on integrating (9.14 and 9.15)

$$P(\alpha) = N \, \exp \left[2\alpha_1 \beta_1 + \frac{2\bar{\gamma}_1 \gamma_2}{\kappa^2} \ln(c^2 - \kappa^2 \alpha_1^2) + 2 \left(\frac{\bar{\gamma}_1 \gamma_2}{\kappa^2} \right)^* \ln(c^{*2} - \kappa^2 \beta_1^2) \right]$$

$$(9.17)$$

where

$$c = \sqrt{2\kappa\varepsilon_2}, \qquad \bar{\gamma}_1 = \gamma_1 - \frac{\kappa^2}{2\gamma_2}.$$

It is clear that this function diverges for the usual integration domain of the complex plane with $\beta_1 = \alpha_1^*$. The observable moments may, however, be obtained by use of the complex P representation. The calculations are described in Appendix 9.A.

The semi-classical solution for the intensity exhibits a threshold behaviour at $\varepsilon_2 = \varepsilon_2^c = \gamma_1 \gamma_2 / \kappa$. This is compared in Fig. 9.1 with the mean intensity $I = \langle \beta_1 \alpha_1 \rangle$ calculated from the solution (9.17), as shown in the Appendix 9.A. For comparison, the mean intensity when thermal fluctuations are dominant (Exercise 9.4) is also plotted. The mean intensity with thermal fluctuations displays the rounding of the transition familiar from classical fluctuation theory. The quantum calculation shows a feature never observed in a classical system where the mean intensity actually drops below the semi-classical intensity. This deviation from the semi-classical behaviour is most significant for small threshold photon numbers. As the parameter $\gamma_1 \gamma_2 / \kappa^2$ is increased the quantum mean approaches the semi-classical value.

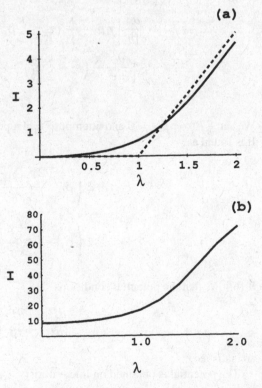

Fig. 9.1 A plot of the mean intensity for the degenerate parametric oscillator versus the scaled driving field λ. (a) The case of zero thermal fluctuations. The *dashed curve* represents the semi-classical intensity, the *solid curve* is the exact quantum result. In both cases $\mu^2 = 2\varepsilon_2^c/\kappa = 5.0$. Note that above threshold the exact quantum result is less than the semi-classical prediction. (b) The case of dominant thermal fluctuations. The mean thermal photon number is 10.0 and $\mu^2 = 2\varepsilon_2^c/\kappa = 100.0$

Fig. 9.2 The log variance of the squeezed (*solid*) and unsqueezed (*dashed*) quadrature in a degenerate parametric amplifier versus the scaled driving field with $\mu^2 = 2\varepsilon_2^c/\kappa = 5.0$

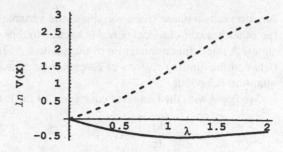

The variance of fluctuations in the quadratures $X_1 = a_1 + a_1^\dagger$ and $X_2 = (a_1 - a_1^\dagger)/\mathrm{i}$ is given by

$$\Delta X_1^2 = [\langle(\alpha_1 + \beta_1)^2\rangle - (\langle\alpha_1 + \beta_1\rangle)^2] + 1\,, \tag{9.18}$$

$$\Delta X_2^2 = -[\langle(\alpha_1 - \beta_1)^2\rangle - (\langle\alpha_1 - \beta_1\rangle)^2] + 1\,. \tag{9.19}$$

The variance in the quadratures is plotted in Fig. 9.2a versus the scaled driving field λ. The variance in the phase quadrature X_2 reaches a minimum at threshold. This minimum approaches $\frac{1}{2}$ as the threshold intensity is increased [10]. The value of one half in the variance of the intracavity field corresponds to zero fluctuations found at the resonance frequency in the external field. The fluctuations in the amplitude quadrature X_1 increase dramatically as the threshold is approached. However, unlike the calculation where the pump is treated classically the fluctuations do not diverge. This is because (9.17) is an exact solution to the nonlinear interaction including pump depletion. As the threshold value increases and therefore the number of pump photons required to reach threshold increases, the fluctuations become larger. In the limit $\gamma_1\gamma_2/\kappa^2 \to \infty$ the fluctuations diverge, as this corresponds to the classical pump (infinite energy). The variance in the amplitude quadrature above threshold continues to increase as the distribution is then double-peaked at the two stable output amplitudes.

The above solution demonstrates the usefulness of the complex P representation. Although the solution obtained for the steady state distribution has no interpretation in terms of a probability distribution, the moments calculated by integrating the distribution on a suitable manifold correspond to the physical moments. We have demonstrated how exact moments of a quantized intracavity field undergoing a nonlinear interaction may be calculated. To calculate the moments of the external field however, we must resort to linearization techniques.

9.2 Optical Parametric Oscillator: Positive P Function

As an alternative to the foregoing description we may consider the use of the positive P representation, following the treatment of *Wolinsky* and *Carmichael* [2]. We can obtain an analytic solution for the steady state positive P function. This solution is a function of two phase space variables; one variable is the *classical field* amplitude, the other is a *non-classical variable* needed to represent superpositions of coherent states. A three-dimensional plot of the positive P function allows one to distinguish between the limiting regions of essentially classical behaviour and predominantly quantum behaviour.

We begin with the Langevin equations for the low frequency mode

$$\frac{d\alpha}{d\tau} = -\alpha - \beta(\lambda - \alpha^2) + g(\lambda - \alpha^2)^{1/2}\eta_1 , \tag{9.20}$$

$$\frac{d\beta}{d\tau} = -\beta - \alpha(\lambda - \beta^2) + g(\lambda - \beta^2)^{1/2}\eta_2 , \tag{9.21}$$

where τ is measured in cavity lifetimes (γ_1^{-1}),

$$g = \frac{\kappa}{(2\gamma_1\gamma_2)^{1/2}} \equiv \frac{1}{\mu} , \tag{9.22}$$

and λ is a dimensionless measure of the pump field amplitude scaled to give the threshold condition $\lambda = 1$, and we have scaled the c-number variables by

$$\alpha = g\alpha_1 , \qquad \beta = g\beta_1 . \tag{9.23}$$

Equations (9.20 and 9.21) describe trajectories in a four-dimensional phase space. The region of phase space satisfying the conjugacy condition $\beta = \alpha^*$ is called the *classical subspace*. Two extra non-classical dimensions are required by the quantum noise. If we neglect the fluctuating forces η_1 and η_2 (9.20 and 9.21) have the stable steady state solution $\alpha = \beta = 0$ below threshold $(\lambda < 1)$, and $\alpha = \beta = \pm(\lambda - 1)^{1/2}$ above threshold $(\lambda > 1)$. In the full phase space there are additional steady states which do not satisfy the conjugacy condition: two steady states $\alpha = \beta = \pm i(1 - \lambda)^{1/2}$ below threshold and two steady states $\alpha = -\beta = \pm(\lambda + 1)^{1/2}$ both below and above threshold.

The variables α and β are restricted to a bounded manifold $\alpha = x$, $\beta = y$ with x and y both real and $|x|, |y| \leq \sqrt{\lambda}$. We denote this manifold by $\Lambda(x, y)$. Trajectories are confined within this manifold by reflecting boundary conditions. If a trajectory starts within this manifold, then it is clear from (9.20 and 9.21) that the drift and noise terms remain real, so a trajectory will remain on the real plane. Furthermore, at the boundary, the trajectory must follow the deterministic flow inwards, as the transverse noise component vanishes. If the initial quantum state is the vacuum state, the entire subsequent evolution will be confined to this manifold.

The manifold $\Lambda(x, y)$ is alternatively denoted by $\Lambda(u, v)$ with $u = \frac{1}{2}(x+y)$, $v = \frac{1}{2}(x-y)$. The line $v = 0$ is a one-dimensional classical subspace, the subspace preserving $\alpha = \beta$. The variable v denotes a transverse, non-classical dimension used by the noise to construct manifestly non-classical states.

We may now construct a pictorial representation of these states which dramatically distinguishes between the quantum and classical regimes.

With $\alpha = x$, $\beta = y$ both real, the solution to the Fokker-Planck equation (9.13) is of the form given by (9.17). With $|x|$, $|y| \leq \sqrt{\lambda}$

$$P_{ss}(x, y) = N[(\lambda - x^2)(\lambda - y^2)]^{1/g^2 - 1} e^{2xy/g^2} . \tag{9.24}$$

For weak noise ($g \ll 1$), $P_{ss}(x, y)$ is illustrated in Fig. 9.3. Below threshold ($\lambda < 1$) $P_{ss}(u, v)$ may be written

$$P_{ss}(u, v) = \frac{(1 - \lambda^2)^{1/2}}{\pi \lambda g / 2} \exp\left(\frac{-(1 - \lambda)u^2 + (1 + \lambda)v^2}{\frac{\lambda g^2}{2}}\right) . \tag{9.25}$$

The normally-ordered field quadrature variances are determined by the quantities

$$\langle : \Delta X_1^2 : \rangle = V\left(\frac{\alpha + \beta}{2}\right) , \tag{9.26}$$

$$\langle : \Delta X_2^2 : \rangle = V\left(\frac{\alpha - \beta}{2}\right) \tag{9.27}$$

where $V(z)$ refers to the variance over the stationary distribution function. As $u = (\alpha + \beta)/2$ and $v = \mathrm{i}(\alpha - \beta)/2$, on the manifold $\Lambda(u, v)$, the quadrature variances are given by

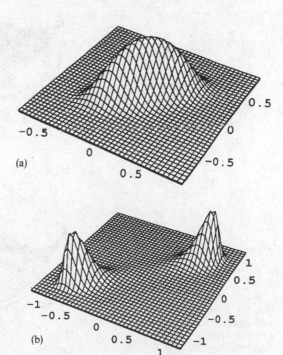

(a)

(b)

Fig. 9.3 A plot of the positive P representation of the steady state of the degenerate parametric amplifier, below and above threshold: (a) $\lambda = 0.8$ (b) $\lambda = 1.5$. In both cases $g = (2\varepsilon_2^c/\kappa)^{-1/2} = 0.25$

$$\langle :\Delta X_1^2: \rangle = V(u)/g^2 , \qquad (9.28)$$

$$\langle :\Delta X_2^2: \rangle = -V(v)/g^2 . \qquad (9.29)$$

The variances $g^{-2}\langle \Delta u^2 \rangle$ and $-g^{-2}\langle \Delta v^2 \rangle$ correspond to the normally ordered variances of the unsqueezed and squeezed quadratures, respectively, of the subharmonic field.

The threshold distribution ($g \ll 1$, $\lambda = 1$) is given by

$$P_{ss}(u,\, v) = \left[4\sqrt{\pi} g^{3/2} \Gamma \left(\frac{1}{4} \right) \right] e^{-(u^4+4v^2)/g^2} . \qquad (9.30)$$

Above threshold the distribution splits into two peaks. We note that in the low-noise regime $P_{ss}(x,\, y)$ is a slightly broadened version of the classical steady state with only a small excursion into the nonclassical space.

Figure 9.4 shows $P_{ss}(x,\, y)$ for the same values of λ as Fig. 9.3 but for the noise strength $g = 1$. The quantum noise has become sufficiently strong to explore

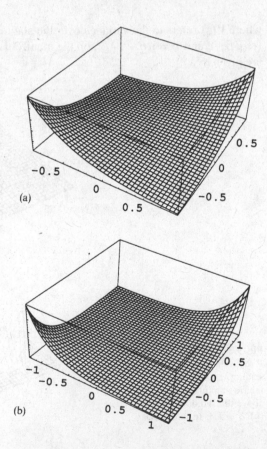

(a)

(b)

Fig. 9.4 As in Fig. 9.3 but with quantum noise parameter $g = 1.0$. **(a)** $\lambda = 0.8$ **(b)** $\lambda = 1.5$

thoroughly the non-classical dimension of the phase space. $P_{ss}(x, y)$ is strongly influenced by the boundary $\Lambda(x, y)$.

As the noise strength g is increased beyond 1, the characteristic threshold behaviour of the parametric oscillator disappears and squeezing is significantly reduced (Fig. 9.5). In the large-g limit the stochastic trajectories are all driven to the boundary of $\Lambda(x, y)$, and then along this boundary to the corners, where both noise terms in (9.20 and 9.21) vanish. $P_{ss}(x, y)$ approaches a sum of δ functions

$$P_{ss}(x,y) = \frac{1}{2}(1 + e^{4\lambda/g^2})^{-1}[\delta(x - \sqrt{\lambda})\delta(y - \sqrt{\lambda})$$
$$+ \delta(x + \sqrt{\lambda})\delta(y + \sqrt{\lambda})] + \frac{1}{2}(1 + e^{4\lambda/g^2})^{-1}$$
$$\times [\delta(x - \sqrt{\lambda})\delta(y + \sqrt{\lambda}) + \delta(x + \sqrt{\lambda})\delta(y - \sqrt{\lambda})]. \quad (9.31)$$

The two δ functions that set $x = -y = \pm\sqrt{\lambda}$ represent off-diagonal or interference terms $e^{-2\sqrt{\lambda}/g}|\sqrt{\lambda}/g\rangle\langle-\sqrt{\lambda}/g|$. Figure 9.6a–c illustrates the behaviour of $P_{ss}(x, y)$ as a function of λ in the strong-noise limit. When $4\lambda/g^2 \ll 1$ all δ functions carry equal weight and the state of the subharmonic field is the coherent state superposition $\frac{1}{2}(|\sqrt{\lambda}/g\rangle + |-\sqrt{\lambda}/g\rangle)$. As λ increases, this superposition state is replaced by a classical mixture of coherent states $|\sqrt{\lambda}/g\rangle$ and $|-\sqrt{\lambda}/g\rangle$ for $4\lambda/g^2 \gg 1$. This is a consequence of the competition between the creation of quantum coherence by the parametric process and the destruction of this coherence by dissipation. It will be shown in Chap. 15, that the decay of quantum coherence in a damped superposition state proceeds at a rate proportional to the phase space separation of the states.

This example has illustrated how quantum dissipative systems can exhibit manifestly quantum behaviour in the limit of large quantum noise. This is outside the realm of linear noise theory where classical states are only slightly perturbed.

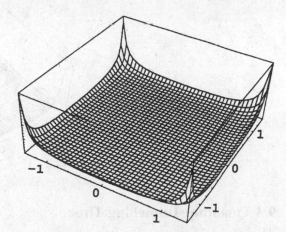

Fig. 9.5 As in Fig. 9.3 but with $\lambda = 1.5$ and $g = 10.0$

Fig. 9.6 As in Fig. 9.3 but
demonstrating the dependence
on λ with $g = 5.0$. (**a**) $\lambda = 1.0$
(**b**) $\lambda = 5.0$ (**c**) $\lambda = 10.0$

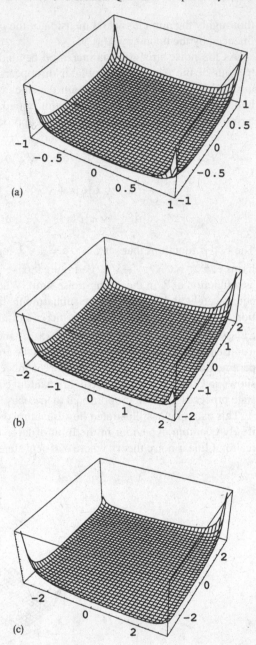

(a)

(b)

(c)

9.3 Quantum Tunnelling Time

We proceed to calculate the quantum tunnelling time between the two stable states.
We shall follow the procedure of *Kinsler* and *Drummond* [3]. In order to calculate

the quantum tunnelling rate, we shall transform the variables α and β to give constant diffusion, or additive stochastic noise.

$$u = \sin^{-1}\left(\frac{g\alpha}{\sqrt{\lambda}}\right) + \sin^{-1}\left(\frac{g\beta}{\sqrt{\lambda}}\right), \tag{9.32}$$

$$v = \sin^{-1}\left(\frac{g\alpha}{\sqrt{\lambda}}\right) - \sin^{-1}\left(\frac{g\beta}{\sqrt{\lambda}}\right). \tag{9.33}$$

These new variables are constrained to have a range such that $|u| + |v| \leq \pi$. Referring back to the variables α and β, it can be seen that the u axis represents the classical subspace of the phase space where $\alpha = \beta$. Thus the variable v is a non-classical dimension which allows for the creation of quantum features. The stochastic equations corresponding to these variables are

$$du = \left\{\lambda \sin(u) - \sigma\left[\tan\left(\frac{u+v}{2}\right) + \tan\left(\frac{u-v}{2}\right)\right]\right\}d\tau + \sqrt{2}g\,dW_u, \tag{9.34}$$

$$dv = \left\{-\lambda \sin(v) - \sigma\left[\tan\left(\frac{u+v}{2}\right) - \tan\left(\frac{u-v}{2}\right)\right]\right\}d\tau + \sqrt{2}g\,dW_v. \tag{9.35}$$

Here $\sigma = 1 - g^2/2$, dW_u, dW_v are Wiener processes.

These Ito equations have a corresponding Fokker–Planck equation and a probability distribution in the limit as $\tau \to \infty$ of

$$P(u, v) = N \exp[-V(u, v)/g^2] \tag{9.36}$$

where the potential $V(u, v)$ is

$$V(u, v) = -2\sigma \ln|\cos u + \cos v| + \lambda \cos u - \lambda \cos v. \tag{9.37}$$

Above threshold the potential has two minima corresponding to the stable states of the oscillator. These minima have equal intensities and amplitudes of opposite sign, and are at classical locations with $\alpha = \alpha^*$

$$(u_0, v_0) = (\pm 2\sin^{-1}[(\lambda - \sigma)^{1/2}/\sqrt{\lambda}], 0) \tag{9.38}$$

or

$$g\alpha_0 = \pm(\lambda - 1 + g^2)^{1/2}. \tag{9.39}$$

There is also a saddle point at $(u_s, v_s) = (0, 0)$.

Along the u axis the second derivative of the potential in the v direction is always positive. The classical subspace $(v = 0)$ is therefore at a minimum of the potential with respect to variations in the non-classical variable v. This valley along the u axis between the two potential wells is the most probable path for a stochastic trajectory in switching from one well to the other. The switching rate between them will be dominated by the rate due to trajectories along this route. Using an extension of Kramer's method, developed by *Landauer* and *Swanson* [4], the mean time taken for the oscillator to switch from one state to the other in the limit of $g^2 \ll 1$ is

$$T_p = \frac{\pi}{\gamma_1} \left(\frac{\lambda + \sigma}{\lambda(\lambda - \sigma)^2} \right)^{1/2} \exp\left\{ \frac{2}{g^2} \left[\lambda - \sigma - \sigma \ln\left(\frac{\lambda}{\sigma} \right) \right] \right\}. \qquad (9.40)$$

The switching time is increased as the pump amplitude λ is increased or the nonlinearity g^2 is reduced.

Previous attempts to compute the tunnelling time for this problem have used the Wigner function [5]. Unfortunately the time-evolution equation for the Wigner function contains third-order derivative terms and is thus not a Fokker–Planck equation. In the case of linear fluctuations around a steady state truncating the evolution equation at second-order derivatives is often a good approximation. However, it is not clear that this procedure will give quantum tunnelling times correctly.

In the limit of large damping in the fundamental mode the truncated Wigner function of the sub-harmonic mode obeys with $\tau = \gamma_1 t$

$$\frac{d}{d\tau} W(\beta, t) = \left\{ \frac{\partial}{\partial \beta} [\beta - \beta^*(\lambda - g^2 \beta^2)] + \frac{\partial}{\partial \beta^*} [\beta^* - \beta(\lambda - g^2 \beta^{*2})] \right.$$

$$\left. + \frac{\partial^2}{\partial \beta \partial \beta^*} (1 + 2g^2 \beta \beta^*) \right\} W(\beta, \tau). \qquad (9.41)$$

This truncated Wigner function equation does not have potential solutions, however an approximate potential solution can be obtained that is valid near threshold. Here, the noise contribution $2g^2 \beta \beta^*$ is small and is neglected leaving only subharmonic noise. Writing $\beta = x + ip$, the solution in the near threshold approximation is

$$W_{\mathrm{NT}} = N_{\mathrm{NT}} \exp[-V_{\mathrm{NT}}(x, p)] \qquad (9.42)$$

where

$$V_{\mathrm{NT}}(x, p) = \frac{2}{g^2} [g^2 x^2 + g^2 p^2 + \frac{1}{2}(g^2 x^2 + g^2 p^2)^2 - \lambda(g^2 x^2 - g^2 p^2)] \qquad (9.43)$$

and N_{NT} is the normalisation constant.

Above threshold this potential has two minima, at $gx = \pm(\lambda - 1)^{1/2}$. In the limit of large-threshold photon numbers, these minima are very close to those obtained in (9.39). The tunneling time has been calculated from the Wigner distribution by *Graham* [6], with the result

$$T_{\mathrm{w}} = \frac{\pi}{\gamma_1} \left(\frac{\lambda + 1}{\lambda(\lambda - 1)^2} \right)^{1/2} \exp\left[\frac{1}{g}(\lambda - 1)^2 \right]. \qquad (9.44)$$

This result is compared with the expression derived using the P function in Fig. 9.7 which shows the variation in the logarithm of the tunnelling rate with the pump amplitude λ. The Wigner function result predicts a slower switching time above threshold. The difference in the two predictions can be many orders of magnitude. The calculations from the exact positive P Fokker–Planck equation represent a true quantum tunnelling rate. Whereas the truncation of the Wigner function equation

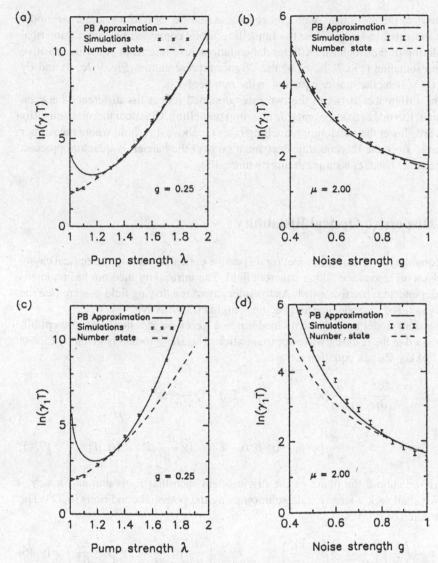

Fig. 9.7 A plot of the log of the tunnelling time for the degenerate parametric amplifier above threshold, versus pump strength or noise strength. In (**a**) and (**b**) we show the results computed by the positive P Representation (PB approximation) while in (**c**) and (**d**) we give the results for the truncated Wigner function model. In all cases we contrast the results obtained by potential methods with the results obtained by direct simulation of the corresponding stochastic differential equations and number state solution of the master equation (*dashed line*) [3]

involves dropping higher order derivatives dependent on the interaction strength g. Thus nonlinear terms in the quantum noise are neglected and the only quantum noise terms included are due to the vacuum fluctuations from the cavity losses. These give a diffusion term in the truncated Wigner Fokker–Planck equation which is identical to classical thermal noise, with an occupation number of half a photon per mode. Also indicated in Fig. 9.7 are the tunnelling times computed by direct numerical simulation of the stochastic differential equations resulting from either the positive P representation (Fig. 9.7a, b) or the Wigner representation (Fig. 9.7c, d) and by directly solving the master equation in the number basis.

The differences between the two rates obtained reflect the difference between classical thermal activation and true quantum tunnelling. Classical thermal-activation rates are slower than quantum tunnelling rates far above threshold where the former are large since the thermal trajectory must go over the barrier. A quantum process, on the other hand, can short cut this by tunnelling.

9.4 Dispersive Optical Bistability

We consider a single mode model for dispersive optical bistability. An optical cavity is driven off resonance with a coherent field. The intracavity medium has an intensity dependent refractive index. As the intensity of the driving field is increased the cavity is tuned to resonance and becomes highly transmissive.

We shall model the intracavity medium as a Kerr type $\chi^{(3)}$ nonlinear susceptibility treated in the rotating wave approximation. The Hamiltonian is given by (5.79), The Fokker–Planck equation is

$$\frac{\partial P}{\partial t} = \left[\frac{\partial}{\partial \alpha}(\kappa \alpha + 2i\chi \alpha^2 \beta - E_0) - i\chi \frac{\partial^2}{\partial \alpha^2}\alpha^2 \right.$$

$$\left. + \frac{\partial}{\partial \beta}(\kappa^* \beta - 2i\chi \beta^2 \alpha - E_0) + i\chi \frac{\partial^2}{\partial \beta^2}\beta^2 \right] P(\alpha, \beta) \qquad (9.45)$$

where we choose the phase of the driving field such that E_0 is real and $\kappa = \gamma + i\delta$. We shall seek a steady state solution using the potential conditions (6.72). The calculation of F gives

$$F_1 = -\left(\frac{i}{\chi} \right) \left(\frac{\bar{\kappa}}{\alpha} + 2\chi \beta - \frac{E_0}{\alpha^2} \right), \qquad F_2 = \left(\frac{i}{\chi^*} \right) \left(\frac{\bar{\kappa}^*}{\alpha} - 2\chi^* \beta - \frac{E_0}{\beta^2} \right), \quad (9.46)$$

where we have defined $\bar{\kappa} = \kappa - 2i\chi$. The cross derivatives

$$\partial_\alpha F_2 = \partial_\beta F_1 = 2 \qquad (9.47)$$

so that the potential conditions are satisfied.

The steady state distribution is given by

$$P_{ss}(\alpha, \beta) = \exp\left[\int^{\alpha} F_\rho(\alpha')d\alpha'_\rho\right]$$

$$= \exp\left\{\int^{\alpha}\left[\frac{1}{i\chi}\left(\frac{\bar{\kappa}}{\alpha_1}+2i\chi\beta_1-\frac{E_0}{\alpha_1^2}\right)d\alpha_1 - \frac{1}{i\chi}\left(\frac{\bar{\kappa}^*}{\beta_1}-2i\chi\alpha_1-\frac{E_0}{\beta_1^2}\right)d\beta_1\right]\right\}$$

$$= \alpha^{c-2}\beta^{d-2}\exp\left[\left(\frac{E_0}{i\chi}\right)\left(\frac{1}{\alpha}+\frac{1}{\beta}\right)+4\alpha\beta\right] \tag{9.48}$$

where $c = \frac{\kappa}{i\chi}$, $d = \left(\frac{\kappa}{i\chi}\right)^*$.

It can be seen immediately that the usual integration domain of the complex plane with $\alpha^* = \beta$ is not possible since the potential diverges for $\alpha\beta \to \infty$. However, the moments may be calculated using the complex P representation. The calculations are described in Appendix 9.A. The results for the mean amplitude $\langle a \rangle$ and correlation function $g^{(2)}(0)$ are plotted in Fig. 9.8 where they are compared with the semi-classical value for the amplitude α_{SS}.

It is seen that, whereas the semi-classical equation predicts a bistability or hysteresis, the exact steady state equation which includes quantum fluctuations does not exhibit bistability or hysteresis. The extent to which bistability is observed in practice will depend on the fluctuations, which in turn determine the time for random switching from one branch to the other. The driving field must be ramped in time intervals shorter than this random switching time in order for bistability to be observed.

The variance of the fluctuations as displayed by $g^{(2)}(0)$ show an increase as the fluctuations are enhanced near the transition point. The dip in the steady state mean at the transition point is due to out-of-phase fluctuations between the upper and lower branches.

Fig. 9.8 The steady state amplitude, and second-order correlation function for optical bistability versus the pump amplitude. The chain curve gives the semi-classical steady state amplitude. The *full curve* gives the exact steady state amplitude. The *broken curve* presents the second-order correlation function $g^{(2)}(0)$. The detuning is chosen so that $\Delta\omega\chi < 0$ with $\Lambda\omega = -10$ and $\chi = 0.5$

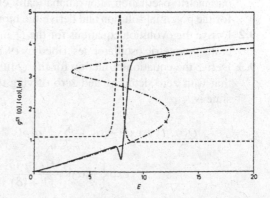

9.5 Comment on the Use of the Q and Wigner Representations

We will compare the above solution we have obtained with the generalised P representation with the equation obtained using the Q and Wigner representations. With the Q representation we obtain the following equation

$$\frac{\partial Q}{\partial t}(\alpha^*,\alpha) = \left[\frac{\partial}{\partial\alpha}(-E_0 + \bar\kappa\alpha + 2i\chi\alpha^2\alpha^*) + i\chi\frac{\partial^2}{\partial\alpha^2}\alpha^2 + (\kappa/2)\frac{\partial^2}{\partial\alpha\partial\alpha^*} + c.c.\right]Q(\alpha^*,\alpha)$$
(9.49)

where $\bar\kappa = \kappa - 4i\chi + i\Delta\omega$.

This equation has a non-positive definite diffusion matrix. Furthermore, it does not satisfy the potential conditions, hence its steady-state solution is not readily obtained.

The equation for the Wigner function may be shown to be as in,

$$\frac{\partial W(\alpha^*,\alpha)}{\partial t} = \left(E_0\frac{\partial}{\partial\alpha} + \kappa\frac{\partial}{\partial\alpha} + \frac{\kappa}{2}\frac{\partial^2}{\partial\alpha^*\partial\alpha} - 2i\chi\frac{\partial}{\partial\alpha} - i\chi\frac{1}{2}\frac{\partial^2}{\partial\alpha^{*2}}\alpha\right.$$
$$\left. + 2i\chi\frac{\partial^3}{\partial\alpha^3}\alpha^*\alpha^2 + c.c.\right)W(\alpha^*,\alpha).$$
(9.50)

This equation is not of a Fokker–Planck form since it contains third-order derivatives. Again a steady-state solution is not readily obtainable. It is clear that for this problem the use of the complex P representation is preferable to the other two representations.

Exercises

9.1 Derive the Fokker–Planck equation for $P(\alpha_1, \alpha_2, t)$ for the non-degenerate parametric oscillation after adiabatically eliminating the pump mode. Solve for the potential solution and derive the moments.

9.2 Derive the evolution equations for the Q and Wigner functions for the degenerate parametric oscillator described by (9.1).

9.3 Derive the equation of motion for the Q function for optical bistability. Show that with zero detuning and zero driving the solution for an initial coherent state is

$$Q(\alpha, t) = \exp(-|\alpha|^2)\sum_{q,\,p=0}^{\infty}(q!p!)^{-1}(\alpha\alpha_0^*)^q(\alpha^*\alpha_0)^p f(t)^{(p+q)/2}$$
$$\times \exp\left\{-|\alpha_0|^2\frac{[f(t)+i\delta]}{(1+i\delta)}\right\}$$

where

$$\delta = (p-q)/\kappa, \quad . \quad f(t) = \exp[-\kappa\nu - i\nu(p-q)], \quad \nu = 2\mu t, \quad \kappa = \frac{\gamma}{2\mu}.$$

9.4 Calculate the steady state distribution $P(\alpha)$ and the mean intensity $\langle \alpha^*\alpha \rangle$ for the degenerate parametric oscillator for the case where the thermal fluctuations dominate the quantum fluctuations.

9.A Appendix

9.A.1 Evaluation of Moments for the Complex P function for Parametric Oscillation (9.17)

It is necessary to integrate on a suitable manifold, chosen so that the distribution (9.17) and all its derivatives vanish at the boundary of integration. If we expand the term $\exp(2\alpha_1\,\beta_1)$ in (9.17) in a power series, the expression for the moment

$$I_{nn'} = \int\int \beta^n \alpha^{n'} P(\alpha) d\alpha\, d\beta \ . \tag{9.A.1}$$

can be written as

$$I_{nn'} = N(2|c|)^{2(j_2-2)} \sum_{m=0}^{\infty} \frac{2^{m+2}}{m!} \left(\frac{-c}{\kappa}\right)^{m+n-1} \left(\frac{-c^*}{\kappa}\right)^{m+n'+1}$$

$$\times \int\int z_1^{j_1-1}(1-z_1)^{j_2-j_1-1}(1-2z_1)^{m+n}(1-2z_2)^{m+n'}$$

$$\times z_2^{j_1-1}(1-z_2)^{j_2} dz_1 dz_2 \tag{9.A.2}$$

where

$$j_1 = \frac{2\gamma_1\gamma_2}{\kappa^2}, \qquad j_2 = \frac{4\gamma_1\gamma_2}{\kappa^2}, \qquad z_1 = \frac{1}{2}\left(1+\frac{\kappa\alpha_1}{c}\right), \qquad z_2 = \frac{1}{2}\left(1+\frac{\kappa\beta_2}{c^*}\right).$$

These integrals are identical to those defining the Gauss' hypergeometric functions. The integration path encircles each pole and traverses the Riemann sheets so that the initial and final values of the integrand are equal, allowing partial integration operations to be defined. The result is [7].

$$I_{nn'} = N' \sum_{m=0}^{\infty} \frac{2^m}{m!} \left(\frac{-c}{\kappa}\right)^{m+n} \left(\frac{-c^*}{\kappa}\right)^{m+n'}$$

$$\times {}_2F_1(-(m+n),j_1,j_2,2)\,{}_2F_1(-(m+n),j_1,j_2,2) \tag{9.A.3}$$

where ${}_2F_1$ are hypergeometric functions.

9.A.2 Evaluation of the Moments for the Complex P Function for Optical Bistability (9.48)

The normalization integral is

$$I(c,\, d) = \int \int_{c} \sum \frac{2^{n}}{n!} x^{-c-n} y^{-d-n} \exp\left[\frac{E_0}{\chi}(x+y)\right] dx\, dy \qquad (9.A.4)$$

where we have made the variable change $x = 1/\alpha$, $y = 1/\beta$, and C is the integration path. $\alpha^{*} = \beta$ since the potential diverges for $|\alpha|^2 \to \infty$. This means no Glauber–Sudarshan P function exists in the steady state (except as a generalised function). Hence, we shall use the complex P function where the paths of integration for α and β are line integrals on the individual $(\alpha,\, \beta)$ complex planes.

The integrand is now in a recognisable form as corresponding to a sum of gamma function integrals. It is therefore appropriate to define each path of integration to be a Hankel path of integration, from $(-\infty)$ on the real axis around the origin in an anticlockwise direction and back to $(-\infty)$. With this definition of the integration domain, the following gamma function identity holds [8]:

$$[\Gamma(c+n)]^{-1} = \left(\frac{t^{1-c-n}}{2\pi i}\right) \int_{c} x^{-c-n} \exp(xt) dx\,. \qquad (9.A.5)$$

Hence, applying this result to both x and y integrations, one obtains with $\tilde{\chi} = i\chi$

$$I(c,\, d) = -4\pi^{2} \sum_{n=0}^{\infty} \frac{2^{n}(E_0/\tilde{\chi})^{c+d+2(n-1)}}{n!\Gamma(c+n)\Gamma(d+n)}\,. \qquad (9.A.6)$$

The series is a transcendental function which can be written in terms of the generalised Gauss hypergeometric series. That is, there is a hypergeometric series called $_0F_2$ which is defined as [9]

$$_0F_2(c,\, d,\, z) = \sum_{n=0}^{\infty} \frac{z^{n}\Gamma(c)\Gamma(d)}{\Gamma(c+n)\Gamma(d+n)n!}\,. \qquad (9.A.7)$$

From now on, for simplicity, we will write just $F()$, instead of $_0F_2()$. Now the normalisation integral can therefore be rewritten in the form

$$I(c,\, d) = \left(\frac{-4\pi^{2}|E_0/\tilde{\chi}|^{c+d-2}}{\Gamma(c)\Gamma(d)}\right) F(c,\, d,\, 2|E_0/\tilde{\chi}|^{2})\,. \qquad (9.A.8)$$

The moments of the distribution function divided by the normalisation factor give all the observable one-time correlation functions. Luckily the moments have exactly the same function form as the normalisation factor [with the replacement of (c, d) by $(c+i,\, d+j)$] so that no new integrals need to be calculated. The ith-order correlation function is

$$G^{(i)} = \langle (a^\dagger)^i (a)^i \rangle = \left(\frac{|E_0/\tilde{\chi}|^{2i} \Gamma(c)\Gamma(d)F(i+c,\, i+d,\, 2|E_0/\tilde{\chi}|^2)}{\Gamma(i+c)\Gamma(i+d)F(c,\, d, 2|E_0/\tilde{\chi}|^2)} \right). \qquad (9.A.9)$$

This is the general expression for the ith-order correlation function of a nonlinear dispersive cavity with a coherent driving field and zero-temperature heat baths.

The results for the mean amplitude $\langle a \rangle$ and correlation function $g^2(0)$ are

$$\langle a \rangle = \frac{1}{c} \frac{|E_0/\tilde{\chi}|F(1+c,\, d,\, 2|E_0/\tilde{\chi}|^2)}{F(c,\, d,\, 2|E_0/\tilde{\chi}|^2)}, \qquad (9.A.10)$$

$$g^{(2)}(0) = \left(\frac{cdF(c,\, d,\, 2|E_0/\tilde{\chi}|^2)F(c+2,\, d+2,\, 2|E_0/\tilde{\chi}|^2)}{(c+1)(d+1)[F(c+1,\, d+1,\, 2|E_0/\tilde{\chi}|^2)]^2} \right). \qquad (9.A.11)$$

References

1. P.D. Drummond, K.J. McNeil, D.F. Walls: Optica Acta **28**, 211 (1981)
2. M. Wolinsky, H.J. Carmichael: Phys. Rev. Lett. **60**, 1836 (1988)
3. P. Kinsler, P.D. Drummond: Phys. Rev. A **43**, 6194 (1991)
4. R. Landauer, S. Swanson: Phys. Rev. **121**, 1668 (1961)
5. P.D. Drummond, D.F. Walls: J. Phys. A**13** 725 (1980)
6. R. Graham: *Quantum Statistics in Optics and Solid State Physics*, Springer Tracts Mod. Phy., Vol. 66 (Springer, Berlin, Heidelberg 1973)
7. T.M. MacRobert: *Function of a Complex Variable* (Macmillan, London 1938)
8. M. Abramowitz, I.A. Stegun: *Handbook of Mathematical Function* (Dover, New York 1965)
9. I.S. Gradsteyn, I.M. Ryzhik: *Tables of Series, Products and Integrals*, (Deutsch, Frankfurt/M 1981)
10. G.J. Milburn, D.F. Walls: Optics Commun. **39**, 401 (1981)

Further Reading

Gibbs, H.: *Optical Bistability*, Academic Press (1985)
Lugiato, L.A. in *Progress in Optics*, p. 69, Ed. E. Wolf, North-Holland (1984)

Chapter 10
Interaction of Radiation with Atoms

Abstract The preceding chapters have been concerned with the properties of the radiation field alone. In this chapter we turn to the interaction between radiation and matter. This is of course the domain of quantum electrodynamics, however in quantum optics we are usually only concerned with low energy systems of bound electrons which simplifies matters considerably. We will use the occupation number representation for bound many-electron systems to quantize the electronic degrees of freedom, following the approach of Haken [1] and also Cohen-Tannoudji et al. [2].

10.1 Quantization of the Many-Electron System

In the full theory of QED, the interaction between the electromagnetic field and charged matter is described by coupling between the vector potential and the Dirac spinor field. In quantum optics we only need the low energy (non relativistic) limit of this interaction. This is given by the minimal coupling Hamiltonian [3]

$$H = \frac{1}{2m}(\vec{p} - e\vec{A})^2 + eV(\vec{x}) + H_{\text{rad}} \qquad (10.1)$$

where $\vec{p}$ is the momentum operator for a particle of charge e moving in a Coulomb potential $V(\vec{x})$. The vector potential is quantised in a box of volume V as

$$\vec{A}(\vec{x},t) = \sum_{n,v} \sqrt{\frac{\hbar}{2\varepsilon_o \omega_n V}} \vec{e}_{n,v} \left[e^{i(\vec{k}_n \cdot \vec{x} - \omega_n t)} a_{n,v} + e^{-i(\vec{k}_n \cdot \vec{x} - \omega_n t)} a_{n,v}^\dagger \right] \qquad (10.2)$$

where $\vec{e}_{n,v}$ are two orthonormal polarisation vectors ($v = 1,2$) which satisfy $\vec{k}_n \cdot \vec{e}_{n,v} = 0$, as required for a transverse field, and the frequency is given by the dispersion relation $\omega_n = c|\vec{k}_n|$. The positive and negative frequency Fourier operators, respectively $a_{n,v}$ and $a_{m,v}^\dagger$, satisfy

$$[a_{n,v}, a_{n',v'}] = \delta_{vv'}\delta_{nn'} \tag{10.3}$$

The last term, H_{rad} is the Hamiltonian of the free radiation field given by

$$H_{\text{rad}} = \sum_k \hbar\omega_k a_k^\dagger a_k \tag{10.4}$$

where we have subsumed polarisation and wave vectors labels into the single sub-script k.

We now use an occupation number representation in the antisymmetric sector of the many body Hilbert space for the electronic system based on a set of single particle states $|\phi_j\rangle$, with position probability amplitudes, $\phi_j(\vec{x})$, which we take as the bound energy eigenstates of the electronic system without radiation. They could for example be the stationary states of a atom, the quasi bound states of a single Cooper pair on a mesoscopic super-conducting metal island, or the bound exciton states of semiconductor quantum dot. We then define the electronic field operators

$$\hat{\psi}(\vec{x}) = \sum_j c_j \phi_j(\vec{x}) \tag{10.5}$$

where the appropriate commutations relations for the antisymmetric sector are the fermionic forms

$$c_k c_l^\dagger + c_l c_k^\dagger = \delta_{kl} \tag{10.6}$$

$$c_k c_l + c_l c_k = c_k^\dagger c_l^\dagger + c_l^\dagger c_k^\dagger = 0 \tag{10.7}$$

In the occupation number representation the Hamiltonian may be written as the sum of three terms, $H = H_{\text{el}} + H_{\text{I}} + H_{\text{rad}}$ where the electronic part is given by

$$H_{\text{el}} = \int d^3\vec{x}\hat{\psi}^\dagger(\vec{x})\left[-\frac{\hbar^2}{2m}\nabla^2 + eV(\vec{x})\right]\hat{\psi}(\vec{x}) = \sum_j E_j c_j^\dagger c_j \tag{10.8}$$

The interaction part may be written as the sum of two terms $H_{\text{I}} = H_{\text{I},1} + H_{\text{I},2}$ where

$$H_{1,1} = \int d^3\vec{x}\hat{\psi}^\dagger(\vec{x})\left(-\frac{e}{2m}(\vec{A}(\vec{x}).\vec{p} + \vec{p}.\vec{A}(\vec{x}))\right)\hat{\psi}(\vec{x}) \tag{10.9}$$

$$H_{1,2} = \int d^3\vec{x}\hat{\psi}^\dagger(\vec{x})\left(\frac{e^2}{2m}(\vec{A}(\vec{x}))^2\right)\hat{\psi}(\vec{x}) \tag{10.10}$$

Unless we are dealing with very intense fields for which multi-photon processes are important, the second term $H_{\text{I},2}$ may be neglected.

The dominant interaction energy may then be written as

$$H_{\text{I}} = \hbar\sum_{\vec{k},n,m} g_{\vec{k},n,m}(b_{\vec{k}} + b_{\vec{k}}^\dagger)c_n^\dagger c_m \tag{10.11}$$

where the interaction coupling constant is

$$g_{\vec{k},n,m} = -\frac{e}{m}\left(\frac{1}{2\varepsilon_0\hbar\omega_k V}\right)^{1/2}\int \mathrm{d}^3\vec{x}\phi_n^*(\vec{x})\left(\mathrm{e}^{\mathrm{i}\vec{k}.\vec{x}}\vec{p}\right)\phi_m(\vec{x}) \tag{10.12}$$

We now proceed by making the *dipole approximation*. The factor $\mathrm{e}^{\mathrm{i}\vec{k}\vec{x}}$ varies on a spatial scale determined by the dominant wavelength scale, λ_0, of the field state. At optical frequencies, $\lambda_0 \approx 10^{-6}$ m. However the atomic wave functions, $\phi_n(\vec{x})$ vary on a scale determined by the Bohr radius, $a_0 \approx 10^{-11}$ m. Thus we may remove the oscillatory exponential from the integral and evaluate it at the position of the atom $\vec{x} = \vec{x}_0$. Using the result

$$[\vec{p}^2,\ \vec{x}] = -\mathrm{i}2\hbar\vec{p} \tag{10.13}$$

we can write the interaction in terms of the atomic dipole moments

$$\int \mathrm{d}^3\vec{x}\phi_n^*(\vec{x})\left(\mathrm{e}^{\mathrm{i}\vec{k}.\vec{x}}\vec{p}\right)\phi_m(\vec{x}) = \mathrm{i}\frac{m}{e}\omega_{nm}\mathrm{e}^{\mathrm{i}\vec{k}.\vec{x}_0}\int \mathrm{d}^3\vec{x}\phi_n^*(\vec{x})(e\vec{k})\phi_m(\vec{x}) \tag{10.14}$$

where $\omega_{nm} = (E_n - E_m)/\hbar$.

In the interaction picture the interaction Hamiltonian becomes explicitly time dependent,

$$\tilde{H}_\mathrm{I}(t) = \hbar\sum_{\vec{k},n,m} g_{\vec{k},n,m}(b_{\vec{k}}\mathrm{e}^{-\mathrm{i}\omega(\vec{k})t} + b_{\vec{k}}^\dagger\mathrm{e}^{\mathrm{i}\omega(\vec{k})t})c_n^\dagger c_m \mathrm{e}^{\mathrm{i}\omega_{nm}t} \tag{10.15}$$

where the tilde indicates that we are in the interaction picture. If the field is in state for which the dominant frequency is such that $\omega(\vec{k}_0) \approx \omega_{nm}$, the field is resonant with a particular atomic transition and we may neglect terms rotating at the very high frequency $\omega(\vec{k}) + \omega_{nm}$. This is known as the *rotating wave approximation*. This assumes that the field strength is not too large and further that the state of the field does not vary rapidly on a time scale of ω_{nm}^{-1} i.e. we ignore fields of very fast strong pulses. As a special case we assume the field is resonant (or near-resonant) with a single pair of levels with $E_2 > E_1$. The interaction picture Hamiltonian in the dipole and rotating wave approximation is then given by

$$\tilde{H}_I = \hbar\sum_{\vec{k}} c_1^\dagger c_2 b_{\vec{k}}^\dagger g_{\vec{k}}\mathrm{e}^{-\mathrm{i}(\omega(\vec{k})-\omega_{21})t} + \mathrm{h.c} \tag{10.16}$$

where

$$g_{\vec{k}} = -\mathrm{i}\left(2\hbar\varepsilon_0\omega(\vec{k})V\right)^{-1/2}\omega_\mathrm{a}\mu_{21}\mathrm{e}^{\mathrm{i}\vec{k}.\vec{x}_0} \tag{10.17}$$

and

$$\mu_{21} = \langle\phi_n|e\vec{x}|\phi_m\rangle \tag{10.18}$$

with $\omega_\mathrm{a} = \omega_2 - \omega_1$.

It is conventional to describe the operator algebra of a two level system in terms of pseudo-spin representation by noting that the Pauli operators may be defined as

$$\sigma_z = c_2^\dagger c_2 - c_1^\dagger c_1 \tag{10.19}$$

$$\sigma_x = c_2^\dagger c_1 + c_1^\dagger c_2 \tag{10.20}$$

$$\sigma_y = -i(c_2^\dagger c_1 - c_1^\dagger c_2) \tag{10.21}$$

$$\sigma_+ = \sigma_-^\dagger = c_2^\dagger c_1 \tag{10.22}$$

The operators $s_\alpha = \sigma_\alpha/2$ (with $\alpha = x, y, z$) then obey the $su(2)$ algebra for a spin half system. In terms of these operators we may write the total Hamiltonian for the system of field plus atom in the dipole and rotating wave approximation as

$$H = \sum_{\vec{k}} \hbar\omega(\vec{k}) b_{\vec{k}}^\dagger b_{\vec{k}} + \frac{\hbar\omega_a}{2}\sigma_z + \hbar\sum_{\vec{k}} g_{\vec{k}} b_{\vec{k}} \sigma_+ + \text{h.c.} \tag{10.23}$$

The free Hamiltonian for the two-level electronic system is

$$\mathcal{H}_{\text{el}} = \frac{\hbar\omega_a}{2}\sigma_z \tag{10.24}$$

Denoting the ground and excited states as $|1\rangle$ and $|2\rangle$ respectively, we see that

$$\mathcal{H}_{\text{el}}|s\rangle = (-1)^s \frac{\hbar\omega_a}{2}|s\rangle \quad s = 1,2 \tag{10.25}$$

The action of the raising and lowering operators on the energy eigenstates is: $\sigma_+|1\rangle = |2\rangle$ and $\sigma_-|2\rangle = |1\rangle$, while $\sigma_\pm^2 = 0$. We now relabel the ground state and excited state respectively as $|1\rangle \equiv |g\rangle$, $|2\rangle \equiv |e\rangle$. If the state of the system at time t is ρ, the probability to find the electronic system in the excited state and ground state are, respectively,

$$p_e(t) = \langle 2|\rho|2\rangle = \langle \sigma_+\sigma_-\rangle \tag{10.26}$$

$$p_g(t) = \langle 1|\rho|1\rangle = \langle \sigma_-\sigma_+\rangle \tag{10.27}$$

The *atomic inversion* is defined as the difference between these two probabilities and is given by

$$p_e(t) - p_g(t) = \langle \sigma_z\rangle \tag{10.28}$$

While the *atomic coherences* are defined by

$$\rho_{12} \equiv \langle 1|\rho|2\rangle = \langle \sigma_+\rangle \tag{10.29}$$

with $\rho_{21} = \rho_{12}^*$.

10.2 Interaction of a Single Two-Level Atom with a Single Mode Field

If we further restrict the state of the field to include only a single mode, with frequency ω_0; perhaps using a high Q optical resonator, we arrive at the *Jaynes–Cummings* hamiltonian,

$$H = \hbar\omega_0 b^\dagger b + \frac{\hbar\omega_a}{2}\sigma_z + \hbar(gb\sigma_+ + g^* b^\dagger \sigma_-) \tag{10.30}$$

coupling a single harmonic oscillator degree of freedom to a two-level system, which might well be called the standard model of quantum optics [4]. The coupling constant g can vary from a few kHz to many MHz. An example is provided by the experiment of Aoki et al. [5] in which a cesium atom interacts with the toroidal whispering gallery mode of a micro-resonator as it falls under the action of gravity from a magneto-optical trap. The atomic resonance at is the $6S_{1/2}$; $F = 4 \to 6P_{3/2}$; $F' = 5$ transition in cesium. A coupling constant as large as $g/2\pi = 50\,\mathrm{MHz}$ was achieved.

On resonance, $\omega_a = \omega_c = \omega$, we see that the interaction Hamiltonian $\mathcal{H}_I = \hbar g(b\sigma_+ + b^\dagger \sigma_-)$ (with g chosen as real), commutes with the free Hamiltonian, $\mathcal{H}_0 = \hbar\omega(b^\dagger b + \frac{1}{2}\sigma_z)$, so that the eigenstates of the full Hamiltonian can be written as a linear combination of the degenerate eigenstates of $\mathcal{H}_0$. Defining $|n,s\rangle = |n\rangle_b \otimes |s\rangle$, where $b^\dagger b|n\rangle = n|n\rangle$, the degenerate eigenstates of the free Hamiltonian are $|n,2\rangle$, $|n+1,1\rangle$. Within this degenerate subspace, the state at time t may be written $|\psi_n(t)\rangle = c_{n,2}(t)|n,2\rangle + c_{n+1,2}(t)|n+1,1\rangle$, and the Schröedinger equation in the interaction picture is

$$\begin{pmatrix} \dot{c}_{n,2} \\ \dot{c}_{n+1,1} \end{pmatrix} = -i\Omega_n \sigma_x \begin{pmatrix} c_{n,2} \\ c_{n+1,1} \end{pmatrix} \tag{10.31}$$

where $\Omega_n = g\sqrt{n+1}$. The eigenvalues of this system of linear equations are $\pm i\Omega_n$, corresponding to the eigenstates of $\mathcal{H}_I$

$$|n,\pm\rangle = \frac{1}{\sqrt{2}}(|n,2\rangle \pm |n+1,1\rangle) \tag{10.32}$$

which are often referred to as the *dressed states*. The splitting of the degeneracy is depicted in Fig. 10.1.

Thus the general solution is

$$c_{n,2}(t) = c_{n,2}(0)\cos\Omega_n t - ic_{n+1,1}(0)\sin\Omega_n t \tag{10.33}$$

$$c_{n+1,1}(t) = c_{n+1,1}(0)\cos\Omega_n t - ic_{n,2}(0)\sin\Omega_n t \tag{10.34}$$

If the atom is initially in the excited state and the cavity field has exactly n photons, the probability for finding the atom in the *same* state at time $t > 0$ is

$$p_e(t) = |\langle n,2|\psi_n(t)\rangle|^2 = \frac{1}{2}(1 + \cos 2\Omega_n t) \tag{10.35}$$

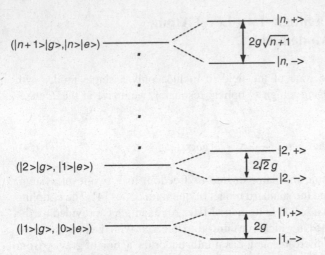

Fig. 10.1 The *dressed states* for the energy eigenstates of the Jaynes–Cummings interaction. On the left are show the degenerate states for zero interaction. When the interaction is turned on the degeneracies are lifted

The excitation oscillates backward and forth between the cavity and the electronic system with frequency Ω_n, the Rabi frequency. Note that for $n = 0$ the separation of these eigenvalues is $2g$, which is known as the *vacuum Rabi splitting*.

If the field is in an arbitrary pure state, $|\phi\rangle = \Sigma_n f_n |n\rangle$ and the atom is initially excited, the probability to find the atom in the excited state at time $t > 0$ may be written

$$p_e(t) = \frac{1}{2}\left[1 + \sum_{n=0}^{\infty} |f_n|^2 \cos(2g\sqrt{n+1}\,t)\right] \tag{10.36}$$

This is a discrete superposition of harmonic oscillations with incommensurate frequencies. Thus it must exhibit quasiperiodic behaviour. If the initial photon number distribution $|f_n|^2$ has narrow support on n, only a few frequencies are involved and there is a beating between these different frequencies leading to what are known as collapses and revivals. The collapse refers to the decay of oscillations at short times due to beating between the incommensurate frequencies. The revival refers to partial re-phasing of the oscillations at later times. In the case of the field initially in a coherent state, $|\alpha\rangle$, the initial number distribution is Poissonian with standard deviation in number given by the root mean, $\bar{n}^{1/2}|\alpha|$. An approximate evaluation of the sum valid for times such that $gt < \bar{n}^{1/2}$ gives [6]

$$p_e(t) = \frac{1}{2}\left[1 + e^{-\frac{g^2 t^2 \bar{n}}{2(n+1)}} \cos(2g\sqrt{\bar{n}+1}\,t)\right] \tag{10.37}$$

There is an average Rabi oscillation frequency under a Gaussian envelope. The characteristic time for the collapse of the oscillation is thus

Fig. 10.2 (a) The experimental observation of collapse and revival of the oscillations in the occupation of the excited state of a two-level atom interacting with a microwave field initially in a coherent state with mean photon number $\bar{n} = 0.85$. In (b) is the Fourier transform of the oscillations with the Rabi frequencies Ω_n, $n = 0, 1, 2, 3$ marked (from [7])

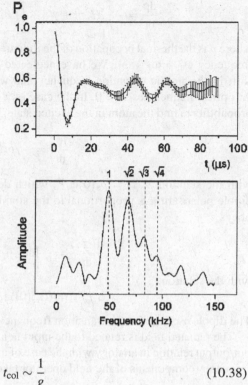

$$t_{\text{col}} \sim \frac{1}{g} \tag{10.38}$$

A more accurate evaluation using the Laplace summation formulae shows that the oscillations first revive at a t time

$$t_{\text{rev}} \sim \frac{2\pi}{g}\bar{n}^{1/2} \tag{10.39}$$

Thus a quasi periodic burst of Rabi oscillations occurs every $\bar{n}$ Rabi periods. The collapse and revival has been seen experimentally using an atom excited to a Ryberg ground state interacting with the microwave field in a superconducting cavity [7]. The results of the experiment are shown in Fig. 10.2.

10.3 Spontaneous Emission from a Two-Level Atom

Spontaneous emission can also be treated using a master equation. In this case the system is a two-level electronic system, with ground state $|g\rangle$ of energy $\hbar\omega_1$ and excited state $|e\rangle$ with energy $\hbar\omega_2$, representing an electric dipole transition, coupled to the many modes of the radiation field in the dipole and rotating wave approximation. The master equation is

$$\frac{d\rho}{dt} = -\frac{i}{\hbar}[H,\rho] + \gamma(\bar{n}+1)\mathscr{D}[\sigma_-]\rho + \gamma\bar{n}\mathscr{D}[\sigma_+]\rho \qquad (10.40)$$

where $\bar{n}$ is the thermal occupation of the radiation field mode at the atomic resonance frequency $\omega_a = \omega_2 - \omega_1$. We have neglected a small term which gives rise to a shift in the atomic transition frequency and which contributes to the Lamb shift. At optical frequencies, $\bar{n} \approx 0$. In the case of a free two-level atom, $H = \frac{\hbar\omega_a}{2}\sigma_z$ the probability to find the atom in the excited state, $p_e(t) = \langle e|\rho|e\rangle$ satisfies the equation

$$\frac{dp_e}{dt} = -\gamma p_e(t) \qquad (10.41)$$

with the solution $p_e(t) = p_e(0)e^{-\gamma t}$, which describes spontaneous emission. The dipole polarisation is proportional to the atomic coherence, $\langle e|\rho|g\rangle = \langle\sigma_-\rangle$ which obeys

$$\frac{d\langle\sigma_-\rangle}{dt} = -\left(i\omega_a + \frac{\gamma}{2}\right)\langle\sigma_-\rangle \qquad (10.42)$$

with the solution

$$\langle\sigma_-(t)\rangle = \langle\sigma_-(0)\rangle e^{-(\gamma/2 + i\omega_a)t} \qquad (10.43)$$

The dipole oscillates at the transition frequency and decays, as it radiates.

The radiated field is related to the input field and the local source through an input/output relation in analogy with the case of a cavity discussed above. The positive frequency components of the field operator takes the form

$$E_o^{(+}(\vec{x},t) = E_i^{(+)}(\vec{x},t) - \frac{\omega_a^2}{4\pi\varepsilon_0 c^2 r}\left(\vec{\mu}\times\frac{\vec{x}}{r}\right)\times\frac{\vec{x}}{r}\sigma_-(t-x/c) \qquad (10.44)$$

where $r = |\vec{x}|$ is the distance from the source to the point $\vec{x}$ and $\vec{\mu}$ is the atomic dipole moment.

10.4 Phase Decay in a Two-Level System

Spontaneous emission is not the only irreversible process involved in the absorption and emission of light. In an atomic vapour, atomic collisions are also a source of decoherence and cause a decay of the atomic polarisation, $\sigma_x + i\sigma_y$, without changing the decay of the inversion, σ_z. We can model this process by a coupling between the inversion and a high temperature heat bath,

$$H_{col} = \sigma_z\Gamma_c(t) \qquad (10.45)$$

where $\Gamma_c(t)$ is a bath operator describing the collisions. This Hamiltonian commutes with σ_z and thus does not contribute to the decay of the inversion. It appears like a fluctuating detuning in the Bloch equations and thus will effect the atomic polarisation. The corresponding master equation, in the interaction picture and including spontaneous emission, is

$$\frac{d\rho}{dt} = \frac{\gamma}{2}(2\sigma_-\rho\sigma_+ - \sigma_+\sigma_+\rho - \rho\sigma_+\sigma_+) - \gamma_p[\sigma_z,[\sigma_z[,\rho]] \tag{10.46}$$

The Bloch equations now become

$$\frac{d\langle\sigma_z\rangle}{dt} = -\gamma(\langle\sigma_z\rangle + 1) \tag{10.47}$$

$$\frac{d\langle\sigma_x\rangle}{dt} = -\left(\frac{\gamma}{2} + \gamma_p\right)\langle\sigma_x\rangle \tag{10.48}$$

$$\frac{d\langle\sigma_y\rangle}{dt} = -\left(\frac{\gamma}{2} + \gamma_p\right)\langle\sigma_y\rangle \tag{10.49}$$

In the presence of collisions the decay time for the polarisation, T_2, is no longer given by twice the decay time for the inversion, $T_1 = \gamma^{-1}$, but rather $T_2 < 2T_1$.

10.5 Resonance Fluorescence

If the atom is driven by a classical radiation field, the Hamiltonian becomes (see (10.30) and replace $b \mapsto \beta$)

$$H = \frac{\hbar\omega_a}{2}\sigma_z + \Omega(\sigma_+e^{-i\omega_L t} + \sigma_-e^{i\omega_L t}) \tag{10.50}$$

where $\Omega = g\beta$ is the Rabi frequency and ω_L is the carrier frequency of the driving field. The master equation in an interaction picture at the frequency ω_L is

$$\frac{d\rho}{dt} = -i\frac{\Delta\omega}{2}[\sigma_z,\rho] - i\Omega[\sigma_+ + \sigma_-,\rho] + \gamma\mathscr{D}[\sigma_-]\rho \tag{10.51}$$

where the detuning is $\Delta\omega = \omega_a - \omega_L$. The resulting *Bloch equations* for the atomic moments are linear

$$\frac{d\langle\sigma_-\rangle}{dt} = -\left(\frac{\gamma}{2} + i\Delta\omega\right)\langle\sigma_-\rangle + i\Omega\langle\sigma_z\rangle \tag{10.52}$$

$$\frac{d\langle\sigma_z\rangle}{dt} = -\gamma(\langle\sigma_z\rangle + 1) - 2i\Omega(\langle\sigma_+\rangle - \langle\sigma_-\rangle) \tag{10.53}$$

These inhomogeneous equations can be written as homogeneous equations as

$$\frac{d}{dt}(\langle\vec{\sigma}(t)\rangle - \langle\vec{\sigma}\rangle_{ss}) = A(\langle\vec{\sigma}(t)\rangle - \langle\vec{\sigma}\rangle_{ss}) \tag{10.54}$$

where

$$A = \begin{pmatrix} -\left(\frac{\gamma}{2} - i\Delta\omega\right) & 0 & -i\Omega \\ 0 & -\left(\frac{\gamma}{2} + i\Delta\omega\right) & i\Omega \\ -2i\Omega & 2i\Omega & -\gamma \end{pmatrix} \qquad (10.55)$$

with $\langle\vec{\sigma}\rangle = (\langle\sigma_+\rangle, \langle\sigma_-\rangle, \langle\sigma_z\rangle)^T$ and the steady state solutions are

$$\langle\sigma_z\rangle_{ss} = -\frac{1+\delta^2}{1+\delta^2+Z^2} \qquad (10.56)$$

$$\langle\sigma_+\rangle_{ss} = \frac{i}{\sqrt{2}}\frac{Z(1+i\delta)}{1+\delta^2+Z^2} \qquad (10.57)$$

with

$$Z = \frac{2\sqrt{2}\Omega}{\gamma}, \qquad \delta = \frac{2\Delta\omega}{\gamma} \qquad (10.58)$$

The solutions for resonance $(\Delta\omega = 0)$, with the atom initially in the ground state, are

$$\langle\sigma_z(t)\rangle = \frac{8\Omega^2}{\gamma^2+8\Omega^2}\left[1 - e^{-3\gamma t/4}\left(\cosh\kappa t + \frac{3\gamma}{4\kappa}\sinh\kappa t\right)\right] - 1 \qquad (10.59)$$

$$\langle\sigma_+(t)\rangle = 2i\Omega\frac{\gamma}{\gamma^2+8\Omega^2}\left[1 - e^{-3\gamma t/4}\left(\cosh\kappa t + \left(\frac{\kappa}{\gamma} + \frac{3\gamma}{16\kappa}\right)\sinh\kappa t\right)\right] \qquad (10.60)$$

where

$$\kappa = \frac{1}{2}\sqrt{\frac{\gamma^2}{4} - 16\Omega^2} \qquad (10.61)$$

Clearly there is a threshold at $\Omega = \gamma/8$ below which the solutions monotonically approach the steady state and above which they are oscillating. A similar threshold occurs in the solutions for the two-time correlation function $\langle\sigma_+(t)\sigma_-(t+\tau)\rangle_{t\to\infty}$ which determines the spectrum of the scattered light.

The stationary spectrum, as measured by a monochromatic detector at the point $\vec{x}$ is defined by [9]

$$S(\vec{x},\omega) = \lim_{t\to\infty}\frac{1}{2\pi}\int_{-\infty}^{\infty}\langle E^{(-)}(\vec{x},t)E^{(+)}(\vec{x},t+\tau)\rangle d\tau \qquad (10.62)$$

the Fourier transform of the stationary two-time correlation function $\langle E^{(-)}(t)E^{(+)}(t+\tau)\rangle$ which using (10.44) is given by

$$S(\vec{x},\omega) = \frac{I_0(\vec{x})}{2\pi}\int_{-\infty}^{\infty}d\tau e^{-i\omega\tau}G(\tau) \qquad (10.63)$$

where

$$I_0(\vec{x}) = \left| \frac{\omega_0^2}{4\pi\varepsilon_0 c^2 r} \left(\vec{\mu} \times \frac{\vec{x}}{r} \right) \times \frac{\vec{x}}{r} \right|^2 \tag{10.64}$$

and

$$G(\tau) = \lim_{t \to \infty} \langle \sigma_+(t)\sigma_-(t+\tau) \rangle \equiv \langle \sigma_+\sigma_-(\tau) \rangle_{ss} \tag{10.65}$$

with

$$\langle \sigma_+(t)\sigma_-(t+\tau) \rangle = \mathrm{tr}\left[\sigma_- e^{\mathscr{L}\tau}\rho(t)\sigma_+ \right] \tag{10.66}$$

The equation of motion for $G(\tau)$ couples in many other moments. If we define the correlation matrix

$$\mathscr{G}(\tau) = \begin{pmatrix} \langle \sigma_+\sigma_+(\tau) \rangle_{ss} & \langle \sigma_-\sigma_+(\tau) \rangle_{ss} & \langle \sigma_z\sigma_+(\tau) \rangle_{ss} \\ \langle \sigma_+\sigma_-(\tau) \rangle_{ss} & \langle \sigma_-\sigma_-(\tau) \rangle_{ss} & \langle \sigma_z\sigma_-(\tau) \rangle_{ss} \\ \langle \sigma_+\sigma_z(\tau) \rangle_{ss} & \langle \sigma_-\sigma_z(\tau) \rangle_{ss} & \langle \sigma_z\sigma_z(\tau) \rangle_{ss} \end{pmatrix} \tag{10.67}$$

The quantum regression theorem indicates that $\mathscr{G}(\tau)$ as a function of τ obeys the same equations of motion as $\langle \vec{\sigma}(\tau) \rangle - \langle \vec{\sigma} \rangle_{ss}$,

$$\frac{\mathscr{G}(\tau)}{d\tau} = A\mathscr{G}(\tau) \tag{10.68}$$

The initial conditions are simplified due to the algebra of the Pauli matrices, for example $\sigma_+\sigma_- = (\sigma_z + 1)/2$ and $\sigma_\pm^2 = 0$, and may thus be written in terms of the stationary solutions in (10.57). On resonance we find that in the Schrödinger picture,

$$\begin{aligned} G(\tau) = \frac{4\Omega^2}{\gamma^2 + 8\Omega^2} \Bigg[& \frac{\gamma^2}{\gamma^2 + 8\Omega^2}e^{-i\omega_a t} + \frac{1}{2}e^{-(\gamma/2 + i\omega_a)\tau} \\ & - \frac{1}{2}\left(\frac{\gamma^2}{\gamma^2 + 8\Omega^2}\frac{3\gamma/4 + \kappa}{\kappa} - \frac{\gamma/2}{\kappa} - \frac{\gamma/4 + \kappa}{2\kappa} \right) \exp\{-(3\gamma/4 - \kappa + i\omega_a)\tau\} \\ & + \frac{1}{2}\left(\frac{\gamma^2}{\gamma^2 + 8\Omega^2}\frac{3\gamma/4 - \kappa}{\kappa} - \frac{\gamma/2}{\kappa} - \frac{\gamma/4 - \kappa}{2\kappa} \right) \exp\{-(3\gamma/4 + \kappa + i\omega_a)\tau\} \Bigg] \end{aligned} \tag{10.69}$$

with $\tau \geq 0$. The corresponding spectrum has a single Lorentzian peak for weak driving fields, $4\Omega \ll \gamma^2/16$,

$$S(\vec{x}, \omega) = I_0(r)\frac{4\Omega^2}{\gamma^2 + 8\Omega^2}\delta(\omega - \omega_a) \tag{10.70}$$

which corresponds to elastic scattering. For very strong driving fields, $\Omega \gg \gamma$ we find that the spectrum acquires three Lorentzian peaks at $\omega = \omega_a$ and $\omega = \omega_a \pm 2\Omega$. The spectrum, including the elastic term, is

$$S(\vec{x},\omega) = \frac{I_0(r)}{2\pi}\left(2\pi\frac{4\Omega^2}{\gamma^2+8\Omega^2}\delta(\omega-\omega_a) + \frac{1}{2}\frac{\gamma/2}{\gamma^2/4+(\omega-\omega_a)^2}\right.$$

$$\left.+\frac{1}{4}\frac{3\gamma/4}{(3\gamma/4)^2+[\omega-(\omega+2\Omega)]^2} + \frac{1}{4}\frac{3\gamma/4}{(3\gamma/4)^2+[\omega-(\omega-2\Omega)]^2}\right)$$

$$(10.71)$$

This is the *Mollow spectrum* [10].

The light scattered by a two-level atom also exhibits *photon anti-bunching*. Consider the conditional probability that given a photon is counted at time t another photon will be counted a time τ later. This is proportional to the second order correlation function

$$G^{(2)}(t,\tau) = \langle a^\dagger(t)a^\dagger(t+\tau)a(t+\tau)a(t)\rangle \qquad (10.72)$$

Usually we are interested in a stationary source so we let $t\to\infty$ and we normalise this by the intensity squared to define

$$g^{(2)}(\tau) = \lim_{t\to\infty}\frac{G^{(2)}(t,\tau)}{\langle a^\dagger(t)a(t)\rangle^2} \qquad (10.73)$$

Using the result in (10.44) we can express this directly in terms of correlation functions for the atomic polarisation. As the equations of motion for the atomic variables are linear, the stationary correlation function $\langle\sigma_+(t)\sigma_+(t+\tau)\sigma_-(t+\tau)\sigma_-(t)\rangle_{t\leftarrow\infty}$ is given by the quantum regression theorem. We then find that

$$g^{(2)}(\tau) = 1 - e^{-3\gamma\tau/4}\left(\cosh\kappa\tau + \frac{3\gamma}{4\kappa}\sinh\kappa\tau\right) \qquad (10.74)$$

The result $g^{(2)}(\tau=0) = 0$ indicates *photon anti-bunching*, as the probability to count a second photon, immediately after a first one has been counted, vanishes. This is a direct result of the emission process of the source. Photons are emitted when an excited atom relaxes back to the ground state. If a photon is counted, the atom is likely to be in ground state and thus a finite time must elapse before it is re-excited and capable of emitting another one. The probability to find the atom in the excited state at time τ given that it starts in the ground state at $\tau=0$ is

$$P_e(\tau) = \frac{4\Omega^2}{\gamma^2+8\Omega^2}\left[1 - e^{-3\gamma\tau/4}\left(\cosh\kappa\tau + \frac{3\gamma}{4\kappa}\sinh\kappa\tau\right)\right] \qquad (10.75)$$

Comparison with (10.74) indicates this interpretation is correct. This prediction, first made by Carmichael and Walls [8], was one of the earliest examples of how quantum optics would differ from a semiclassical description of light. In Fig. 10.3 we plot $g^{(2)}(\tau)$ for two values of the Rabi frequency.

The first observation of photon antibunching was made by Kimble et al. in 1977 on atomic beams [11]. They saw a positive slope for $g^{(2)}(\tau)$ which is consistent with the predictions of the theory, however fluctuations from atomic numbers in the beam made a detailed comparison with the single atom result impossible. Ion traps

Fig. 10.3 The second order correlation function of the fluorescent light, given by (10.74) versus delay time τ. The *solid line* corresponds to $\Omega = 2.5$, while the *dashed line* corresponds to $\Omega = 0.25$. In both cases $\gamma = 1.0$

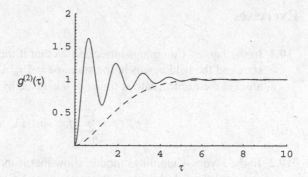

(see Chap. 17) provided a means to observe photon antibunching from a single atom [12]. In Fig. 10.4 we show the results of a measurement of the second order correlation function performed on a single trapped mercury ion by Walther's Garching group [13].

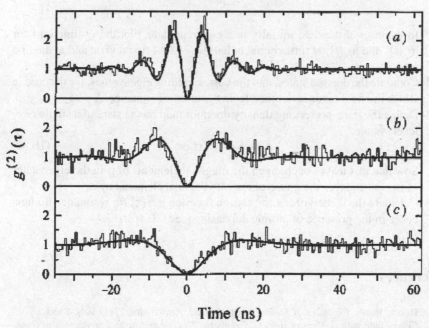

Fig. 10.4 The second order correlation function of the fluorescent light form a single mercury ion in a trap versus delay, τ. (a)$\Delta = -2.3\gamma$, $\Omega = 2.8\gamma$. (b)$\Delta = -1.1\gamma$, $\Omega = \gamma$, (c)$\Delta = -0.5\gamma$, $\Omega = 0.6\gamma$ (from [13])

Exercises

10.1 In the Jaynes–Cummings model, show that if the atom begins in the ground state and the field begins in the state $|\phi\rangle = \Sigma f_n |n\rangle$, the probability to find the atom in the excited state at time $t > 0$ is given by

$$p_e(t) = \sum_{n=1}^{\infty} |f_n|^2 \sin^2(\Omega_{n-1} t) \qquad (10.76)$$

10.2 In the Jaynes–Cummings model, show that if the atom begins in the ground state and the field begins in the state $|\phi\rangle = \Sigma f_n |n\rangle$, the state at time $t > 0$ is the entangled state

$$|\psi(t)\rangle = |\phi_g(t)\rangle|g\rangle + |\phi_e(t)\rangle|e\rangle \qquad (10.77)$$

where

$$|\phi_g(t)\rangle = \sum_n f_n \cos(\Omega_{n-1} t) \qquad (10.78)$$

$$|\phi_e(t)\rangle = i \sum_n f_n \sin(\Omega_{n-1} t) \qquad (10.79)$$

In the case of the field initially in a coherent state, plot the Q-functions for $|\phi_e(t)\rangle$ and $|\phi_e(t)\rangle$ at times equal to half way to the first revival and at the first revival.

10.3 Compute the dressed states, and the corresponding eigenvalues, for the case in which the field mode is detuned from the atomic resonance, $\Delta = \omega_a - \omega_c \neq 0$.

10.4 Define the trace-preserving density operator map on the state of a single two-level system,

$$\rho \mapsto \mathcal{E}(\rho) = \sigma_z \rho \sigma_z \qquad (10.80)$$

Sow that this leaves unchanged the diagonal elements of ρ in the eigenstates of σ_z, but changes the phase of the off-diagonal elements by π.

10.5 Calculate the second order correlation function $g^{(2)}(\tau)$ for resonance fluorescence in the presence of atomic dephasing (see (10.46).

References

1. H. Haken: *Waves, Photons and Atoms* (North Holland, Amsterdam, 1981) Vols. 1 and 2.
2. C. Cohen-Tannoudji, J. Dupont-Roc, G. Grynberg: *Photons and Atom-Introduction to Quantum Electrodynamics* (Wiley-Interscience, 1997).
3. R. Loudon: *Quantum Theory of Light* (Oxford University Press, Oxford 1973).
4. B.W. Shore, P.L. Knight: *The Jaynes–Cummings Model*, J. Mod. Opt. **40**, 1195–1238 (1993).
5. T. Aoki, B. Dayan, E. Wilcut, W.P. Bowen, A.S. Parkins, T.J. Kippenberg, K.J. Vahala, H.J. Kimble: *Observation of Strong Coupling Between One Atom and a Monolithic Microresonator*, Nature **443**, 671 (2006).

6. S. Haroche: In *New Trends in Atomic Physics*, ed. by G. Grynberg, R. Stora, Les Houche Session XXXVIII (Elsevier, Amsterdam 1984).
7. J.M. Raimond, M. Brune, S. Haroche: Rev. Mod. Phys. **73**, 565–582 (2001).
8. H.J. Carmichael, D.F. Walls: *Proposal for the Measurement of the Resonant Stark Effect by Photon Correlation Techniques*, J. Phys. B **9**, L43 (1976).
9. R.J. Glauber: Phys. Rev. **131**, 2766 (1963).
10. B.R. Mollow: Phys. Rev. **188**, 1969 (1969).
11. H.J. Kimble, M. Degenais, L. Mandel: Phys. Rev. Lett. **39**, 691 (1977).
12. F. Diedrich, H. Walther: Phys. Rev. Lett. **58**, 203 (1987).
13. H. Walther: Proc. R. Soc. Lond. A *454*, 431 (1998).

Further Reading

Allen, L., J. Eberly: *Optical Resonance and Two-Level Atoms* (Wiley, New York 1975)

Cohen-Tannoudji, C., J. Duppont, G. Grynberg: *Photons and Atoms: Introduction to Quantum Electrodynamics* (Wiley, New York 1989)

Mollow, B.R.: *Progress in Optics* **19** (North-Holland, Amsterdam 1987). H.J. (Springer, Berlin 1999)

Carmichael, H.J. "Statistical methods in quantum optics 1: Master equations and Fokker-Planck equations" (Springer, Berlin, 1999)

Chapter 11
CQED

Abstract In this chapter we will discuss two different physical systems in which a single mode of the electromagnetic field in a cavity interacts with a two-level dipole emitter. In the first example, the system is comprised of a single two-level atom inside an optical cavity. The study of this system is often known as cavity quantum electrodynamics (cavity QED) as it may be described using the techniques of the previous chapter.

The second example is comprised of a superconducting Cooper pair box inside a co-planar microwave resonator. The description of this system is given in terms of the quantisation of an equivalent electronic circuit and thus goes by the name of circuit quantum electrodynamics (circuit QED).

In both cases we are typically interested in the *strong coupling regime* in which the single photon Rabi frequency g (the coupling constant in the Jaynes–Cummings model) is lager than both the spontaneous decay rate, γ, of the two-level emitter and the rate, κ, at which photons are lost from the cavity.

11.1 Cavity QED

The primary difficulty we face in cavity QED is finding a way to localise a single two-level atom in the cavity mode for long time intervals. One approach, pioneered by the Caltech group of Kimble [1], is to first trap and cool two-level atoms in a magneto-optical trap (MOT) (see Chap. 18) and then let them fall into a high finesse cavity placed directly below the MOT. If the geometry is correctly arranged then at most one atom will slowly fall through the cavity at a time. Another approach is to use constraining forces to trap a single atom in the optical cavity. This can be done using the light shift forces of a far off resonant laser field on a two-level atom[2, 3] (see Chap. 18), or it can be done using an ion trap scheme [4] (see Chap. 17). A very novel way to get atoms from the MOT into the cavity deterministically has been pioneered by the Chapman group in Georgia [5]. They use an optical dipole standing wave trap as a kind of atomic conveyor belt to move atoms from the MOT into the cavity.

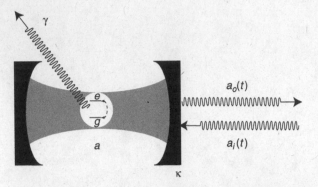

Fig. 11.1 A cavity QED scheme: a single two-level dipole emitter is fixed at a particular location inside a Fabry-Perot cavity. The dipole is strongly coupled to a single cavity mode, a, but can emit photons at rate γ into external modes. Photons are emitted from the end mirror of the cavity at rate κ

Consider the scheme in Fig. 11.1. The interaction Hamiltonian between a single two-level atom at the point $\vec{x}$ in a Fabry–Perot cavity is given by

$$H_{\mathrm{I}} = g(\vec{x})a^{\dagger}\sigma_{+} + g^{*}(\vec{x})a\sigma_{-} \tag{11.1}$$

where

$$g(\vec{x}) = \left(\frac{\mu^{2}\omega_{c}}{2\hbar\varepsilon_{0}V}\right)^{1/2} U(\vec{x}) \equiv g_{0}U(\vec{x}) \tag{11.2}$$

This is obtained from (10.17) with the traveling wave mode function replaced by a cavity standing wave mode function, $U(\vec{x})$. Here μ is the dipole moment for the two-level system and V is the cavity mode volume defined by $V = \int \sin |U(\vec{x})|^{2}\mathrm{d}^{3}x$.

Let us consider the interaction between a single cavity mode and a two-level system. For the present we neglect the spatial dependance of $g(\vec{x})$. The master equation, in the interaction picture, for a single two-level atom interacting with a single cavity mode, at optical frequencies, is

$$\begin{aligned}
\frac{\mathrm{d}\rho}{\mathrm{d}t} &= -\mathrm{i}\delta[a^{\dagger}a,\rho] - \mathrm{i}\frac{\Delta}{2}[\sigma_{z},\rho] - \mathrm{i}[\varepsilon^{*}a + \varepsilon a^{\dagger},\rho] - \mathrm{i}g[a\sigma_{+} + a^{\dagger}\sigma_{-},\rho] \\
&\quad + \frac{\kappa}{2}(2a\rho a^{\dagger} - a^{\dagger}a\rho - \rho a^{\dagger}a) + \frac{\gamma}{2}(2\sigma_{-}\rho\sigma_{+} - \sigma_{+}\sigma_{-}\rho - \rho\sigma_{+}\sigma_{-})
\end{aligned} \tag{11.3}$$

where ε represents a classical coherent laser field driving the cavity mode at frequency ω_{L}, the detuning between the cavity field and the driving field is $\delta = \omega_{c} - \omega_{\mathrm{L}}$ and $\Delta = \omega_{a} - \omega_{\mathrm{L}}$ is the detuning between the two-level system and the driving field. From this equation we can derive equations for first order field/atom moments;

$$\frac{\mathrm{d}\langle a\rangle}{\mathrm{d}t} = -\left(\frac{\kappa}{2} + \mathrm{i}\delta\right)\langle a\rangle - \mathrm{i}\varepsilon - \mathrm{i}g\langle\sigma_{-}\rangle \tag{11.4}$$

$$\frac{\mathrm{d}\langle\sigma_{-}\rangle}{\mathrm{d}t} = -\left(\frac{\gamma}{2} + \mathrm{i}\Delta\right)\langle\sigma_{-}\rangle + \mathrm{i}g\langle a\sigma_{z}\rangle \tag{11.5}$$

$$\frac{d\langle\sigma_z\rangle}{dt} = -\gamma(\langle\sigma_z\rangle+1)-2ig(\langle a\sigma_+\rangle - \langle a^\dagger\sigma_-\rangle) \tag{11.6}$$

Looking at these equations we see that we do not get a closed set of equations for the first order moments, for example the equation for $\langle\sigma_-\rangle$ is coupled to $\langle a\sigma_z\rangle$. A number of procedures have been developed to deal with this. If there are many atoms interacting with a single mode field, an expansion in the inverse atomic number can be undertaken and we will describe this approach in Sect. 11.1.3. However a good idea of the behaviour expected can be obtained simply by factorising all higher order moments. This of course neglects quantum correlations and is thus not expected to be able to give correct expressions for, say, the noise power spectrum of light emitted from the cavity. Nonetheless it is often a good pace to start as it captures the underlying dynamical structure of the problem. We thus define the *semiclassical* equations as

$$\dot{\alpha} = -\frac{\tilde{\kappa}}{2}\alpha - i\varepsilon - igv \tag{11.7}$$

$$\dot{v} = -\frac{\tilde{\gamma}}{2}v + ig\alpha z \tag{11.8}$$

$$\dot{z} = -2ig(\alpha v^* - \alpha^* v) - \gamma(z+1) \tag{11.9}$$

where the dot signifies differentiation with respect to time and

$$\tilde{\kappa} = \kappa + 2i\delta \tag{11.10}$$

$$\tilde{\gamma} = \gamma + 2i\Delta \tag{11.11}$$

The first thing to consider is the steady state solutions, α_s, z_s, v_s which are given implicitly by

$$z_s = -\left[1 + \frac{n}{n_0(1+\Delta_1^2)}\right]^{-1} \tag{11.12}$$

$$\alpha_s = -\frac{2i\varepsilon}{\tilde{\kappa}}\left[1 + \frac{2C(1+i\phi)^{-1}(1+i\Delta_1)^{-1}}{1+\frac{n}{n_0(1+\Delta_1^2)}}\right]^{-1} \tag{11.13}$$

$$v_s = \frac{2ig}{\tilde{\gamma}}\alpha_s z_s \tag{11.14}$$

where

$$n = |\alpha_s|^2 \tag{11.15}$$

is the steady state intracavity intensity and

$$\phi = 2\delta/\kappa \tag{11.16}$$

$$\Delta_1 = 2\Delta/\gamma \tag{11.17}$$

$$n_0 = \frac{\gamma^2}{8g^2} \tag{11.18}$$

$$C = \frac{2g^2}{\kappa\gamma} \qquad (11.19)$$

The parameter n_0 sets the scale for the intracavity intensity to saturate the atomic inversion and is know as the *critical* photon number. The parameter C is sometimes defined in terms of the critical atomic number, N_0, as $C = N_0^{-1}$. Why this name is appropriate is explained in Sect. 11.1.3 where we consider cavity QED with many atoms.

We can now determine how the steady state intracavity intensity depends on the driving intensity. We first define the scaled driving intensity and intracavity intensity by

$$I_d = \frac{4\varepsilon^2}{\kappa^2 n_0} \qquad (11.20)$$

$$I_c = \frac{n}{n_0} \qquad (11.21)$$

The driving intensity and the intracavity intensity are then related by

$$I_d = I_c \left[\left(1 + \frac{2C}{1 + \Delta_1^2 + I_c}\right)^2 + \left(\phi - \frac{2C\Delta_1}{1 + \Delta_1^2 + I_c}\right)^2 \right] \qquad (11.22)$$

The phase θ_s of the steady state cavity field is shifted from the phase of the driving field (here taken as real) where

$$\tan\theta_s = -\frac{\phi - 2\Delta_1 C/(1 + \Delta_1^2 + I_c)}{1 + 2C/(1 + \Delta_1^2 + I_c)} \qquad (11.23)$$

Equation (11.22) is known as the *bistability state equation*, a name that makes sense when we plot the intracavity intensity versus the driving intensity, see Fig. 11.2. It can be shown that the steady state corresponding to those parts of the curve with

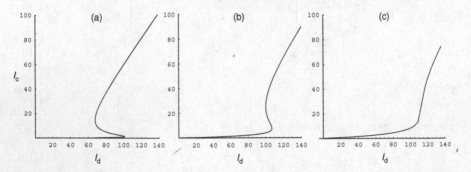

Fig. 11.2 The intracavity intensity versus the driving intensity, as given implicitly by (11.22), for various values of the detuning between the atom and the driving field. In all cases we assume the driving is on resonance with the cavity so that $\phi = 0$ and $C = 9$. (a) $\Delta_1 = 0$, (b) $\Delta_1 = 2$, (c) $\Delta_1 = 3$

negative slope are unstable. Clearly there are regions for which two stable steady states coexist for a given driving intensity.

Cavity QED requires that we are in the *strong coupling* limit in which $g_0 > \gamma, \kappa$. Furthermore a necessary condition for strong coupling is that $(n_0, N_0) << 1$. In this limit a single photon in the photon can lead to significant dynamics. One way to make g_0 large is to use a very small mode volume V and a large dipole moment. In recent years, with optical Fabry-Perot cavities, it has been possible to achieve $n_0 \approx 10^{-3}$–10^{-4} and $N_0 \approx 10^{-2}$–10^{-3}. The mirrors of these cavities are highly reflective, with reflectivity coefficients greater than 0.999998. This means that the inter mirror spacing can be made very small giving a small mode volume. Typical parameters for the Caltech group of Kimble, using atomic cesium, are [1]

$$(g_0, \kappa, \gamma) = (34, 2, 1.25)\text{MHz} \tag{11.24}$$

which give critical parameters $n_0 = 0.0029$ and $N_0 = 0.018$. Even better performance is possible using microtoroidal resonators, again implemented by the Caltech group [6], or excitonic dipoles in quantum dots integrated into photonic band gap materials, implemented by the Imamoglu group in Zurich [7]. A very different approach is to use Rydberg atoms, which have very large dipole moments, in superconducting microwave cavities. This approach has been pioneered by the group of Haroche in Paris [8].

11.1.1 Vacuum Rabi Splitting

With the ability to trap a single atom in the cavity and cool it to very low kinetic energies, it becomes possible to measure the vacuum Rabi splitting. This is the splitting energy, induced by the interaction in (11.1), of the degenerate states $|n = 0\rangle|e\rangle, |n = 1\rangle|g\rangle$ where $a^\dagger a |n\rangle = n|n\rangle$ is a photon number eigenstate for the intracavity field. As we saw in Chap. 10, Sect. 10.2, these states are split in energy by $2g$. If g is large enough an excited atom is likely to emit a single photon into the cavity mode and periodically reabsorb and remit before the excitation is lost.

Boca et al. [9] observed the vacuum Rabi splitting using a single Cs atom trapped inside an optical Fabry-Perot cavity using a far off-resonance optical dipole trap. An important breakthrough that enabled this experiment was the ability to cool the atom (see Chap. 18) using a Raman cooling scheme for motion of the atom along the cavity axis. The inferred uncertainties in the axial and transverse position of the atom in the trap were $\Delta z_{ax} \approx 33$ nm and $\Delta r_{trans} \approx 5.5\,\mu$m. The two electronic levels used were the $6S_{1/2}, F = 4 \to 6P_{3/2}, F' = 5$ transition of the $D2$ with a maximum single photon Rabi frequency of $2g_0/2\pi = 68$ MHz. The transverse atomic decay rate is $\gamma/2\pi = 1.3$ MHz and the cavity decay rate is $\kappa/2\pi = 2.05$ MHz. Clearly this is in the strong coupling regime.

A weak probe laser beam is incident on the cavity with a frequency ω_p that can be tuned through the atomic resonance frequency. The transmitted light is detected

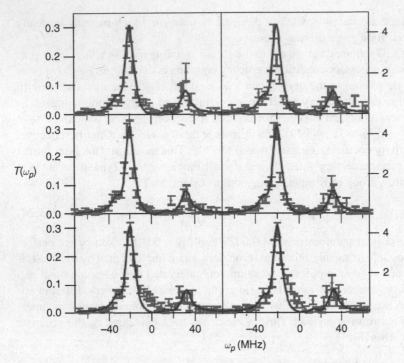

$T(\omega_p)$

ω_p (MHz)

Fig. 11.3 The results of a measurement of the vacuum Rabi splitting performed by the Caltech group. Six different studies are shown, together with a comparison to theory (*solid line*). From [9]

at a photodetector and thus the transmission coefficient $T(\omega_p)$ can be measured. The results for six cases in which one atom was present in the cavity are shown in Fig. 11.3. Also shown as a solid line is the theoretical prediction based on the steady state solution to the master equation. The asymmetry of the peaks is due to the different Stark shifts for the Zeeman sub-levels of the excited state and optical pumping.

11.1.2 Single Photon Sources

In Chap. 16 we discuss how single photons can be used to encode and process information in a fully quantum coherent fashion. To realise such scheme however requires a very special kind of light source that produces a train of transform limited pulses each containing one and only one photon with high probability. Cavity QED schemes can be used to generate such states. If an atom, coupled to a single mode cavity field, was prepared in the excited state at time $t = t_0$, it will emit a photon into the cavity on a time scale determined by g^{-1}. If we are in the strong coupling regime, this photon will be reabsorbed by the atom on the same time scale. This is not what we want for a single photon source. In order to ensure the photon is emitted from the cavity a time $t = t_0 + \tau$, we will need a bad cavity, i.e. one for which $\kappa >> g, \gamma$. We also need to ensure that the atom does not emit a photon in any mode other than the cavity mode, so that we require $g >> \gamma$. The net effect is that the rate for the photon to be emitted preferentially into the output mode from the cavity is much greater than its free-space spontaneous emission rate. This is know as the Purcell effect.

Of course there is still some uncertainty in the emission of the photon from the cavity as this is a Poisson process at rate κ and T is a random variable. The probability density for T is

$$p(T) = \kappa e^{-\kappa T} \tag{11.25}$$

This has a mean given by κ^{-1}, which also happens to be the uncertainty in the emission time. Such a system necessarily has some "time jitter" in the single photon pulses emitted from the cavity. Once the photon is emitted, we need to re-excite the atom to generate another pulse. Let us suppose that the repetition time for this is T. If we can arrange things so that $T >> \kappa^{-1}$, the relative time jitter is small.

Of course the excitation itself is a dynamical process and takes some time. There is some time scale associated with this excitation and the excitation pulse itself may have some non trivial time dependence. There are two models of interest for the excitation process. In the first model, a strong classical pump pulse excites a multi level atomic system which then decays *non radiatively* into the excited state of the dipole coupled to the cavity mode, see Fig. 11.4 (a). The problem with this scheme is that the entry of the system into the excited state $|e\rangle$ is a random process (likely a Poisson process). One might think this will cause no problems so long as Γ is large enough. However the condition $g >> \gamma$ means that these fluctuations are important when the Purcell effect becomes large [10]. In the second model, (b), the classical pulsed field together with the cavity field excites a two photon Raman transition to state $|2\rangle$. The effective interaction Hamiltonian between the field and the atom is

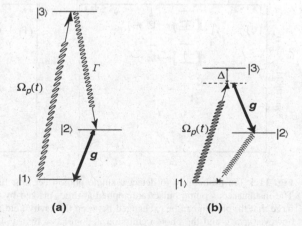

Fig. 11.4 Two schemes for a single photon source via cavity QED. In (**a**), a pulsed optical field excites the system to an auxiliary excited state (or states), which then decays non radiatively into the excited state $|e\rangle$. The transition $|1\rangle \leftrightarrow |2\rangle$ is coupled to a cavity mode with coupling constant g. In (**b**) the pulsed field together with the cavity field excites a two photon Raman transition to state $|a\rangle$

$$H_1(t) = \hbar \frac{\Omega_p(t)g}{\Delta}(a^\dagger|2\rangle\langle1| + a|1\rangle\langle2|) \qquad (11.26)$$

In effect, the coupling of the atom to the cavity field has become time dependent through the dependance on the pump field, $\omega_p(t)$.

In Fig. 11.5 we show the results of simulation for scheme (a) [11]. We assume that the pump pulse excitation is instantaneous. The quality of such a single

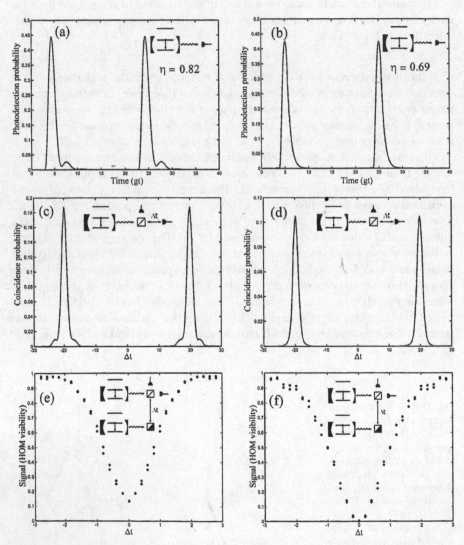

Fig. 11.5 The probability to detect a single photon per unit time for scheme (**a**) in Fig. 11.4. The instantaneous pump pulses are applied at times marked by the arrow. In (**a**) g is sufficiently large that the photon can be exchanged between the cavity field and atom before emission. In (**b**) however $\kappa > g$ and the photon emission is dominated by a Poisson decay process. Also shown are the corresponding results for a simulated $g^{(2)}(\tau)$ experiment (**c**, **d**) and a HOM coincidence experiment (**e**, **f**)

photon source can be operationally determined by two experiments: a Hanbury-Brown/Twiss experiment to measure $g^{(2)}(\tau)$ and a Hong-Ou-Mandel two-photon interference experiment (see Sect. 16.4.2). If we do indeed have a source of single photon pulses with one and only one photon per pulse $g^{(2)}(\tau)$ should be zero at zero delay and peak at the pulse repetition rate. Like wise if we mode match two identical single photon pulses on a 50/50 beam splitter as in the Hong-Ou-Mandel two-photon interference experiment, the coincidence rate for zero delays between the pulses should go to zero. Clearly we do not want to be in a regime where significant Rabi oscillations can occur before photon emission.

The second scheme, (b), has been implemented by *Keller* et al. [4] using a ion trap CQED scheme. They used a laser cooled Ca^+ ion confined to the centre of an optical cavity with a linear RF trap. The levels are given as $|1\rangle \rightarrow 4^2S_{1/2}$, $|2\rangle \rightarrow 3^2D_{3/2}$, $|3\rangle \rightarrow 4^2P_{1/2}$. The pump pulse had a carrier wavelength of 397 nm while the Raman resonance was tuned to the cavity resonant wavelength of 866 nm. The single photon Rabi frequency was $g/2\pi = 0.92$ MHz. The spontaneous decay rate of the $P_{1/2} \rightarrow D_{3/2}$ transition was $\gamma/2\pi = 1.69$ MHz and the cavity decay rate was $\kappa/2\pi = 1.2$ MHz. The Raman pump pulse had a predefined intensity profile of up to 6 ms duration and was repeated at a rate of 100 kHz. An interesting feature of the experiment was that the temporal shape of the pump pulse could be controlled to some extent. The output from the cavity is then directed towards a avalanche photodiode detector with overall detection efficiency of about $\eta = 0.05$. The time of each photodetection event is recorded with a 2 ns resolution and the resulting arrival time distribution gives the detection probability per unit time, $n(t)$ for the single photon (with $n(t) = \eta\langle a^\dagger(t)a(t)\rangle$, see Sect. 16.4.2). The results are shown in Fig. 11.6 for various choices for the temporal structure of the pump pulse. Also shown is a full simulation of $n(t)$ based on the master equation. Keller et al. also measured the cross correlation events in a Hanbury-Brown/Twiss experiment. The peak suppression at zero delay was by a factor of the order of 10^5 compared to the counts in all the other peaks.

11.1.3 Cavity QED with N Atoms

There are a large class of experiments in which more than one atom (for example in an atomic vapour) interacts with a single cavity mode. In that case a single cavity photon is *shared* over many atomic excitations, and so the effect on any single atom is reduced. In this case we might expect that an approximate scheme based on an expansion in N^{-1} where N is the number of atoms involved, might be possible. We will present one approach to this problem developed by *Drummond* and *Walls* [12].

The interaction Hamiltonian in (11.1) becomes

$$H_I = g_0 \sum_{j=1}^{N} a^\dagger \sigma_-^{(j)} e^{i\vec{k}.\vec{x}_j} + a\sigma_+^{(j)} e^{-i\vec{k}.\vec{x}_j} \qquad (11.27)$$

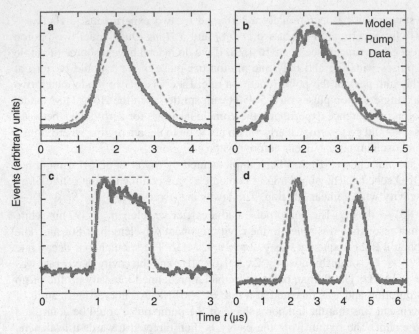

Fig. 11.6 The probability to detect a single photon per unit time, versus time, for scheme (b) in the experiment of *Keller* et al. [5], with various pump pulse shapes (*dashed line*): (**a**), Strong Gaussian pump; (**b**), Weak Gaussian pump; (**c**), Square-wave pump; (**d**), Double-peaked pump

$$= g(a^{\dagger}S + aS^{\dagger}) \tag{11.28}$$

where we have used plane wave modes to be specific, and where we have defined the collective atomic polarisation operators as

$$S = \sum_{j=1}^{N} \sigma_{-}^{(j)} e^{i\vec{k}.\vec{x}_j} \tag{11.29}$$

This immediately suggests that we should also define the collective atomic inversion operator

$$D = \sum_{j=1}^{N} \sigma_{z}^{(j)} \tag{11.30}$$

These operators obey the following commutation relations,

$$[S^{\dagger}, S] = D \tag{11.31}$$

$$[D, S] = -2S \tag{11.32}$$

$$[D, S^{\dagger}] = 2S^{\dagger} \tag{11.33}$$

We assume that each atom undergoes independent spontaneous emission out of the cavity mode and that such emitted photons are not reabsorbed. The master equation then takes the form

$$\frac{d\rho}{dt} = -i\delta[a^\dagger a,\rho] - i\frac{\Delta}{2}[D,\rho] - i[\varepsilon^* a + \varepsilon a^\dagger,\rho] - ig[a^\dagger S + aS^\dagger,\rho]$$

$$+ \frac{\kappa}{2}(2a\rho a^\dagger - a^\dagger a\rho - \rho a^\dagger a)$$

$$+ \frac{\gamma}{2}\sum_{j=1}^{N}(2\sigma_-^{(j)}\rho\sigma_+^{(j)} - \sigma_+^{(j)}\sigma_-^{(j)}\rho - \rho\sigma_+^{(j)}\sigma_-^{(j)}) \qquad (11.34)$$

We next define the normally ordered characteristic function

$$\chi(\vec{\lambda}) = \text{Tr}\{\rho\,\Xi(\vec{\lambda})\} \qquad (11.35)$$

where

$$\Xi(\vec{\lambda}) = e^{i\lambda_5 S^\dagger}e^{i\lambda_4 D}e^{i\lambda_3 S}e^{i\lambda_2 a^\dagger}e^{i\lambda_1 a} \qquad (11.36)$$

with $\vec{\lambda}^T = (\lambda_1,\lambda_2,\lambda_3,\lambda_4,\lambda_5)$. The positive P-representation for the atom-field state is then defined as the multi-dimensional Fourier transform of the characteristic function,

$$P(\vec{\alpha}) = \frac{1}{(2\pi)^5}\int \chi(\vec{\lambda})e^{-i\vec{\lambda}\vec{\alpha}}d^5\lambda \qquad (11.37)$$

where $\vec{\alpha}^T = (\alpha,\beta,v,D,u)$.

To obtain an equation of motion for $P(\vec{\alpha})$ we first need to find an equation of motion for the characteristic function, $\chi(\vec{\lambda})$ and then integrate by parts. Consider, for example, the term arising from the interaction with the cavity field mode,

$$\left(\frac{d\chi}{dt}\right)_I = -ig\langle[\Xi,a^\dagger S]\rangle - \langle[\Xi,aS^\dagger]\rangle \qquad (11.38)$$

We can use the commutation relations for these operators to show that, for example,

$$\langle[\Xi,aS^\dagger]\rangle = \left(\lambda_1\frac{\partial}{\partial\lambda_3} - (1 - e^{2i\lambda_4})\frac{\partial^2}{\partial\lambda_2\partial\lambda_3} - i\lambda_5\frac{\partial^2}{\partial\lambda_2\partial\lambda_4} + \lambda_5^2\frac{\partial^2}{\partial\lambda_2\partial\lambda_5}\right)\chi(\vec{\lambda})$$
$$(11.39)$$

The corresponding term in the equation of motion for $P(\vec{\alpha})$ is

$$\left(\frac{\partial P}{\partial t}\right)_I = -ig\left(-\frac{\partial}{\partial\alpha}v + (1 - e^{-2\partial_D})\beta v + \frac{\partial^2}{\partial u^2}\beta u - \frac{\partial}{\partial u}\beta D\right)P \quad (11.40)$$

$$+ ig\left(-\frac{\partial}{\partial\beta}u + (1 - e^{-2\partial_D})\alpha u + \frac{\partial^2}{\partial v^2}\alpha v - \frac{\partial}{\partial v}\alpha D\right)P \quad (11.41)$$

where $\partial_D = \frac{\partial}{\partial D}$.

A problem is immediately apparent: this term contains infinite order derivatives and thus the resulting equation is not of Fokker–Planck form and does not define a stochastic process. However a careful analysis shows that the exponential may be truncated to second order in an asymptotic expansion in N^{-1}, which for $N \gg 1$ is a reasonable approximation. The result after truncating is

$$
\begin{aligned}
\frac{\partial P}{\partial t} = [&-\partial_\alpha[-i\varepsilon - \tilde{\kappa}\alpha/2 - igv] - \partial_v[ig\alpha D - \tilde{\gamma}v/2] \\
&-\partial_D[-\gamma(D+N) - 2ig(u\alpha - v\beta)] \\
&+\partial_v^2[ig\alpha v] + \partial_D^2[\gamma(D+N) - 2ig(u\alpha - v\beta)] + \text{CC}]\, P
\end{aligned}
\tag{11.42}
$$

where CC stands for complex conjugate under the condition that $v^* \mapsto u$, $u^* \mapsto v$, $\alpha^* \mapsto \beta$, $\beta^* \mapsto \alpha$. The corresponding stochastic differential equations are

$$
\begin{aligned}
\dot{\alpha} &= -\tilde{\kappa}\alpha/2 - i\varepsilon - igv + \Gamma_\alpha \\
\dot{\beta} &= -\tilde{\kappa}^*\alpha/2 + i\varepsilon + igv + \Gamma_\beta \\
\dot{v} &= -\tilde{\gamma}v/2 + ig\alpha D + \Gamma_v \\
\dot{u} &= -\tilde{\gamma}^*u/2 - ig\beta D + \Gamma_u \\
\dot{D} &= -\gamma(D+N) - 2ig(\alpha u - \beta v) + \Gamma_D
\end{aligned}
\tag{11.43}
$$

which should be compared to the single atom semiclassical equations to which they reduce when noise is neglected and we make the replacements $\beta \mapsto \alpha^*$, $u \mapsto v^*\, N \mapsto 1$. The only non-zero noise correlation functions are

$$
\langle \Gamma_v(t)\Gamma_v(t') \rangle = 2ig\alpha v\, \delta(t - t')
\tag{11.44}
$$

$$
\langle \Gamma_u(t)\Gamma_v(t') \rangle = 2ig\alpha u\, \delta(t - t')
\tag{11.45}
$$

$$
\langle \Gamma_D(t)\Gamma_D(t') \rangle = 2\gamma(D+N) - 4ig(\alpha u - \beta v)\delta(t - t')
\tag{11.46}
$$

These equations now provide a basis to obtain a phenomenological description of optical bistability in terms of an intensity dependent cavity detuning. To proceed we will assume that $\gamma \gg \kappa$ so that the atomic variables can be assumed to reach a steady state slaved to the instantaneous values of the field variables. This is called *adiabatic elimination*. The next step is to approximate the noise correlation functions for the atomic variables by replacing u, v by the steady state values of the deterministic equations for these variables. This is equivalent to a linearisation around the deterministic steady state. The validity of this approximation rests on $N \gg 1$. The resulting atomic variables are then substituted into the field equations to give

$$
\dot{\alpha} = -i\varepsilon - \tilde{\kappa}\alpha/2 - \frac{2g^2 N\alpha}{\tilde{\gamma}\Pi(\alpha\beta)} + \Gamma(t)
\tag{11.47}
$$

where

$$
\Pi(\alpha\beta) = 1 + \frac{\alpha\beta}{n_0(1 + \Delta_1^2)}
\tag{11.48}
$$

with n_0 the saturation photon number and Δ_1 as previously defined in (11.17 and 11.18).

$$\langle \Gamma(t)\Gamma(t') \rangle = \frac{\kappa C x^2}{(1+X+\Delta_1^2)^3}\left[(1-i\Delta_1)^3 + \frac{X^2}{2}\right]\delta(t-t')$$

$$\langle \Gamma^\dagger(t)\Gamma(t') \rangle = \frac{\kappa C x^2}{(1+X+\Delta_1^2)^3}\left[2X + \frac{X^2}{2}\right]\delta(t-t') \qquad (11.49)$$

with $x = \alpha/\sqrt{n_0}$ and $X = \alpha\beta/n_0$.

11.2 Circuit QED

Superconducting coplanar microwave cavities [13] enable a new class of experiments in circuit quantum electrodynamics in the strong coupling regime [14]. The dipole emitter in this case is a single superconducting metalic island separated by tunnel junctions form a Cooper pair reservoir. Under appropriate conditions it is possible for the charge on the island to be restricted to at most a single Cooper pair. This Cooper pair tunneling on an off the island constitutes a single large electric dipole system.

A possible experimental implementation is shown in Fig. 11.7.

The coupling between a Cooper pair box charge system and the microwave field of circuit QED is given by [13]

$$H = 4E_c \sum_N (N - n_g(t))^2 |N\rangle\langle N| - \frac{E_J}{2}\sum_{N=0}(|N\rangle\langle N+1| + |N+1\rangle\langle N|) \qquad (11.50)$$

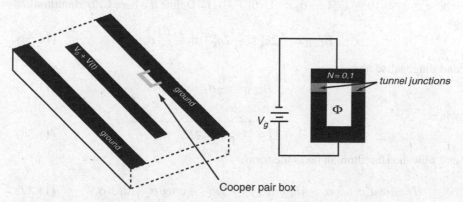

Fig. 11.7 A co-planar microwave resonator is coupled to a Cooper pair box electric dipole

where

$$E_C = \frac{e^2}{2C_\Sigma}$$

$$n_g(t) = \frac{C_g V_g(t)}{2e}$$

with C_Σ the capacitance between the island and the rest of the circuit, C_g is the capacitance between the CPB island and the bias gate for the island, and $V_g(t)$ is the total voltage applied to the island by the bias gate composed of a DC field, $V_g^{(0)}$ and microwave field in the cavity, $\hat{v}(t)$. Thus we can write $V_g(t) = V_g^{(0)} + \hat{v}(t)$, where the hat indicates a quantisation of the cavity field. Including the time dependent cavity field we can write

$$n_g(t) = n_g^{(0)} + \delta\hat{n}_g(t) \tag{11.51}$$

where

$$\delta\hat{n}_g(t) = \frac{C_g}{2e}\hat{v}(t) \tag{11.52}$$

The Hamiltonian in (11.50) is written in the Cooper pair number basis. In this basis the electrostatic energy of the first term is quite clear. The Josephson energy term describes tunneling of single Cooper pairs across the junction. This term is more traditionally (i.e. in mean-field theory) written in the phase representation as $E_J \cos\theta$. The connection between these two representations is discussed in [15].

If the Cooper pair box is sufficiently small, the electrostatic charging energy is so large that it is very unlikely that there will be more than a single Cooper pair on the island at any time. We can then usual restrict the CPB Hilbert space to $N = 0, 1$, we can write the Hamiltonian as

$$H = -2E_C(1 - 2n_g^{(0)})\bar{\sigma}_z - \frac{E_J}{2}\bar{\sigma}_x - 4E_C\delta\hat{n}_g(t)(1 - 2n_g^{(0)} - \bar{\sigma}_z) \tag{11.53}$$

where $\bar{\sigma}_z = |0\rangle\langle 0| - |1\rangle\langle 1|$, $\bar{\sigma}_x = |1\rangle\langle 0| + |0\rangle\langle 1|$. Define the bare CPB Hamiltonian as

$$H_{CPB} = -2E_C(1 - 2n_g^{(0)})\bar{\sigma}_z - \frac{E_J}{2}\bar{\sigma}_x \tag{11.54}$$

and diagonalise it as

$$H_{CPB} = \frac{\varepsilon}{2}\sigma_z \tag{11.55}$$

where

$$\varepsilon = \sqrt{E_J^2 + [4E_C(1 - 2N_g^{(0)})]^2} \tag{11.56}$$

and now the Hamiltonian takes the form,

$$H = \hbar\omega_r a^\dagger a + \frac{\varepsilon}{2}\sigma_z - 4E_C\delta\hat{n}_g(t)[1 - 2n_g^{(0)} - \cos\theta\sigma_z + \sin\theta\sigma_x] \tag{11.57}$$

where we have now included the free Hamiltonian for the microwave cavity field, and

$$\theta = \arctan\left(\frac{E_J}{4E_C(1 - 2n_g^{(0)})}\right) \tag{11.58}$$

Operating at the charge degeneracy point, $n_g^{(0)} = 1/2$ so that $\theta = \pi/2$, the Hamiltonian becomes

$$H = \hbar\omega_c a^\dagger a + \frac{\varepsilon}{2}\sigma_z - 4E_C\frac{C_g}{2e}\hat{v}(t)\sigma_x \tag{11.59}$$

$$= \hbar\omega_c a^\dagger a + \frac{\varepsilon}{2}\sigma_z - \hbar g(a + a^\dagger)\sigma_x \tag{11.60}$$

where the coupling constant is

$$\hbar g = e\frac{C_g}{C_\Sigma}\sqrt{\frac{\hbar\omega_r}{Lc}} \tag{11.61}$$

This can be as large as 50 MHz[13]. The circuit resonance is typically at $\omega_c \approx$ 1 GHz. However, we can detune the qubit from this resonant frequency by a few MHz [14]. We can then make the rotating wave approximation and take the Hamiltonian in the interaction picture to be the usual Jaynes–Cummings form

$$H_I = \hbar\delta_a a^\dagger a + \hbar g(a\sigma_+ + a^\dagger\sigma_-) \tag{11.62}$$

with $\delta_a = \omega_c - \omega_b$ is the detuning between the cavity resonance and the CPB and $\hbar\omega_b = \varepsilon$.

Exercises

11.1 A laser may be modeled by the master equation in (11.34) with $\varepsilon = 0$ and with the addition of the incoherent pump term

$$\mathcal{L}_p\rho = \frac{r}{2}\sum_{j=1}^{N}(2\sigma_+^{(j)}\rho\sigma_-^{(j)} - \sigma_-^{(j)}\sigma_+^{(j)}\rho - \rho\sigma_-^{(j)}\sigma_+^{(j)}) \tag{11.63}$$

We will also assume that $\omega_A = \omega_c$

(a) Show that the Fokker-Planck equation is now given by

$$\begin{aligned}
\frac{\partial P}{\partial t} = &[-\partial_\alpha[-\kappa\alpha/2 - igv] - \partial_v[ig\alpha D - \gamma_\parallel v/2] \\
&-\partial_D[-\gamma_\parallel(D - D_0) - 2ig(u\alpha - v\beta)] \\
&+\partial_v^2[ig\alpha v] - 2r\frac{\partial^2}{\partial v\partial D} + NP\frac{\partial^2}{\partial v\partial u} \\
&+\partial_D^2[r(N - D) + \gamma(D + N) - 2ig(u\alpha - v\beta) + CC]P \tag{11.64}
\end{aligned}$$

where $\gamma_\parallel = \gamma + r$ and

$$D_0 = N\left(\frac{r-\gamma}{r+\gamma}\right) \tag{11.65}$$

(b) By considering the deterministic equations of motion show that laser action requires $D_0 > 0$.

(c) Show that the steady state field amplitude obeys the equation

$$x\left(1 - \frac{C_1}{1+|x|^2}\right) = 0 \tag{11.66}$$

where $x = \alpha/\sqrt{n_0}$ with

$$n_0 = \frac{\gamma\gamma_\parallel}{8g^2} \tag{11.67}$$

$$C_1 = \frac{4g^2 D_0}{\gamma_\parallel \kappa} \tag{11.68}$$

(d) Define $I = |x|^2$. Show that the stable solutions are

$$I = \begin{cases} 0 & \text{if } C_1 < 1 \\ C_1 - 1 & \text{if } C_1 > 1 \end{cases} \tag{11.69}$$

References

1. R. Miller, T.E. Norththrop, K.M. Birnbaum, A. Boca, A.D. Boozer, H.J. Kimble: J. Phys. B: At. Mol. Opt. Phys. **38**, s551 (2005)
2. J. Ye, D.W. Vernooy, H.J. Kimble: Phys. Rev. Lett. **83**, 4987 (1999)
3. T. Puppe, I. Schuster, A. Grothe, A. Kubanek, K. Murr, P.W.H. Pinske, G. Rempe: quant-ph/072162 (2007)
4. M. Keller, B. Lange, K. Hayasaka, W. Lange, H. Walther: Nature **431**, 1075 (2004)
5. K. Fortier, S.Y. kim, M.J. Gibbons, P. Ahmadi, M.S. Chapman: Phys. Rev. Letts. **98**, 233601 (2007)
6. T. Aoki, B. Dayan, E. Wilcut, W.P. Bowen, A.S. Parkins, H.J. Kimble: Nature, **443**, 671 (2006)
7. K. Hennessy, A. Badolato, M. Winger, D. Gerace, M. Attüre, S. Gulde, S. Fält, E.L. Hu, A. Imamoglu: Nature **445**, 896 (2007)
8. J.M. Raimond, M. Brune, S. Haroche: Rev. Mod. Phys. **73**, 565 (2001)
9. A. Boca, R. Miller, K.M. Birnbaum, A.D. Boozer, J. McKeever, H.J. Kimble: Phys. Rev. Lett. **93**, 233603 (2004)
10. A. Kirag, M. Attatüre, A. Imamoğlu: Phys. Rev. A., **69**, 032305 (2004)
11. M.J. Fernee, H. Rubinsztein-Dunlop, G.J. Milburn, Phys. Rev. A., **75**, 043815 (2007)
12. P.D. Drummond, D.F. Walls: Phys. Rev. A **23**, 2563 (1981)
13. A. Blais, R. Huang, A. Walraff, S. Girvin, R.J. Schoelkopf: Phys. Rev. A **69**, 062320 (2004)

14. D.I. Schuster et al.: Nature **445**, 515 (2007)
15. J.F. Annett, B.L. Gyorffy, T.P. Spiller: *Superconducting Devices for Quantum Computation*. In 'Exotic States in Quantum Nanostructures', S. Sarkar (ed.) (Kluwer Academic Publishers 2002)

Further Reading

Haroche, S., J.M. Raimond: *Exploring the quantum* (Oxford University Press, Oxford, UK 2006)

Chapter 12
Quantum Theory of the Laser

Abstract The quantum theory of the laser was developed in the 1960s principally by the schools associated with H. Haken, W.E. Lamb and M. Lax, see [1, 2, 3, 4]. Haken and Lax independently developed sophisticated techniques to convert operator master equations into c-number Fokker–Planck equations or equivalent Langevin equations.

In this chapter we shall follow the approach of *Scully* and *Lamb* [3] to compute photon statistics and the linewidth of the laser. In the *Scully–Lamb* treatment the pumping is modelled by the injection of a sequence of inverted atoms into the laser cavity. In a usual laser, with a thermal pumping mechanism, a Poisson distributed sequence of inverted atoms is assumed. Introduction of a Bernoulli distribution enables a more general class of pumping mechanisms to be considered, including the case of the regularly pumped laser. Diode lasers with more regular pumping than usual lasers have recently been shown to give rise to sub-shot-noise photocurrent fluctuations.

12.1 Master Equation

A single mode cavity field is excited by a sequence of atoms injected into the cavity. Let t_i be the arrival time of the atom i in the cavity and τ the time spent by each atom in the cavity. The change in the density operator for the field due to the interaction with the ith atom may be represented by

$$\rho(t_i + \tau) = \mathscr{P}(\tau)\rho(t_i) . \tag{12.1}$$

The explicit form of $\mathscr{P}(\tau)$ depends on the particular atomic system used in the excitation process. The model we will employ is indicated in Fig. 12.1.

Of the four levels, only levels $|1\rangle$ and $|2\rangle$ are coupled to the intracavity field, which thus are referred to as the lasing levels. Each of these levels may then decay. Level $|1\rangle$ decays to level $|3\rangle$ at a rate γ_1 while level $|2\rangle$ decays to level $|4\rangle$ at a rate γ_2.

Fig. 12.1 Schematic representation of the four-level atomic model of a laser. Only levels 1 and 2 at coupled to the laser field

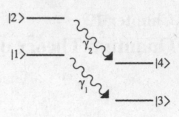

We will assume that these decay rates are very much greater than the spontaneous decay rate of level $|2\rangle$ to level $|1\rangle$, and thus we neglect spontaneous emission in the lasing levels. Each atom is prepared in the excited state $|2\rangle$ prior to interaction with the cavity field. In the usual laser system the lifetimes γ_1^{-1} and γ_2^{-1} are much shorter than the time τ spent by each atom in the cavity. This means that each atom rapidly attains a steady state in passing through the cavity and the pump operation $\mathscr{P}(\tau)$ is effectively independent of the time τ. The effect of a single atom on the state of the field may then be written as

$$\rho' = \mathscr{P}\rho ,\qquad(12.2)$$

where we have dropped the time dependence in ρ for simplicity, the prime serving to indicate the state of the field after the passage of a single atom through the cavity. We may represent the initial state of the field quite generally as

$$\rho = \sum_{n,m=0}^{\infty} \rho_{n,m}(0)|n\rangle\langle m| .\qquad(12.3)$$

In Appendix [12.A] we solve the master equation for the system over the time τ under the assumptions discussed above. The result is

$$\rho' = \sum_{n,m=0}^{\infty} \rho_{n,m}(0)(A_{nm}|n\rangle\langle m| + B_{nm}|n+1\rangle\langle m+1|) ,\qquad(12.4)$$

where the explicit expressions for A_{nm}, B_{nm} are given in the appendix.

We now assume that each atom contributes independently to the field. (This assumption remains valid even if there is more than one atom in the cavity at any time, provided that they are sufficiently dilute.) Thus, if k atoms are passed through the cavity from time 0 to time t the field density operator at time t is given by

$$\rho(t) = \mathscr{P}^k\rho(0) .\qquad(12.5)$$

More generally, however, not all atoms entering the cavity are prepared in the excited state. Let the probability for an excited atom to enter the cavity between t and $t+\Delta t$ be $r\Delta t$, r being the average injection rate. This defines a Poisson excitation process. Thus the field at time $t+\Delta t$ is made up of a mixture of states corresponding to atomic excitation and no atomic excitation, thus

$$\rho(t+\Delta t) = r\Delta t \mathscr{P}\rho(t) + (1-r\Delta t)\rho(t) .\qquad(12.6)$$

In the limit $\Delta t \to 0$ we have

$$\frac{d\rho(t)}{dt} = r\mathscr{U}\rho(t) \tag{12.7}$$

where

$$\mathscr{U} = \mathscr{P} - 1 . \tag{12.8}$$

We must now include the decay of the cavity field through the end mirrors. This is modelled in the usual way by coupling the field to a zero temperature heat bath. Thus the total master equation for the field density operator is

$$\frac{d\rho}{dt} = r\mathscr{U}\rho + \frac{\kappa}{2}(2a\rho a^\dagger - a^\dagger a\rho - \rho a^\dagger a) , \tag{12.9}$$

where κ is the cavity decay rate. This is the usual *Scully–Lamb* laser master equation.

In the special case that $\gamma_1 = \gamma_2 = \gamma$ the matrix elements of $\mathscr{U}$ in the number basis are greatly simplified. In this case the master equation in the number basis may be written as

$$\frac{d\rho_{nm}}{dt} = G\left(\frac{\sqrt{nm}}{1 + (n+m)/2n_s} \rho_{n-1,m-1} \right.$$
$$\left. - \frac{(m+n+2)/2 + (m-n)^2/8n_s}{1 + (n+m+2)/2n_s} \rho_{nm} \right)$$
$$+ \frac{\kappa}{2}[2\sqrt{(n+1)(m+1)}\rho_{n+1,m+1} - (n+m)\rho_{nm}] , \tag{12.10}$$

where

$$G = \frac{r}{2n_s} \tag{12.11}$$

and

$$n_s = \frac{\gamma^2}{4g^2} . \tag{12.12}$$

where g is the coupling strength between the cavity and the levels 1 and 2.

we have neglected terms $\propto n_s^{-2}$ in the denominators of the first two coefficients.

12.2 Photon Statistics

The photon number distribution obeys the equation

$$\frac{dp_n}{dt} = -G\left(\frac{n+1}{1 + (n+1)/n_s} p_n - \frac{n}{1 + (n/n_s)} p_{n-1} \right)$$
$$+ \kappa(n+1)p_{n+1} - \kappa n p_n . \tag{12.13}$$

The gain coefficient G is defined by

$$G = \frac{r\gamma_1}{2\gamma_+ + n_s},$$
(12.14)

where $\gamma_+ = (\gamma_1 + \gamma_2)/2$.

If we expand the denominators in (12.13) to first-order an approximate equation for the mean photon number may be obtained, namely

$$\frac{d\bar{n}}{dt} = (G - \kappa)\bar{n} - \frac{G}{n_s}(\overline{n^2} + 2\bar{n} + 1) + G.$$
(12.15)

If $G > \kappa$ there will be an initial exponential increase in the mean photon number. Thus $G = \kappa$ is the threshold condition for the laser.

The steady state photon number distribution may be deduced directly from (12.13), using the condition of detailed balance. It may be written in the form

$$p_n^{ss} = \mathcal{N} \frac{(G n_s/\kappa)^{n+n_s}}{(n+n_s)!},$$
(12.16)

where $\mathcal{N}$ is a normalisation constant. Below threshold $(G < \kappa)$ this distribution may be approximated by a chaotic (thermal) distribution with the mean $\bar{n} = G/(\kappa - G)$ (Exercise 12.1). Above threshold $(G > \kappa)$ the mean and variance are given, to a good approximation, by (Exercise 12.2),

$$\bar{n} = n_s\left(\frac{G}{\kappa} - 1\right),$$
(12.17)

$$V(n) = \bar{n} + n_s.$$
(12.18)

Well above threshold $\bar{n} \gg n_s$ and thus $V(n) \approx \bar{n}$, indicating an approach to Poisson statistics. In Fig. 12.2 we show the exact photon number distribution for below and above threshold. The transition from power law to the Poisson distribution is quite evident.

Photon counting experiments by *Arecchi* [5], *Johnson* et al. [6], and *Morgan* and *Mandel* [7], demonstrated that the photon statistics of a laser well above threshold, approaches a Poisson distribution. In Fig. 12.3 we present the results of photon counting measurements by *Arecchi* on both thermal and laser light. A comparison of the experimental data with the thermal and Poisson distributions is also shown.

12.2.1 Spectrum of Intensity Fluctuations

Equations (12.17 and 12.18) give the photon number fluctuations for the internal cavity mode. This quantity, however, is not directly observable. We must now determine how the photon number fluctuations inside the cavity determine the intensity

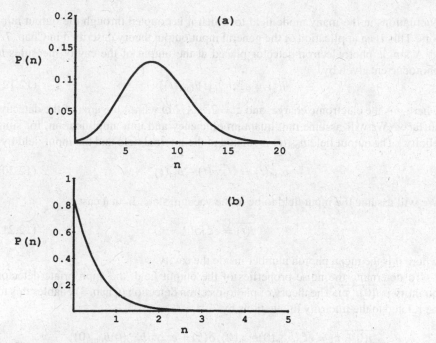

Fig. 12.2 The steady state photon number distribution of a laser operating above and below threshold. In **(a)** $G/\kappa = 5.0$, in **(b)** $G/\kappa = 0.25$. In both cases $n_{\mathrm{s}} = 2$

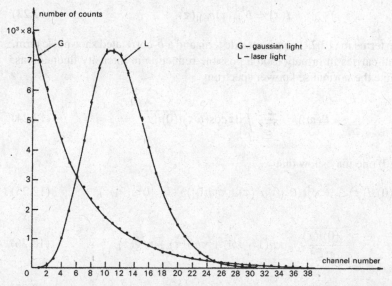

Fig. 12.3 Experimental results for the steady state photon number distribution for a thermal (i.e. Gaussian) light source and a laser operator above threshold. The laser exhibits Poissonian photon number statistics [5]

fluctuations in the many mode field to which it is coupled through the output mirrors. This is an application of the general input/output theory described in Chap. 7.

A single photoelectron detector placed at the output of the cavity measures a photocurrent given by

$$\overline{i(t)} = e\xi \langle b_{\text{out}}^\dagger(t)b_{\text{out}}(t)\rangle \,, \tag{12.19}$$

where e is the electronic charge, and $\xi = 2\varepsilon_0 cA/\hbar\omega$ with A the area of the detector surface. (We will assume unit quantum efficiency and unit amplification, for simplicity.) The output field $b_{\text{out}}(t)$ is related to the internal field and the input field by

$$b_{\text{out}}(t) = \sqrt{\kappa}a(t) - b_{\text{in}}(t) \,. \tag{12.20}$$

We will assume the input field to be in the vacuum state. In that case

$$\overline{i(t)} = e\xi\kappa\bar{n} \,, \tag{12.21}$$

where $\bar{n}$ is the mean photon number inside the cavity.

To determine the noise properties of the output field, the appropriate detector quantity is $\overline{i(0)i(\tau)}$. The theory of photo-electron detection (Chap. 3) enables this to be related to the intensity fluctuations by

$$\overline{i(0)i(\tau)} = e\xi \langle b_{\text{out}}^\dagger(0)b_{\text{out}}(0)\rangle\delta(\tau) + e^2\xi^2\langle b_{\text{out}}^\dagger(0)b_{\text{out}}(0)\rangle^2$$
$$+ e^2\xi^2\langle : I(0),I(\tau): \rangle \tag{12.22}$$

where : : denotes normal and time ordering and

$$I(\tau) = b_{\text{out}}^\dagger(\tau)b_{\text{out}}(\tau). \tag{12.23}$$

The first two terms in (12.22) represent a dc term and a δ-correlated shot-noise term. The last term carries information on a possible reduction in intensity fluctuations. We now define the normalised power spectrum

$$P(\omega) = \frac{2}{e^2\xi^2}\int_0^\infty d\tau \cos(\omega\tau)\overline{i(0)i(\tau)} \,. \tag{12.24}$$

Using (12.20) one may show that

$$\langle : I(0),I(\tau): \rangle = \kappa^2(\langle a^\dagger(0)a^\dagger(\tau)a(\tau)a(0)\rangle - \langle a^\dagger(0)a(0)\rangle^2) \,. \tag{12.25}$$

Thus

$$\frac{\overline{i(0)i(\tau)}}{e^2\xi^2} = \kappa\bar{n}(1-\kappa\bar{n}) + \kappa\bar{n}\delta(\tau) + \kappa^2 g(\tau) \tag{12.26}$$

where

$$g(\tau) = \langle a^\dagger(0)a^\dagger(\tau)a(\tau)a(0)\rangle \,. \tag{12.27}$$

We are only interested in the steady state fluctuations of the output field. In which case we can show that $g(\tau) = \bar{n}^2$ and thus

$$P(\omega) = \kappa\bar{n} .$$ (12.28)

This flat photocurrent spectrum is the shot-noise limit of the laser.

12.3 Laser Linewidth

Well above threshold the laser produces Poisson photon statistics. A coherent state has the same photon statistics, and this suggests that well above threshold the laser might be producing a coherent state. This is not the case. While the intensity of the laser is stabilised with a Poissonian distribution the phase of the laser undergoes a diffusion process. The effect of this phase diffusion is to cause a decay in the mean amplitude of the laser field, as the phase becomes uniformly distributed over 2π. The rate of amplitude decay Γ is thus a direct measure of the phase diffusion rate.

We will only discuss the case $\gamma_1 = \gamma_2 = \gamma$. The mean amplitude is defined by

$$\langle a(t) \rangle = \sum_{n=0}^{\infty} n^{1/2} \rho_{n,n-1}(t) .$$ (12.29)

Using (12.10) we find

$$\frac{d\langle a \rangle}{dt} = -\frac{G}{2} \sum_{n=0}^{\infty} \frac{(1/4n_s) - 1}{1 + (2n+1)/2n_s} \sqrt{n}\rho_{n,n-1} .$$ (12.30)

Assuming the laser operates well above threshold we can replace n by $\bar{n}$ in the denominator of each coefficient. Then as $\bar{n} \gg n_s$

$$\frac{d\langle a \rangle}{dt} = -\frac{G}{8\bar{n}}\langle a \rangle .$$ (12.31)

Thus the phase diffusion rate is inversely proportional to the intensity of the laser. All second-order phase dependent correlation functions will decay at a similar rate. In particular, the two-time correlation function

$$F(\tau) = \langle a^\dagger(\tau)a(0) \rangle$$ (12.32)

will decay at the rate $\Gamma = G/8\bar{n}$, i.e.

$$F(\tau) = \bar{n}e^{-\Gamma\tau} .$$ (12.33)

The Fourier transform of this function defines the laser spectrum

$$S(\omega) = \frac{\bar{n}}{\omega^2 + \Gamma^2} ,$$ (12.34)

and thus the laser linewidth is simply

$$\Gamma = \frac{G}{8\bar{n}} \, . \tag{12.35}$$

It must be emphasised, however, that these results only apply well above threshold.

12.4 Regularly Pumped Laser

The Poissonian photon statistics of a laser reflect the contributions from the random pumping mechanism and spontaneous emission, which lead to an irregular photo-emission sequence. By suppressing the pump fluctuations, sub-Poissonian photon statistics and thus sub-shot-noise photo-current fluctuations, are possible. This has been demonstrated in recent experiments by *Machida* et al. [8] and also *Richardson* and *Shelby* [9] with semiconductor lasers. The pump amplitude fluctuations were reduced by high impedance suppression of the electron injection rate.

We shall demonstrate how the *Scully–Lamb* laser theory can readily be modified to incorporate regular pumping. Regular pumped lasers have been considered theoretically by a number of researchers [10, 11, 12, 13, 14]. We shall follow the approach of *Golubov* and *Sokolov* [10], with some modifications.

Consider a time interval Δt short compared to the time scale on which the field is changing due to damping through the end mirrors. However, the time Δt is very long compared to the time interval between successive pumping atoms entering the cavity. Divide the interval Δt into N steps of length τ. The probability for an excited atom to enter the cavity at time $t_j = j\tau$ is defined to be p. The fundamental probabilities of interest are then the probability of r excited atoms to enter the cavity at any of the N time steps, over the interval Δt. These probabilities are

$$P_r(\Delta t) = \frac{\Delta t (\Delta t - \tau)(\Delta t - 2\tau) \ldots (\Delta t - r\tau)}{\tau^r r!} \left(\frac{p}{1-p} \right)^r$$

$$\times \exp \left[\frac{\Delta t}{\tau} \ln(1-p) \right] . \tag{12.36}$$

To first order in Δt this is

$$P_r(t) = (-1)^{r+1} \left(\frac{p}{1-p} \right)^r \frac{\Delta t}{r\tau} \, . \tag{12.37}$$

Between each atom entering the cavity the field evolves freely according to

$$\mathscr{T}(\tau) = e^{-i\omega_a \tau a^\dagger a} \rho \, e^{+i\omega_a \tau a^\dagger a} \, . \tag{12.38}$$

The change in the state of the field due to the passage of a single atom is given by (12.4). It is a simple matter to prove that the operation describing the effect of the pump atoms $\mathscr{P}$ commutes with the free evolution operator $\mathscr{T}$ (if this is not the case a simple master equation for the field state cannot be obtained in general). Thus the change in the state of the field over a time Δt is

$$\rho(t+\Delta t) = \mathscr{T}(\Delta t) \left[\sum_{n=0}^{N} P_n(\Delta t) \mathscr{P}^n \rho(t) \right] \tag{12.39}$$

(i.e., we can factor out the free evolution between each time step). We henceforth assume we are working in the interaction picture and drop the free evolution term. As we assume $\tau \ll \Delta t$ we extend the upper limit on the sum to ∞, then

$$\rho(t+\Delta t) = \rho(t) + \frac{\Delta t}{\tau} \ln(1-p) \rho(t)$$

$$+ \frac{\Delta t}{\tau} \left[\frac{p}{1-p} \mathscr{P} - \frac{1}{2} \left(\frac{p}{1-p} \right)^2 \mathscr{P}^2 + \dots \right] \rho(t) . \tag{12.40}$$

From which we obtain

$$\frac{d\rho}{dt} = \frac{1}{\tau} \ln(1-p) \rho(t) + \frac{1}{\tau} \ln \left(1 + \frac{p}{1-p} \mathscr{P} \right) \rho(t) \tag{12.41}$$

$$= R \ln(1 + p\mathscr{U}) \rho(t) , \tag{12.42}$$

where $R = \tau^{-1}$ is the pumping rate for a perfectly regular process $(p = 1)$ and

$$\mathscr{U} = \mathscr{P} - 1 . \tag{12.43}$$

We can define an average injection rate $r = pR$, then

$$\frac{d\rho}{dt} = \frac{r}{p} \ln(1 + p\mathscr{U}) \rho(t) . \tag{12.44}$$

In this form we can take the Poisson limit defined by $p \to 0$, $R \to \infty$, such that $pR = \text{constant} = r$. In this limit the equation reduces to that for a normal laser.

The difficulty in discussing the regularly pumped laser is the logarithm term in (12.42). As $\mathscr{U}$ represents the change in the state of the field due to a single atom we might expect $\mathscr{U}$ to be in some sense small. With this in mind we expand the logarithm to second order. Unfortunately this leads to a rather pathological master equation. However, the procedure does give accurate results for the first-and second-order moments of the photon number.

The photon number distribution now obeys the equation

$$\frac{dp_n}{dt} = \kappa[-np_n + (n+1)p_{n+1}] + r(-a_{n+1}P_n + a_n p_{n-1})$$

$$+ \frac{pr}{2} [-a_{n+1}^2 P_n + a_n(a_n + a_{n+1})p_{n-1} - a_n a_{n-1} p_{n-2}] , \tag{12.45}$$

where

$$a_n = \frac{Gn}{r(1 + (n/n_s))} \, . \tag{12.46}$$

To obtain the stationary state variances is a rather more difficult process than for the Poisson pumped case. The mean photon number above threshold is not changed, however the variance is given by

$$V(n) = \bar{n}\left(1 - \frac{p\gamma_1}{2(\gamma_1 + \gamma_2)}\right) + n_s \, . \tag{12.47}$$

We consider some special cases of this result for regular pumping, $p = 1$. Far above threshold $\bar{n} \gg n_s$ so we may neglect the last term in (12.47). When the decay rates are equal, the photon number variance is

$$V(n) = \frac{3\bar{n}}{4}(\gamma_1 = \gamma_2) \, . \tag{12.48}$$

In this case spontaneous emission from level $|2\rangle$ is contributing to the noise. This effect may be reduced by increasing the decay rate of the lower level with respect to the upper level, $\gamma_1 \gg \gamma_2$. In this case

$$V(n) = \frac{\bar{n}}{2}(\gamma_1 \gg \gamma_2) \, . \tag{12.49}$$

Thus the width of the photon number distribution inside the cavity is reduced by half.

We now consider intensity fluctuations of the light emerging from the cavity. We may obtain a solution for the normally ordered two-time correlation function $g(\tau)$, from the master equation (12.45) assuming a Gaussian steady state distribution. The result is

$$g(\tau) = \bar{n}^2 + [V(n) - \bar{n}]e^{-\delta\tau} \, , \tag{12.50}$$

where

$$\delta = \kappa\frac{\bar{n}/n_s}{1 + (\bar{n}/n_s)} \, . \tag{12.51}$$

Substituting (12.47 and 12.50) into (12.26) the spectrum of the photocurrent fluctuations is given by

$$P(\omega) = \kappa\bar{n}\left(1 + \frac{2\kappa Q\delta}{\omega^2 + \delta^2}\right) \, , \tag{12.52}$$

where

$$Q = \frac{V(n) - \bar{n}}{\bar{n}} \tag{12.53}$$

$$= -\frac{p\gamma_1}{2(\gamma_1 + \gamma_2)} + \frac{n_s}{\bar{n}} \, . \tag{12.54}$$

The Q parameter measures the deviation of the intracavity field from Poisson statistics. In the limit of regular pumping ($p = 1$) and far above threshold

$$P(\omega) = \kappa\bar{n}\left(1 - \frac{\gamma_1\kappa^2}{(\gamma_1+\gamma_2)(\kappa^2+\omega^2)}\right). \tag{12.55}$$

Thus at cavity resonance ($\omega = 0$),

$$\begin{aligned} P(0) &= \kappa\bar{n}(1+2Q) \\ &= \kappa\bar{n}\frac{\gamma_1}{\gamma_1+\gamma_2}. \end{aligned} \tag{12.56}$$

The first term in the first equation represents the shot-noise contribution. A negative value of Q leads to a reduction below the shot-noise limit. If the decay rates are equal ($\gamma_1 = \gamma_2$), spontaneous emission is not suppressed and

$$P(0) = \frac{\kappa\bar{n}}{2}, \tag{12.57}$$

which represents a 50% reduction below the shot-noise level. However, in the limit $\gamma_1 \gg \gamma_2$, Q approaches -0.5 far above threshold and

$$P(\omega) = \kappa\bar{n}\left(1 - \frac{\kappa^2}{\kappa^2+\omega^2}\right). \tag{12.58}$$

Then at cavity resonance, the fluctuation spectrum is zero. This may be compared with the light inside the cavity where the photon number fluctuations were only reduced by one half. This result has the same interpretation as the limit to the intracavity squeezing in a parametric oscillator; there is a destructive interference from the vacuum fluctuations reflected from the cavity mirror and the reduced noise light emerging from the cavity. This results in no fluctuations in the output light on resonance.

In Fig. 12.4 we show the results of the experiment by *Machida* et al. [8] for a regularly pumped semiconductor laser.

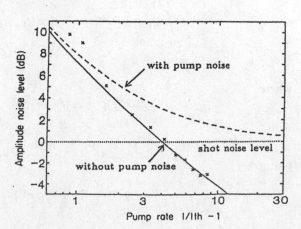

Fig. 12.4 The normalised amplitude noise level versus the pump rate for a laser with pump noise (*dashed*) and with pump noise suppressed (*solid*) [11]

12.A Appendix: Derivation of the Single-Atom Increment

Consider a single multilevel atom (Fig. 12.1) prepared in the state $|2\rangle$. Level $|1\rangle$ is damped at the rate γ_1 to level $|3\rangle$ and level $|2\rangle$ is damped at the rate γ_2 to level $|4\rangle$. Only levels $|1\rangle$ and $|2\rangle$ interact with the cavity field. The master equation describing the dynamics of this system is

$$\frac{d\rho}{dt} = ig[a^\dagger \sigma_-^{12} + a\sigma_+^{12}, \rho]$$

$$+ \frac{\gamma_1}{2}(2\sigma_-^{13}\rho\sigma_+^{13} - \sigma_+^{13}\sigma_-^{13}\rho - \rho\sigma_+^{13}\sigma_-^{13})$$

$$+ \frac{\gamma_2}{2}(2\sigma_-^{24}\rho\sigma_+^{24} - \sigma_+^{24}\sigma_-^{24}\rho - \rho\sigma_+^{24}\sigma_-^{24}) \qquad (12.59)$$

(we ignore spontaneous emission on the lasing levels $|1\rangle, |2\rangle$). We will present a complete operator solution to the master equation over the time τ and then consider the limit $\gamma_i \tau \gg 1$.

Define the operation

$$\mathscr{J}\rho = \gamma_1 \sigma_-^{13}\rho\sigma_+^{13} + \gamma_2 \sigma_-^{24}\rho\sigma_+^{24} \qquad (12.60)$$

and the rate operator

$$R = \gamma_1 \sigma_+^{13}\sigma_-^{13} + \gamma_2 \sigma_+^{24}\sigma_-^{24} \qquad (12.61)$$

$$= \gamma_1|1\rangle\langle 1| + \gamma_2|2\rangle\langle 2| . \qquad (12.62)$$

The solution to the master equation may then be written formally

$$\rho(t) = \mathscr{S}(t)\rho(0) + \int_0^t dt_1 \mathscr{S}(t-t_1)\, \mathscr{J}\, \mathscr{S}(t_1)\rho(0)$$

$$+ \int_0^t dt_1 \int_0^{t_1} dt_2 \mathscr{S}(t-t_1)\, \mathscr{J}\, \mathscr{S}(t_1-t_2)\, \mathscr{J}\, \mathscr{S}(t_2)\rho(0)$$

$$+ \cdots , \qquad (12.63)$$

where

$$\mathscr{S}(t)\rho = \mathscr{B}(t)\rho\mathscr{B}^\dagger(t) , \qquad (12.64)$$

with

$$\mathscr{B}(t) = \exp[-ig(a^\dagger \sigma_-^{12} + a\sigma_+^{12}) - \gamma_1 t|1\rangle\langle 1| - \gamma_2 t|2\rangle\langle 2|] . \qquad (12.65)$$

For

$$\rho(0) = |2\rangle\langle 2| \otimes \rho_F(0) \qquad (12.66)$$

$$= \sum_{n,\,m=0}^{\infty} \rho_{nm}(0)|n,2\rangle\langle m,2| \; ; \tag{12.67}$$

with

$$|n,2\rangle = |n\rangle_{\mathrm{F}} \otimes |2\rangle \; . \tag{12.68}$$

Note that after the action of $\mathscr{J}$ the atom is in a mixture of the states $|3\rangle$ and $|4\rangle$ and is then decoupled from the field. All further action of $\mathscr{S}(t)$ is just the identity, and $\mathscr{J}$ destroys the state. The series thus truncates at first order.

Now one may use the eigenstates of the free Hamiltonian (Chap. 10)

$$|n,+\rangle = \frac{1}{\sqrt{2}}(|n,2\rangle + |n+1,1\rangle) \; , \tag{12.69}$$

$$|n,-\rangle = \frac{1}{\sqrt{2}}(|n,2\rangle - |n+1,1\rangle) \; , \tag{12.70}$$

to show that

$$\mathscr{S}(t)(|n,2\rangle\langle m,2|) = (c_n^+(t)|n,+\rangle + c_n^-(t)|n,-\rangle)$$

$$\times (\langle m,+|c_m^+(t)^* + \langle m,-|c_m^-(t)^*) \tag{12.71}$$

where

$$c_n^+(t) = \frac{-\mathrm{i}\exp\left(-\frac{\gamma_+ t}{2}\right)}{2\sqrt{2}\Delta\Omega(n)} \left\{ \left[-\mathrm{i}\Omega(n)(1-\Delta) + \frac{\gamma_-}{2} \right] \mathrm{e}^{\mathrm{i}\Delta\Omega(n)t} \right.$$

$$\left. + \left[\mathrm{i}\Omega(n)(1+\Delta) - \frac{\gamma_-}{2} \right] \mathrm{e}^{-\mathrm{i}\Delta\Omega(n)t} \right\} \tag{12.72}$$

and

$$c_n^-(t) = \frac{-\mathrm{i}\exp\left(-\frac{\gamma_+ t}{2}\right)}{2\sqrt{2}\Delta\Omega(n)} \left\{ \left[\mathrm{i}\Omega(n)(1+\Delta) + \frac{\gamma_-}{2} \right] \mathrm{e}^{\mathrm{i}\Delta\Omega(n)t} \right.$$

$$\left. + \left[-\mathrm{i}\Omega(n)(1-\Delta) - \frac{\gamma_-}{2} \right] \mathrm{e}^{-\mathrm{i}\Delta\Omega(n)t} \right\} \tag{12.73}$$

where

$$\gamma_{\pm} = \frac{1}{2}(\gamma_1 \pm \gamma_2) \; , \tag{12.74}$$

$$\Delta = \left(1 - \frac{\gamma^2}{4\Omega(n)^2} \right)^{1/2} \; , \tag{12.75}$$

$$\Omega(n) = g\sqrt{n+1} \; . \tag{12.76}$$

We now assume $\gamma_{1,2} t \gg 1$. The first term in (12.63) may be ignored as it simply decays to zero.

We are interested in the state of the field alone which is obtained by tracing out over the atomic states. We use

$$\mathrm{Tr}_A(\mathscr{J}\mathscr{S}(t)|n,\,2\rangle\langle m,\,2|) = \frac{\gamma_2}{2}|n\rangle\langle m|[c_n^+(t) + c_n^-(t)][c_m^+(t)^* + c_m^-(t)^*]$$

$$+ \frac{\gamma_1}{2}|n+1\rangle\langle m+1|[c_n^+(t) - c_n^-(t)]$$

$$\times [c_m^+(t)^* - c_m^-(t)^*] \,. \qquad (12.77)$$

Thus we obtain in the steady state, the single atom increment

$$\rho' = \sum_{n,\,m=0}^{\infty} \rho_{nm}(0)(A_{nm}|n\rangle\langle m| + B_{nm}|n+1\rangle\langle m+1|)\,, \qquad (12.78)$$

where

$$A_{nm} = \frac{\gamma_2}{2}\int_0^{\infty} dt\,[c_n^+(t) + c_n^-(t)][c_m^+(t)^* + c_m^-(t)^*]\,, \qquad (12.79)$$

$$B_{nm} = \frac{\gamma_1}{2}\int_0^{\infty} dt\,[c_n^+(t) - c_n^-(t)][c_m^+(t)^* - c_m^-(t)^*]\,. \qquad (12.80)$$

Note that $\mathrm{Tr}(\rho) = 1$ requires that $A_{nn} + B_{nn} = 1$. We quote only the results for the diagonal matrix elements,

$$A_{nn} = \left(\frac{\gamma_2}{2\gamma_+}\right)\frac{4\Omega(n)^2 + 2\gamma_1\gamma_+}{4\Omega(n)^2 + \gamma_1\gamma_2} \qquad (12.81)$$

and

$$B_{nn} \stackrel{!}{=} 1 - A_{nn} = \left(\frac{\gamma_2}{2\gamma_+}\right)\frac{4\Omega(n)^2}{4\Omega(n)^2 + \gamma_1\gamma_2}\,. \qquad (12.82)$$

To compute the change in the state we write

$$\rho' = (1 + \mathscr{U})\rho = \mathscr{P}\rho\,. \qquad (12.83)$$

The diagonal matrix elements of $\mathscr{U}\rho$ are then found to be

$$\langle n|\mathscr{U}\rho|n\rangle = -a_{n+1}p_n + a_n p_{n+1}\,, \qquad (12.84)$$

where

$$a_{n+1} = A_{nn} - 1\,. \qquad (12.85)$$

Exercises

12.1 Show that below threshold ($G < \kappa$) the master equation may be approximated by

$$\frac{d\rho}{dt} = \frac{G}{2}(2a^\dagger \rho a - aa^\dagger \rho - \rho aa^\dagger) + \kappa(2a\rho a^\dagger - a^\dagger a\rho - \rho a^\dagger a) \,.$$

Thus demonstrate that the steady state density operator is

$$\rho^{ss} = \left(1 - \frac{G}{\kappa}\right) \sum_{n=0}^{\infty} \left(\frac{G}{\kappa}\right)^n |n\rangle\langle n| \qquad (12.86)$$

which is equivalent to a chaotic state.

12.2 Show that well above threshold the laser master equation may be approximated by

$$\frac{d\rho}{dt} = \frac{Gn_s}{2}(2n^{-1/2}a^\dagger \rho an^{-1/2} - an^{-1}a^\dagger \rho - \rho an^{-1}a^\dagger)$$

$$+ \frac{\kappa}{2}(2a\rho a^\dagger - a^\dagger \rho a - \rho a^\dagger a)$$

where $n = a^\dagger a$. Show that the steady-state solution is

$$\rho^{ss} = \exp\left(-\frac{Gn_s}{\kappa}\right) \sum_{n=0}^{\infty} \frac{(Gn_s/\kappa)^n}{n!} |n\rangle\langle n| \,. \qquad (12.87)$$

12.3 Show that the contours of the Q-function for the laser steady states in Exercises 12.1, 12.2 are: (a) Circles centred on the origin for below threshold, (b) annulli centered at the radius $r = (Gn_s/\kappa)^{1/2}$, for the above threshold state. Thus in both cases the phase of the field is random.

References

1. H. Haken: *Laser Theory*, Reproduction from *Handbuch der Physik* (Springer, Berlin, Heidelberg 1984)
2. M. Sargent III, M.O. Scully, W.E. Lamb: *Laser Physics* (Addison Wesley 1974)
3. M.O. Scully, W.E. Lamb: Phys. Rev. **159**, 208 (1967)
4. M. Lax, W.H. Louisell: Phys. Rev. **185**, 568 (1969)
5. F.T. Arecchi: Phys. Rev. Lett. **15**, 912 (1965)
6. E.A. Johnson, R. Jones, T.P. McLean, E.R. Pike: Phys. Rev. Lett. **16**, 589 (1966)
7. B.L. Morgan, L. Mandel: Phys. Rev. Lett. **16**, 1012 (1966)
8. S. Machida, Y. Yamamoto, Y. Itaya: Phys. Rev. Lett. **58**, 1000 (1987)

9. W. Richardson, R.E. Shelby: Phys. Rev. Lett. **64**, 400 (1990)
10. Y.M. Golubov, I.V. Sokolov: Sov. Phys. JETP **60**, 234 (1984)
11. Y. Yamamoto, S. Machida, O. Nilsson: Phys. Rev. A **34**, 4025 (1986)
12. F. Haake, S.M. Tan, D.F. Walls: Phys. Rev. A **40**, 7121 (1989)
13. J. Bergou, L. Davidovich, M. Orszag, C. Berkert, M. Hillery, M.O. Scully: Opts. Commun. **72**, 82 (1989)
14. M. Marte, P. Zoller: Quantum Optics **2**, 229 (1990)

Chapter 13
Bells Inequalities in Quantum Optics

Abstract The early days of quantum mechanics were characterised by debates over the applicability of established classical concepts, such as position and momentum, to the new formulation of mechanics. The issues became quite distinct in the protracted exchange between A. Einstein and N. Bohr, culminating in the paper of *Einstein, Podolsky* and *Rosen* (EPR) in 1935 [1]. *Bohr*, in his response to this paper, [2] expanded upon his concept of complementarity and showed that the EPR argument did not establish the incompleteness of quantum mechanics, as EPR had claimed, but rather highlighted the inapplicability of classical modes of description in the quantum domain. A. Einstein, however, did not accept this position and the two sides of the debate remained unreconciled, while most physicists generally believed that N. Bohr's argument carried the day.

Thus the matter rested until 1964 when J.S. Bell opened up the possibility of directly testing the consequences of the EPR premises. We will discuss the EPR argument and the analysis of Bell in the context of correlated photon states.

13.1 The Einstein–Podolsky–Rosen (EPR) Argument

The essential step in the EPR argument is to introduce correlated pure states of two particles (or photons) of the form

$$|\Psi\rangle = \sum_n |a_n\rangle_1 \otimes |b_n\rangle_2 , \tag{13.1}$$

where $\{|a_n\rangle_1\}$ and $\{|b_n\rangle_2\}$ are ortho-normal eigenstates for some operators $\hat{A}_1$ and $\hat{B}_2$ of particles 1 and 2, respectively. The correlations between the particles persist even if in the course of the experiment the particles become spatially separated after the interaction responsible for the correlated state.

Now suppose one were to measure the operator $\hat{A}_1$ on particle one long after the interaction between the particles has ended, and the two particles are far apart. If the result is some eigenvalue a_n, particle 1 must thence-forth be considered to be in the state $|a_n\rangle_1$, while particle 2 must be in the state $|b_n\rangle_2$. As the state of particle 2 is now an eigenstate of $\hat{B}_2$ we can predict with probability one that the physical quantity represented by $\hat{B}_2$ if measured will give the result b_n. Thus we can predict the value of this physical quantity for particle 2 without in any way interacting with it.

Suppose, however, that instead of measuring $\hat{A}_1$ on particle 1 we measured some other quantity, $\hat{C}_1$, with eigenstates $|c_n\rangle_1$. We then rewrite the state in (13.1) as

$$|\Psi\rangle = \sum_n |c_n\rangle_1 \otimes |d_n\rangle_2 , \tag{13.2}$$

where $|d_n\rangle_2$ is an eigenstate of some other operator $\hat{D}_2$ for particle 2. If the result c_n is obtained for the measurement on particle 1, particle 2 must be in the state $|d_n\rangle_2$ for which a measurement of $\hat{D}_2$ must give the result d_n. Thus depending on what we choose to measure on particle 1 the state of particle 2 after the measurement, can be an eigenstate of two quite different operators. This is another example of the measurement ambiguity discussed in the previous chapter. However, the EPR argument now raises one very important question. Is it possible that the two operators on particle 2, $\hat{B}_2$ and $\hat{D}_2$, do not commute? If this were the case the EPR argument establishes that, depending on what is measured on particle 1, we can predict with certainty the values of physical quantities, represented by noncommuting operators without in anyway interacting with this particle. By explicit construction *Einstein, Podolsky* and *Rosen* showed that this is indeed possible.

EPR claimed that "if without in anyway disturbing a system, we can predict with certainty (i.e., with probability equal to unity), the value of a physical quantity, then there exists an element of physical reality corresponding to that quantity".

Assuming that the wave function does contain a complete description of the two-particle system it would seem that the argument of EPR establishes that it is possible to assign two different states ($|b_n\rangle_2$ and $|d_n\rangle_2$) to the same reality. However, in the language of EPR, two physical quantities represented by operators which do not commute cannot have simultaneous reality. The conclusion of EPR was that the quantum mechanical description of physical reality given by the wave function is not complete.

13.2 Bell Inequalities and the Aspect Experiment

Were one to adopt the conclusion of EPR it would seem necessary to search for a more complete physical theory than quantum mechanics. To obtain such a theory, quantum mechanics should be supplemented by additional, perhaps inaccessible, variables. As *Bell* [3] showed, attempting to complete the theory in this way and maintain the locality condition (that measurements on particle 1 carried out when the particles are spatially separated should have no effect on particle 2) leads to statistical predictions which differ from those of standard quantum theory.

To elucidate *Bell*'s argument we consider a system in which correlated photon polarisation states are produced. Such a system is the $(J = 0) \rightarrow (J = 1) \rightarrow (J = 0)$

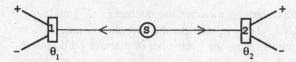

Fig. 13.1 Schematic representation of the experiment of *Aspect* et al. [4] to test quantum mechanics against the Bell inequality. S is a source of two polarised photons. 1 and 2 label *polarisation analysers*, with orthogonal output channels labelled $+$ and $-$. The polarisation analysers are set at angles θ_1, θ_2

cascade two-photon transition in calcium-40 (Fig. 13.1). The two photons are emitted in opposite directions (by conservation of momentum) with correlated polarisation states. Each photon passes through separate polarisation analysers, emerging in either the horizontal $(+)$ channel, or the vertical channel $(-)$ of each analyser. Initially let us assume that the horizontal polarisation is chosen to be orthogonal to the plane of the experiment and that both analysers are so aligned. However, we are free to rotate the polarisers in the plane orthogonal to the propagation direction of the photons. We follow the treatment of *Reid* and *Walls* [5].

Let $a_\pm(b_\pm)$ be the annihilation operator for the horizontally $(+)$ or vertically $(-)$ polarised mode for the field travelling to analyser 1 (labelled 1) or analyser 2 (labelled 2). The state of the two photons may be written as

$$|\Psi\rangle = \frac{1}{\sqrt{2}}(a_+^\dagger b_+^\dagger + a_-^\dagger b_-^\dagger)|0\rangle , \qquad (13.3)$$

where $|0\rangle$ is the vacuum state. Using the notation $|n_1, n_2, n_3, n_4\rangle$ to denote n_1 photons in mode a_+, n_2 photons in mode a_-, n_3 photons in mode b_+ and n_4 photons in mode b_-, the state may be expressed as

$$|\psi\rangle = \frac{1}{\sqrt{2}}(|1,0,1,0\rangle + |0,1,0,1\rangle) . \qquad (13.4)$$

If the photon in analyser 1 is detected in the $(+)$ channel, the state of the photon directed towards 2 must be polarised in the horizontal direction. This correlation is thus precisely of the kind required for the EPR experiment.

We are free to measure the polarisation in any direction by rotating the analysers through the angles θ_1 and θ_2 for detector 1 and 2, respectively. The detected modes in this case are orthogonal transformations of the modes $a_\pm$ and $b_\pm$;

$$c_+ = a_+ \cos\theta_1 + a_- \sin\theta_1 , \qquad (13.5a)$$

$$c_- = -a_+ \sin\theta_1 + a_- \cos\theta_1 , \qquad (13.5b)$$

$$d_+ = b_+ \cos\theta_2 + b_- \sin\theta_2 , \qquad (13.5c)$$

$$d_- = -b_+ \sin\theta_2 + b_- \sin\theta_2 . \qquad (13.5d)$$

The detectors placed after the polarisers measure the intensities $\langle I_1^{\pm} \rangle$ and $\langle I_2^{\pm} \rangle$, while the correlators measure $\langle I_1^+ I_2^+ \rangle$, etc. In fact, for the two-photon state $\langle I_i^{\pm} \rangle = P_i^{\pm}$, is the probability for one count in the $+$ or $-$ channel of detector i. Of course, these moments depend on θ_1 and θ_2. Let us further suppose that in a complete theory these functions also depend on the variable λ which remains hidden from direct determination and for which only a statistical description is available. This variable is distributed according to some density $\rho(\lambda)$. In general, we may then write

$$\langle I_1^{\pm} I_2^{\pm} \rangle_{\theta_1 \theta_2} = \int \rho(\lambda) I_1^{\pm}(\lambda, \theta_1, \theta_2) I_2^{\pm}(\lambda, \theta_1, \theta_2) d\lambda \,, \tag{13.6}$$

where I_1^+ denotes the expected intensity at detector 1 given a value for λ, namely

$$I_1^+(\lambda, \theta_1, \theta_2) = \int I_1^+ \rho(I_1^+ | \lambda, \theta_1, \theta_2) dI_1^+ \,. \tag{13.7}$$

It is reasonable to assume, as in EPR, that for a given value of λ the results at 1 cannot depend on the angle θ_2 chosen at 2, (and conversely). This is the "locality assumption", it is formally represented by

$$I_1^{\pm}(\lambda, \theta_1, \theta_2) = I_1^{\pm}(\lambda, \theta_1) \,, \tag{13.8a}$$

$$I_2^{\pm}(\lambda, \theta_1, \theta_2) = I_2^{\pm}(\lambda, \theta_2) \,. \tag{13.8b}$$

Consider the following correlation functions:

$$E(\theta_1, \theta_2) = \frac{\langle (I_1^+ - I_1^-)(I_2^+ - I_2^-) \rangle}{\langle (I_1^+ + I_1^-)(I_2^+ + I_2^-) \rangle} \,. \tag{13.9}$$

In terms of the detected mode operators this may be written in the form

$$E(\theta_1, \theta_2) = \frac{\langle : (c_+^{\dagger} c_+ - c_-^{\dagger} c_-)(d_+^{\dagger} d_+ - d_-^{\dagger} d_-) : \rangle}{\langle : (c_+^{\dagger} c_+ + c_-^{\dagger} c_-)(d_+^{\dagger} d_+ + d_-^{\dagger} d_-) : \rangle} \tag{13.10}$$

where $:\,:$ denotes normal ordering.

Assuming a local hidden variable theory we may write

$$E(\theta_1, \theta_2) = N^{-1} \int f(\lambda) S_1(\lambda, \theta_1) S_2(\lambda, \theta_2) d\lambda \,, \tag{13.11}$$

where

$$S_1(\lambda, \theta_1) = \frac{I_1^+(\lambda, \theta_1) - I_1^-(\lambda, \theta_1)}{I_1(\lambda)} \,, \tag{13.12}$$

$$S_2(\lambda, \theta_1) = \frac{I_2^+(\lambda, \theta_2) - I_2^-(\lambda, \theta_2)}{I_2(\lambda)} \,, \tag{13.13}$$

$$f(\lambda) = \rho(\lambda)I_1(\lambda)I_2(\lambda) \tag{13.14}$$

with

$$I_1(\lambda) = I_1^+(\lambda, \theta_1) + I_1^-(\lambda, \theta_2) \,, \tag{13.15a}$$

$$I_2(\lambda) = I_2^+(\lambda, \theta_2) + I_2^-(\lambda, \theta_2) \,. \tag{13.15b}$$

The latter equations correspond to the intensity of light measured at 1 or 2 with the polarisers removed. The normalisation N is

$$N = \int f(\lambda)\mathrm{d}\lambda \,. \tag{13.16}$$

The functions $|S_1(\lambda, \theta_1)|$ and $|S_2(\lambda, \theta_2)|$ are bounded by unity:

$$|S_1(\lambda, \theta_1)| \le 1 \,, \tag{13.17a}$$

$$|S_2(\lambda, \theta_2)| \le 1 \,. \tag{13.17b}$$

To obtain a testable statistical quantity we need to consider how $E(\theta_1, \theta_2)$ changes as the orientation of the polarisers are changed. With this in mind, consider $E(\theta_1, \theta_2)$ $E(\theta_1, 0_2')$. This quantity may be expressed as

$$E(\theta_1, \theta_2) - E(\theta_1, \theta_2') = N^{-1} \int \mathrm{d}\lambda f(\lambda) S_1(\lambda, \theta_1) S_2(\lambda, \theta_2)[1 \pm S_1(\lambda, \theta_1') S_2(\lambda, \theta_2')]$$
$$- N^{-1} \int \mathrm{d}\lambda f(\lambda) S_1(\lambda, \theta_1) S_2(\lambda, \theta_2')$$
$$\times [1 \pm S_1(\lambda, \theta_1') S_2(\lambda, \theta_2)] \,. \tag{13.18}$$

Then using (13.17a and b)

$$|E(\theta_1, \theta_2) - E(\theta_1, \theta_2')| \le N^{-1} \int \mathrm{d}\lambda f(\lambda)[1 \pm S_1(\lambda, \theta_1') S_2(\lambda, \theta_2')]$$
$$+ N^{-1} \int \mathrm{d}\lambda f(\lambda)[1 \pm S_1(\lambda, \theta_1') S_2(\lambda, \theta_2)]$$
$$= 2 \pm [E(\theta_1', \theta_2') + E(\theta_1', \theta_2)] \,.$$

Finally, we obtain the Bell inequality

$$|B| \le 2 \,, \tag{13.19}$$

where
$$B = E(\theta_1, \theta_2) - E(\theta_1, \theta_2') + E(\theta_1', \theta_2') + E(\theta_1', \theta_2) \,.$$

This particular Bell inequality is known as the Clauser–Horne–Shimony–Holt (CHSH) inequality.

As we shall see, there are states of the field which violate the inequality equation (13.19) [for example, the state given in (5)]. We note firstly, however, that if the state of the field can be represented by a positive, normalisable Glauber–Sudarshan P-representation no violation of this inequality is possible. Let $\alpha = (\alpha_+, \alpha_-, \beta_+, \beta_-)$ be the c-number corresponding to the modes $a_\pm, b_\pm$. If we define the following 'transformation' variables for the modes $c_\pm, d_\pm$,

$$\gamma_+ = \alpha_+ \cos\theta_1 + \alpha_- \sin\theta_1, \qquad \delta_+ = \beta_+ \cos\theta_2 + \beta_- \sin\theta_2,$$
$$\gamma_- = -\alpha_+ \sin\theta_1 + \alpha_- \cos\theta_1, \qquad \delta_- = -\beta_+ \sin\theta_2 + \beta_- \cos\theta_2 , \qquad (13.20)$$

the correlation function $E(\theta_1, \theta_2)$ becomes

$$E(\theta_1, \theta_2) = N^{-1} \int P(\alpha)(|\gamma_+|^2 - |\gamma_-|^2)(|\delta_+|^2 - |\delta_-|^2) d^2\alpha \qquad (13.21)$$

with

$$N = \int P(\alpha)(|\gamma_+|^2 + |\gamma_-|^2)(|\delta_+|^2 + |\delta_-|^2) d^2\alpha .$$

Recalling that the transformations in (13.20) are orthogonal we note that

$$|\gamma_+|^2 + |\gamma_-|^2 = |\alpha_+|^2 + |\alpha_-|^2 \text{ and } |\delta_+|^2 + |\delta_-|^2 = |\beta_+|^2 + |\beta_-|^2$$

the normalisation may be written

$$N = \int P(\alpha)(|\alpha_+|^2 + |\alpha_-|^2)(|\beta_+|^2 + |\beta_-|^2) d^2\alpha , \qquad (13.22)$$

where the integrand does not depend on θ_1 or θ_2. Then

$$E(\theta_1, \theta_2) = N^{-1} \int P(\alpha)(|\alpha_+|^2 + |\alpha_-|^2)(|\beta_+|^2 + |\beta_-|^2) S(\gamma) S(\delta) , \qquad (13.23)$$

where

$$S(\gamma) = \frac{|\gamma_+|^2 - |\gamma_-|^2}{|\alpha_+|^2 + |\alpha_-|^2} \qquad (13.24)$$

and

$$S(\delta) = \frac{|\delta_+|^2 - |\delta_-|^2}{|\beta_+|^2 + |\beta_-|^2} . \qquad (13.25)$$

As $S(\gamma)$ is a function of θ_1 and not θ_2 while $S(\delta)$ is a function of θ_2 and not θ_1, the Glauber–Sudarshan representation is local. It then follows immediately that provided $P(\alpha)$ is positive and normalisable, the Bell inequality in (13.19) must hold.

The correlation function $E(\theta_1, \theta_2)$ may be evaluated directly for the state in (13.4) using the normally-ordered moment in (13.9). One finds

$$E(\theta_1, \theta_2) = \cos 2\psi , \qquad (13.26)$$

where

$$\psi \equiv \theta_1 - \theta_2 \,.$$

If we choose

$$\psi = \theta_2 - \theta_1 = \theta_1' - \theta_2 = \theta_1' - \theta_2' = \frac{1}{3}(\theta_1 - \theta_2')\,,$$

one finds

$$B = 3\cos 2\psi - \cos 6\psi \,. \tag{13.27}$$

When $\psi = 22.5°$, $B = 2\sqrt{2}$ showing a clear violation of the Bell inequality $|B| \leq 2$.

This violation has convincingly been demonstrated in the experiment of Aspect [4]. In this experiment the polarisation analysers were essentially beam splitters with polarisation-dependent transmittivity. Ideally, one would like to have the transmittivity (T^+) for the modes a_+ and b_+ equal to one, and the reflectivity (R^-) for the modes a_- and b_- also equal to one. However, in the experiment the measured values were $T_1^+ = R_1^- = 0.950$, $T_1^- = R_1^+ = 0.007$ and $T_2^+ = T_2^- = 0.930$, $T_2^- = R_2^+ = 0.007$.

The expression for $E(\theta_1, \theta_2)$ is then modified:

$$E(\theta_1, \theta_2) = F \frac{(T_1^+ - T_1^-)(T_2^+ - T_2^-)}{(T_1^+ + T_1^-)(T_2^+ + T_2^-)} \cos 2\psi \,, \tag{13.28}$$

where F is a geometrical factor accounting for finite solid angles of detection. In this experiment $F = 0.984$, and quantum mechanics would give for $\psi = 22.5°$, $B = 2.7$.

The observed value was 2.697 ± 0.015, in quite good agreement with quantum theory and a clear violation of the Bell inequality. In Fig. 13.2 is shown a plot of the theoretical and experimental results as a function of ψ. The agreement with quantum mechanics is better than 1%. It would appear in the light of this experiment that realistic local theories for completing quantum mechanics are untenable.

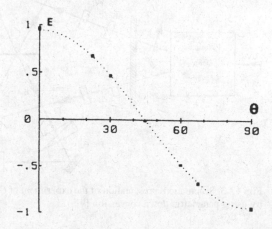

Fig. 13.2 Correlation of polarisations as a function of the relative angle of the polarisation analysers. The indicated errors are ± 2 standard deviations. The *dotted curve* is the quantum-mechanical prediction for the experiment. For ideal polarisers the curves woudl reach the values ± 1. (From Aspect et al. Phys. Rev. Letts. **49**, 92 (1982))

13.3 Violations of Bell's Inequalities Using a Parametric Amplifier Source

The Bell inequality presented in (13.19) is only one of a large class of inequalities violated by quantum mechanics. Another inequality has recently been tested by *Ou* and *Mandel* [6] based on an experiment first suggested by *Reid* and *Walls* [5]. It has now been realised in a number of configurations [7, 8]. We shall discuss the *Ou* and *Mandel* experiment presented schematically in Fig. 13.3. A parametric down converter produces two beams of linearly polarized signal and idler photons. Phase matching conditions give a relative angle of 4° between the propagation direction of the two beams. The idler photons pass through a 90° polarization rotator. The signal and idler beams are then incident from opposite sides onto a beam splitter. After the beam splitter, the two beams now consisting of mixed signal and idler photons pass through linear polarizers set at adjustable angles θ_1 and θ_2 before falling on two photodetectors. The coincidence counting rate of the two detectors is then measured with a time-to-digital converter. This provides a measure of the joint probability of detecting two photons for various settings θ_1 and θ_2 of the two polarizers.

In this experiment the polarisation analysers used have only a single output channel. However, we can still derive a Bell inequality violated by quantum mechanics. If we define the correlation function $P(\theta_1, \theta_2)$ by

$$P(\theta_1, \theta_2) = \langle I_1 I_2 \rangle_{\theta_1 \theta_2} \tag{13.29}$$

the following Bell inequality may be derived as

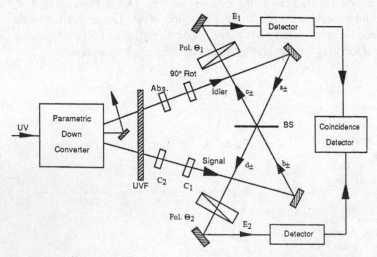

Fig. 13.3 Schematic representation of the experiment of *Ou* and *Mandel* to test the CHSH inequality using parametric down conversion [9]

$$S = P(\theta_1, \theta_2) - P(\theta_1, \theta_2') + P(\theta_1', \theta_2') + P(\theta_1', \theta_2)$$
$$- P(\theta_1, -) - P(-, \theta_2) \leq 0, \tag{13.30}$$

where

$$P(\theta_1, -) = \langle I_1 I_2 \rangle_{\theta_1}, \tag{13.31}$$

$$P(-, \theta_2) = \langle I_1 I_2 \rangle_{\theta_2} \tag{13.32}$$

are the intensity correlation functions with one or the other polariser removed. The inequality in (13.30) is known as the Clauser–Horne inequality. Just as in the case of the CHSH inequality (13.19) this inequality is satisfied for states of the field which can be represented by a positive, normalisable Glauber–Sudarshan P-representation [5]. It may, however, be violated for certain quantum fields.

We follow closely the treatment given by *Tan* and *Walls* [9]. We now proceed to calculate $P(\theta_1, \theta_2)$ and S for the experiment of *Ou* and *Mandel* [6]. We include the possibility of placing an attenuator in the idler beam. Let a_+ and a_- denote the annihilation operators for the x and y polarized modes in the idler beam, and let b_+ and b_- denote the operators for the corresponding modes in the signal beam. The outgoing modes from the beam splitter are described by the operators $c_\pm$ and $d_\pm$, and obey the following relationships:

$$c_+ = \sqrt{T_+} b_+ + i\sqrt{R_+} a_+,$$
$$c_- = \sqrt{T_-} b_- - i\sqrt{R_-} a_-,$$
$$d_+ = \sqrt{T_+} a_+ + i\sqrt{R_+} b_+,$$
$$d_- = \sqrt{T_-} a_- - i\sqrt{R_-} b_-, \tag{13.33}$$

where $T_\pm$ and $R_\pm$ are the intensity transmission and reflection coefficients for the x and y polarizations. The phase relationships arise from the Fresnel formula. Since the signal beam is polarized in the x direction, and the idler beam is polarized in the y direction, the modes associated with the operators a_+ and b_+ are the annihilation operators for the output modes of the parametric down converter. When an attenuator with the intensity transmission coefficient η is placed in the idler beam, the operators a_- in the above equations is replaced by

$$\sqrt{\eta} a_- + \sqrt{1 - \eta} v, \tag{13.34}$$

where the vacuum mode operator v is included to give the correct level of fluctuations in the attenuated beam. Photodetectors 1 and 2 respond to the fields $E_1^{(+)}$ and $E_2^{(+)}$, respectively, where

$$E_1^{(+)} = c_+ \cos \theta_1 + c_- \sin \theta_1 ,$$
$$E_2^{(+)} = d_+ \cos \theta_2 + d_- \sin \theta_2 , \tag{13.35}$$

The joint two-photon detection probability (for perfect detector efficiency) is

$$P(\theta_1, \theta_2) = \langle \psi | E_1^{(-)} E_2^{(-)} E_2^{(+)} E_1^{(+)} | \psi \rangle . \tag{13.36}$$

For low conversion efficiencies the output of the parametric down converter is a pair of photons, one in each of the signal and idler modes a_- and b_+. Thus $|\psi\rangle = |1,1\rangle$, this yields

$$P(\theta_1, \theta_2) = \eta (\sqrt{R_+ R_-} \sin \theta_1 \cos \theta_2 + \sqrt{T_+ T_-} \cos \theta_1 \sin \theta_2)^2 . \tag{13.37}$$

Taking a 50/50 beam splitter ($R_+ = R_- = T_+ = T_- = \frac{1}{2}$),

$$P(\theta_1, \theta_2) = \frac{1}{4} \eta \sin^2(\theta_1 + \theta_2) . \tag{13.38}$$

Removing one polarizer, we must calculate

$$P(-, \theta_2) = \langle \psi | : (c_+^\dagger c_+ + c_-^\dagger c_-) E_2^{(-)} E_2^{(+)} : | \psi \rangle , \tag{13.39}$$

where : : represents normal ordering. For the input state $|1,1\rangle$ we find

$$P(-, \theta_2) = \frac{1}{4} \eta , \tag{13.40}$$

and similarly

$$P(\theta_1, -) = \frac{1}{4} \eta , \tag{13.41}$$

for a 50/50 beam splitter.

Substituting (13.38, 13.40 and 13.41) into the Clauser–Horne–Bell inequality (13.2) gives

$$S = \frac{1}{4} \eta [\sin^2(\theta_1 + \theta_2) - \sin^2(\theta_1 + \theta_2') + \sin^2(\theta_1' + \theta_2') + \sin^2(\theta_1' + \theta_2) - 2] .$$

Choosing the angles such that $\theta_1 = \pi/8, \theta_2 = \pi/4, \theta_1' = 3\pi/8$ and $\theta_2' = 0$,

$$S = \frac{1}{4} \eta (\sqrt{2} - 1) > 0 , \tag{13.42}$$

which violates the inequality.

In a classical wave analysis of the parametric down converter, we represent the signal and idler fields incident on the beam splitter by the complex numbers E_s and E_i. The beam splitter combines these fields to produce E_1 and E_2 at the detectors, where

$$E_1 = \cos\theta_1 \sqrt{T_+} E_s - i\sin\theta_1 \sqrt{R_-} E_i,$$

$$E_2 = i\cos\theta_2 \sqrt{R_+} E_s + \sin\theta_2 \sqrt{T_-} E_i . \qquad (13.43)$$

The joint detection probability $P(\theta_1, \theta_2)$ is proportional to the intensity correlation $\langle |E_1|^2 E_2|^2 \rangle$. Using the above forms for E_1 and E_2, and assuming that the difference in the phases of the signal and idler fields is random, we find that for a 50/50 beam splitter

$$P(\theta_1, \theta_2) \propto \langle I_s I_i \rangle \sin^2(\theta_1 + \theta_2)$$
$$+ \langle I_s^2 \rangle \cos^2\theta_1 \cos^2\theta_2 + \langle I_i^2 \rangle \sin^2\theta_1 \sin^2\theta_2 . \qquad (13.44)$$

where we have written I_s for $|E_s|^2$ and for I_i for $|E_i|^2$. With the attenuator in the idler beam, $I_i = \eta I_s$, and if we assume that the intensity fluctuations are such that $\langle I^2 \rangle \propto \langle I \rangle^2$ and $\langle I_i I_s \rangle \propto \langle I_i \rangle \langle I_s \rangle$, then

$$P(\theta_1, \theta_2) \propto \eta \sin^2(\theta_1 + \theta_2) + \cos^2\theta_1 \cos^2\theta_2 + \eta^2 \sin^2\theta_1 \sin^2\theta_2 . \qquad (13.45)$$

In order to compare the quantum and classical result we consider $P(\theta, \pi/4)$ with $R_+ = R_- = T_+ = T_- = \frac{1}{2}$. Then

$$P(\theta, \pi/4) = \frac{\eta}{8}(1 + \sin 2\theta) , \qquad (13.46)$$

which exhibits a sinusoidal modulation with respect to the angle 2θ. The visibility of the resulting modulation is unity. However, the classical result gives

$$P(\theta, \pi/4) \propto \frac{\eta}{2}(1 + \sin 2\theta) + \frac{1}{2}\cos^2\theta + \frac{\eta^2}{2}\sin^2\theta , \qquad (13.47)$$

which in the absence of the absorber ($\eta = 1$) gives

$$P(\theta, \pi/4) \propto (1 + \frac{1}{2}\sin 2\theta) . \qquad (13.48)$$

In the classical case the modulation is not 100%, in fact the visibility is only one half.

In the experiment of *Ou* and *Mandel* the value of S was found to be positive with an accuracy of six standard deviations, in clear violation of the Bell inequality (13.30). The experiment also distinguished between the different quantum and classical predictions for the phase dependence of $P(\theta, \pi/4)$. These results are shown in Fig. 13.4. The solid and dashed-dotted lines correspond to the quantum and classical wave predictions, respectively, with constants of proportionality adjusted to fit. Clearly, $P(\theta, \pi/4)$ does exhibit the phase dependence predicted by quantum mechanics. The observed visibility obtained from a best fit was 0.76; greater than the classical prediction of 0.5.

Instead of the correlated two-photon state discussed above we can also use the output state of a parametric down converter. We assume that the pump field in this

Fig. 13.4 The coincidence counting rate as a function of polariser angle θ_1 with θ_2 fixed at 45°. The *solid curve* is the quantum prediction based on (13.46), and the *dash-dot curve* is the classical prediction based on (13.48). The *dashed* and *dotted curves* are the quantum and classical predictions, respectively, including a detector inefficiency of 0.76

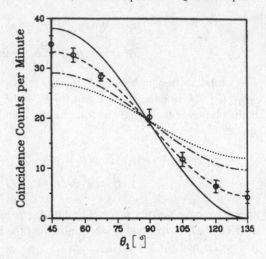

device may be treated classically. The solutions for the output modes of the device are

$$a_- = a_0 \cosh \kappa + b_0^\dagger \sinh \kappa \,, \tag{13.49}$$

$$b_+ = b_0 \cosh \kappa + a_0^\dagger \sinh \kappa \,, \tag{13.50}$$

where κ is proportional to the second-order nonlinear susceptibility of the crystal, and a_0, b_0 are the input modes. We assume that the input state is a vacuum. With a 50:50 beam splitter $\eta = 1$ and with $\theta_2' = 0, \theta_1 = \psi, \theta_2 = 2\psi, \theta_1' = 3\psi$ the quantity S which occurs in the Bell inequality (13.30) is given by (Exercise 13.2)

$$S = \frac{1}{4} \sinh^2 \kappa \{ F(\psi) + 2 \sinh^2 \kappa [F(\psi) + 2G(\psi)] \} \,, \tag{13.51}$$

where

$$F(\psi) = 2\sin^2 3\psi - \sin^2 \psi + \sin^2 5\psi - 2,$$

$$G(\psi) = \sin^2 \psi \sin^2 2\psi + \sin^2 3\psi \sin^2 2\psi - \sin^2 3\psi - \sin^2 2\psi \,.$$

When $\kappa \ll 1$ this may be approximated by

$$S \approx \frac{1}{4} \kappa^2 F(\psi) \,. \tag{13.52}$$

For purposes of comparison the two-photon state, with the same choice of angles would give

$$S = \frac{1}{4} F(\psi) \,. \tag{13.53}$$

Up to a scale constant in this limit the parametric down converter gives the same result as for a correlated photon pair.

Fig. 13.5 The correlation
function in (13.51) normalised
by $\sinh^2 \kappa$, versus ψ for various values of κ. The violation
of the classical inequality is
evident for small κ

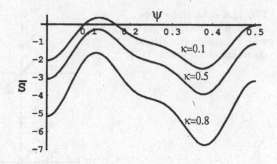

In the limit $\kappa \gg 1$ we find

$$S \propto F(\psi) - 2G(\psi) . \tag{13.54}$$

As the function on the right-hand side is always nonpositive no violation of the
Clauser–Horne inequality is possible. In Fig. 13.5 we plot S normalised by the intensity $I = \sinh^2 k$ versus ψ for various values of κ. We see that the maximum
violation for $\kappa \ll 1$ occurs when $\theta = \pi/8$ (solid curve).

We note that the form of the intensity correlation function for the parametric
down-converter in the limit of $\kappa \gg 1$ coincides with that of the classical analysis (13.44).

13.4 One-Photon Interference

In all the schemes discussed above the states which lead to a violation of the Bell
inequalities are correlated two-photon states. We now consider a scheme which
demonstrates the non-local nature of quantum mechanics, which does not rely on
two-photon states. This experiment illustrates on the nonlocal behaviour of a single
photon.

The scheme is illustrated in Fig. 13.6. A field is split at a 50:50 beam splitter,
and each of the two output fields directed to homodyne detectors. Each of the homodyne detectors mix the output field from the first beam splitter with a coherent
local oscillator of amplitude $\alpha_k = \alpha e^{i\theta_k}$, and the final intensities at the two output
channels of each of the homodyne detectors are measured using photodetectors. We
follow closely the treatment of *Tan* et al. [10].

Referring to Fig. 13.6 we see that the homodyne detector k may be regarded as
making a measurement of b_k with a local parameter θ_k. This parameter is analogous
to the angle of the polarisation analysers in the two-photon schemes. We wish to
determine the probabilities with which the individual photodetectors respond, and
the coincidence probabilities for pairs of photodetectors, one in each homodyne
detector.

Fig. 13.6 Schematic representation of an experiment with a single photon state to demonstrate non-locality [10]

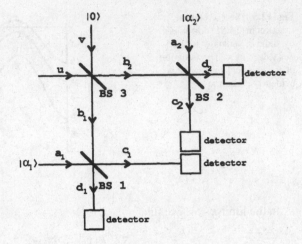

The transformation between the mode operators shown in Fig. 13.6 are given by

$$\begin{pmatrix} c_k \\ d_k \end{pmatrix} = \frac{1}{\sqrt{2}} \begin{pmatrix} 1 & i \\ i & 1 \end{pmatrix} \begin{pmatrix} a_k \\ b_k \end{pmatrix},$$

$$\begin{pmatrix} b_1 \\ b_2 \end{pmatrix} = \frac{1}{\sqrt{2}} \begin{pmatrix} 1 & i \\ i & 1 \end{pmatrix} \begin{pmatrix} v \\ u \end{pmatrix}. \tag{13.55}$$

Thus the modes input into the detectors may be expressed in terms of the input mode operators by

$$\begin{pmatrix} c_1 \\ d_1 \\ c_2 \\ d_2 \end{pmatrix} = \begin{pmatrix} \frac{1}{\sqrt{2}} & \frac{i}{2} & 0 & -\frac{1}{2} \\ \frac{i}{\sqrt{2}} & \frac{1}{2} & 0 & \frac{i}{2} \\ 0 & -\frac{1}{2} & \frac{1}{\sqrt{2}} & \frac{i}{2} \\ 0 & \frac{i}{2} & \frac{i}{\sqrt{2}} & \frac{1}{2} \end{pmatrix} \begin{pmatrix} a_1 \\ v \\ a_2 \\ u \end{pmatrix}. \tag{13.56}$$

This enables us to calculate the coincidence probabilities between the detectors directly in terms of the input field.

We begin by considering vacuum inputs to the modes u and v. The local oscillators are assumed to be in coherent states $|\alpha e^{i\theta_1}\rangle$, $|\alpha e^{i\theta_2}\rangle$. The intensities at all detectors are found to be equal

$$\langle I_{c_1} \rangle = \langle I_{c_2} \rangle = \langle I_{d_1} \rangle = \langle I_{d_2} \rangle = \frac{1}{2}\alpha^2. \tag{13.57}$$

The two-photon coincidence rates due to rare chance coincidences between the local oscillators are also equal between the pairs of detectors

$$\langle I_{c_1} I_{c_2} \rangle = \langle I_{d_1} I_{d_2} \rangle = \langle I_{c_1} I_{d_2} \rangle = \langle I_{d_1} I_{c_2} \rangle = \frac{1}{4}\alpha^4. \tag{13.58}$$

We now consider the input of a single photon in mode u while the mode v is the vacuum. The state of the two-mode field (b_1 and b_2) after the first beam-splitter is then an entangled state of a one-photon state and the vacuum

$$|\psi\rangle = \frac{1}{\sqrt{2}}(i|1\rangle|0\rangle + |0\rangle|1\rangle) \tag{13.59}$$

which is precisely the same state as one gets (except for a phase factor) for a one-photon state incident on the two slits in Young's interference experiment.

The photon count probabilities at the individual detectors are now

$$\langle I_{c_1}\rangle = \langle I_{c_2}\rangle = \langle I_{d_1}\rangle = \langle I_{d_2}\rangle = \frac{1}{2}\alpha^2 + \frac{1}{4}. \tag{13.60}$$

Thus the intensities at each detector are increased by $\frac{1}{4}$, being the probability that the one-photon input is detected by any given detector. The coincidence count probabilities between the pairs of detectors differ, now depending on which is considered. We find

$$\langle I_{c_1}I_{c_2}\rangle = \langle I_{d_1}I_{d_2}\rangle = \frac{1}{4}\{\alpha^4 + \alpha^2[1 + \sin(\theta_1 - \theta_2)]\} \tag{13.61}$$

and

$$\langle I_{c_1}I_{d_2}\rangle = \langle I_{d_1}I_{c_2}\rangle = \frac{1}{4}\{\alpha^4 + \alpha^2[1 - \sin(\theta_1 - \theta_2)]\}. \tag{13.62}$$

The coincidence probabilities depend on the phase difference between the local oscillators $\theta_1 - \theta_2$. If this is set to $-\pi/2$, we get the minimum possible coincidence probability of $\frac{1}{4}\alpha^4$ between detector pairs (c_1, c_2) and (d_1, d_2) and the maximum coincidence probability of $\frac{1}{4}\alpha^4 + \frac{1}{2}\alpha^2$ between the pairs (c_1, d_2) and (d_1, d_2). We shall be most interested in the situation where α is small compared to one.

Let us first try to interpret these results from a naïve particle viewpoint. The great enhancement of the single count probability over that with vacuum inputs is easily understood by the above argument. On the other hand, a coincidence between two detectors is expected to be a rare event since there is only one incident photon, and a coincidence can only occur if an additional photon is generated by the (weak) local oscillator of the homodyne detector which the photon does *not* reach. Since these two photons are detected at two spatially separated detectors and have apparently arisen from independent sources, we would not expect any correlation between the paths of these photons within each homodyne detector. Nevertheless, the quantum mechanical analysis reveals that such a correlation is present. In fact, this correlation is so great that for the choice of phases given above, no additional coincidence (above the vacuum level) occur for particular detector pairs, whereas there is a relatively large coincidence probability (proportional to the local oscillator intensity) for the other pairs.

Non-local intensity correlation and their dependence on the local oscillator phases are not unexpected from a classical wave description of light. A classical analogue to the single photon input is a wave of low amplitude and unspecified phase. We may formally obtain the results for the classical wave theory from the quantum-mechanical calculation by substituting the wave amplitude $\beta e^{\pm i\phi}$ for b and $b^\dagger$, respectively, and averaging over the random phase ϕ. It is easy to check that the predicted average intensities and intensity correlations are given by

$$\langle I_{c_1}\rangle = \langle I_{c_2}\rangle = \langle I_{d_1}\rangle = \langle I_{d_2}\rangle = \frac{1}{2}\alpha^2 + \frac{1}{4}\beta^2 , \tag{13.63}$$

$$\langle I_{c_1}I_{c_2}\rangle = \langle I_{d_1}I_{d_2}\rangle = \frac{1}{4}\{\alpha^4 + \alpha^2\beta^2[1+\sin(\theta_1-\theta_2)] + \frac{1}{4}\beta^4\} , \tag{13.64}$$

$$\langle I_{c_1}I_{d_2}\rangle = \langle I_{d_1}I_{c_2}\rangle = \frac{1}{4}\{\alpha^4 + \alpha^2\beta^2[1-\sin(\theta_1-\theta_2)] + \frac{1}{4}\beta^4\} . \tag{13.65}$$

If we consider the coincidence probabilities as a function of $(\theta_1-\theta_2)$, we see that they can vary between $\frac{1}{4}(\alpha^4+\frac{1}{4}\beta^4)$ to $\frac{1}{4}(\alpha^4+2\alpha^2\beta^2+\frac{1}{4}\beta^4)$. This corresponds to a "visibility" of

$$v = \frac{\rho}{\rho^2+\rho+\frac{1}{4}} \tag{13.66}$$

where $\rho = (\alpha/\beta)^2$. The visibility attains a maximum value of $\frac{1}{2}$ when $\rho = \frac{1}{2}$. By contrast, the visibility as calculated from the quantum-mechanical result is

$$v = \frac{1}{\alpha^2+1} . \tag{13.67}$$

This can be made arbitrarily close to unity by choosing a sufficiently small value of α. Figure 13.7 shows the coincidence probabilities $\langle I_{c_1}I_{c_2}\rangle = \langle I_{d_1}I_{d_2}\rangle$ as a function of the local oscillator phase difference for the quantum mechanical and classical results with $\beta = 1$ and $\alpha = 1/\sqrt{2}$. This gives the same single count probability of $\frac{1}{2}$ in each detector, and the local oscillator amplitudes are optimized for maximum visibility

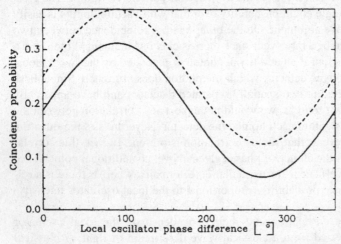

Fig. 13.7 Coincidence probability for the single photon non-locality experiment. The *solid line* is the quantum mechanical model, the *dashed line* is the prediction for a classical wave model [10]

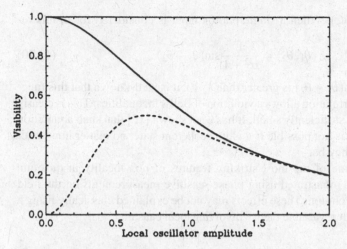

Fig. 13.8 Variation of visibility with the amplitude of the local oscillator for the quantum model (*solid line*) and a classical wave model (*dashed line*) [10]

in the classical result. However, the quantum mechanical visibility is considerably larger than that expected classically. This is clearly seen in Fig. 13.8 where the visibility v is plotted as a function of the coherent local oscillator amplitude α for the quantum mechanical single state and the classical wave mode with $\beta = 1$.

We thus see that by measuring the coincidence probability in a pair of detectors, it is possible to distinguish between the classical and quantum mechanical models. If the detector efficiencies are less than unity, coincidences will be missed, but this does not affect the measurement of the visibility of the effect.

Preparation of a single photon state may be achieved experimentally by using the signal beam of a parametric amplifier while monitoring photons in the idler beam [11]. *Hong* and *Mandel* [12] described an experiment in which a nearly pure single photon state was produced using this method. If the pump for the parametric amplifier is derived by frequency doubling a coherent beam, this provides a convenient source for the local oscillator required in the experiment under discussion.

In order to rigorously rule out classical explanations for the quantum mechanical result, it is necessary to show that Bell's inequality may be violated.

An intensity correlation coefficient is used which involves all four photo-detectors

$$E(\theta_1 - \theta_2) = \frac{\langle (I_{d_1} - I_{c_1})(I_{d_2} - I_{c_2}) \rangle}{\langle (I_{d_1} + I_{c_1})(I_{d_2} + I_{c_2}) \rangle} . \tag{13.68}$$

Evaluating this in terms of the statistics of the input mode u, where v is the vacuum yields

$$E(\theta_1 - \theta_2) = -\frac{\alpha^2 \{ \langle u^\dagger u \rangle \sin(\theta_2 - \theta_1) + |\langle u^2 \rangle| \sin(\theta_2 + \theta_1 - \xi) \}}{\alpha^4 + \langle u^\dagger u \rangle \alpha^2 + \frac{1}{4} \langle u^{\dagger 2} u^2 \rangle} \tag{13.69}$$

where $\langle u^2 \rangle = R \exp(\mathrm{i}\xi)$. When a single photon input is considered for u, this reduces to

$$E(\theta_1, \theta_2) = \frac{1}{\alpha^2 + 1} \sin(\theta_1 - \theta_2) . \tag{13.70}$$

If th coefficient of $\sin(\theta_1 - \theta_2)$ is greater than $1/\sqrt{2}$ it is well-known that this functional form for the correlation allows a violation of Bell's inequalities. This is clearly possible if α is made sufficiently small. It has been shown [13] that such a violation of Bell's inequalities is not possible if u is in a coherent state, no matter how small the input amplitude may be.

In conclusion, some of the most striking features of non-locality in quantum mechanics may be demonstrated using phase-sensitive measurements on the field produced by a single photon. These effects may not be explained classically using a particle, wave or hidden-variable theory involving local causality.

Exercises

13.1 Derive (13.26) for the correlation function $E(\theta_1, \theta_2)$. Show that with the choice $\psi = \theta_2 - \theta_1 = \theta_1' - \theta_2 = \theta_1' - \theta_2' = \frac{1}{3}(\theta_1 - \theta_2')$ one obtains (13.27) for **B**.

13.2 Derive (13.51) for the Bell parameter S for the parametric amplifier .

13.3 The state going from the beam splitter in the one-photon interference experiment is the linear superposition state

$$|\psi\rangle = \frac{1}{\sqrt{2}} (\mathrm{i}|1\rangle|0\rangle + |0\rangle|1\rangle) .$$

Compute the intensity correlations were this state replaced by the mixed state

$$\rho = \frac{1}{2} (|1\rangle_u \langle 1| \otimes |0\rangle_v \langle 0| + |0\rangle_u \langle 0| \otimes |1\rangle_v \langle 1|)$$

and show that no violation of the Bell inequality can occur.

References

1. A. Einstein, B. Podolsky, N. Rosen: Phys. Rev. **47**, 777 (1935)
2. N. Bohr: Phys. Rev. **48**, 696 (1935)
3. J.S. Bell: Physics **1**, 105 (1964); Rev. Mod. Phys. **38**, 447 (1966)
4. A. Aspect, P. Grangier, G. Roger: Phys. Rev. Lett. **49**, 91 (1982). A. Aspect, J. Dalibard, G. Roger: Phys. Rev. Lett. **49**, 1804 (1982)
5. M.D. Reid, D.F. Walls: Phys. Rev. A **34**, 1260 (1986)
6. Z.Y. Ou, L. Mandel: Phys. Rev. Lett. **61**, 50 (1988)
7. Y.H. Shih, C. Alley: Phys. Rev. Lett. **61**, 2921 (1988)

8. P.G. Kwiat, W.A. Vareka, C.K. Hong, H. Nathel, R.Y. Chiao: Phys. Rev. A **41**, 2910 (1990)
9. S.M. Tan, D.F. Walls: Optics Commun. **71**, 235 (1989)
10. S.M. Tan, D.F. Walls, M.J. Collet: Phys. Rev. Lett. **66**, 252 (1991)
11. C.A. Holmes, G.J. Milburn, D.F. Walls: Phys. Rev. Lett. A **39**, 2493 (1989)
12. C.K. Hong, L. Mandel: Phys. Rev. Lett. **56**, 58 (1986)
13. S.M. Tan, M.J. Holland, D.F. Walls: Opt. Commun. **77**, 285 (1990)

Chapter 14
Quantum Nondemolition Measurements

Abstract Current attempts to detect gravitational radiation have to take into account the quantum uncertainties in the measurement process. Considering that the detectors are macroscopic objects in some cases as large as a 10-ton bar, the fact that quantum fluctuations in the detector must be taken into account seems surprising. However, as discussed in Chap. 8, gravitational waves interact so weakly with terrestrial detectors that a displacement of the order of 10^{-19} cm is expected. To illustrate how the measurement process may introduce uncertainties which obscure the signal we consider the simple example of a free mass. A measurement of the position of a free mass with a precision $\Delta x_i \approx 10^{-19}$ cm will disturb the momentum by an amount given by the uncertainty principles as $\Delta p \geq \hbar (2\Delta x_i)^{-1}$. The period of the gravitational waves is expected to be about 10^{-3} s, hence a second measurement of the position should be made after this time. During this period, however, the position uncertainty will grow under free evolution by an amount $\Delta x^2(\tau) = \Delta x^2(0) + [\Delta p^2(0)\tau^2/m^2]$. The following inequality then holds

$$\Delta x^2(\tau) \geq 2\Delta x(0)\Delta p(0)\frac{\tau}{m}. \tag{14.1}$$

Using the uncertainty principle we then find $\Delta x^2(\tau) \geq \hbar\tau/m$. Taking the detector mass equal to 10 tons, we find $\Delta x \geq 5 \times 10^{-19}$ cm. That is, the uncertainty introduced by the first measurement has made it impossible for a second measurement to determine with certainty whether a gravitational wave has acted or not. This is the standard quantum limit.

It is instructive to consider measurements of momentum instead of position. The first measurement of momentum causes an uncertainty in position. This however does not feed back to disturb the momentum as the momentum is a constant of motion for a free mass. Hence, subsequent determination of the momentum may be made with great predictability. The momentum of a free mass is an example of a quantum nondemolition (QND) variable. The concept of quantum nondemolition measurements has been introduced over the past few years to allow the detection, in principle, of very weak forces below the level of quantum noise in the detector. In the next section we will give a brief review of the concept of a quantum nondemolition measurement.

We mention here another way in which the standard quantum limit might be overcome. Quantum nondemolition measurements generally presume that nothing at all is known about the state of the system to be measured. The standard quantum limit for a free mass, for example, was derived by assuming no correlation between position and momentum. If however we are permitted to prepare the state of the system to be measured, the accuracy of a measurement can be improved without resort to a QND scheme. For example, in the case of a free particle the position variance at time τ is given by

$$\Delta x^2(\tau) = \Delta x^2(0) + \frac{\Delta p^2(0)\tau^2}{m^2} + \langle \Delta x(0)\Delta p(0) + \Delta p(0)\Delta x(0) \rangle \frac{\tau}{m} \qquad (14.2)$$

where the possibility of nonzero correlation between position and momentum has been included. In fact, this correlation may be negative if the initial state of the particle is chosen to be a 'contractive state'. If this is the case it is clear that at a later time τ it is possible that $\Delta x^2(\tau) < \hbar\tau/m$, thus allowing a greater accuracy than the standard quantum limit.

14.1 Concept of a QND Measurement

The basic requirement of a QND measurement is the availability of a variable which may be measured repeatedly giving predictable results in the absence of a gravitational wave [1]. Clearly this requires that the act of measurement itself does not degrade the predictability of subsequent measurements. Then in a sufficiently long sequence of measurements the output becomes predictable.

This requirement is satisfied if for an observable $A^I(t)$ (in the interaction picture)

$$[A^I(t), A^I(t')] = 0 . \qquad (14.3)$$

The condition ensures that if the system is in an eigenstate of $A^I(t_0)$ it remains in this eigenstate for all subsequent times although the eigenvalues may change. Such observables are called *QND observables*. Clearly constants of motion will be QND observables. Thus for a free particle, energy and momentum are QND observables while the position is not as

$$x(t + \tau) = x(t) + p\frac{\tau}{m} \qquad (14.4)$$

and

$$[x(t), x(t + \tau)] = \frac{i\hbar\tau}{m} . \qquad (14.5)$$

For a harmonic oscillator of unit mass

$$[x(t), x(t + \tau)] = \frac{i\hbar}{\omega}\sin \omega\tau . \qquad (14.6)$$

and

$$[p(t),\ p(t+\tau)] = i\hbar\omega\ \sin\ \omega\tau\,, \tag{14.7}$$

thus position and momentum are not QND observables for the harmonic oscillator.

There are, however, QND observables for the harmonic oscillator. We define the explicitly time dependent quadrature phase amplitudes for the oscillator as follows.

$$X_1(t) = ae^{i\omega t} + a^\dagger e^{-i\omega t} \tag{14.8}$$

and

$$X_2(t) = -i(ae^{i\omega t} - a^\dagger e^{-i\omega t})\,. \tag{14.9}$$

In the Heisenberg picture the quadrature phase operators are given by

$$X_1 = a + a^\dagger\,, \tag{14.10}$$

$$X_2 = -i(a - a^\dagger)\,, \tag{14.11}$$

which clearly shows that the quadrature phase operators are constants of the motion. In terms of the position and momentum the quadrature phase operators are

$$X_1(t) = \left(\frac{2\omega}{\hbar}\right)^{1/2}\left[x(t)\cos\ \omega t - \frac{p(t)}{\omega}\sin\ \omega t\right] \tag{14.12}$$

and

$$X_2(t) = \left(\frac{2\omega}{\hbar}\right)^{1/2}\left[x(t)\sin\ \omega t + \frac{p(t)}{\omega}\cos\ \omega t\right]\,. \tag{14.13}$$

Thus the X_1 and X_2 axes rotate with respect to the position and momentum axes of phase space, at frequency ω.

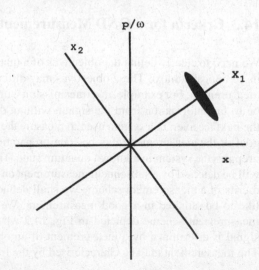

Fig. 14.1 Error box in the phase plane for a harmonic oscillator. The error box rotates with respect to the x and p/ω axes but is stationary with to the X_1, X_2 axes

The behaviour of X_1 and X_2 are most easily discussed with reference to an amplitude and phase diagram. In such a diagram the state of the system is represented by a set of points centred on the mean and contained within an error ellipse determined by the variance of the quadrature phases. Alternatively the error box may be regarded as a contour of the Wigner function. In Fig. 14.1 an error ellipse for the oscillator is shown. The error ellipse is stationary with respect to the X_1 and X_2 axes but rotates with respect to the x and p axes. This clearly illustrates how uncertainties in momentum feed back into position.

14.2 Back Action Evasion

Having first determined the QND variables of the detector it is necessary to couple the detector to a readout system or meter. It is essential that the coupling to the meter does not feed back fluctuations into the QND variable of the detector. In order to avoid this it is sufficient if the QND variable A commutes with the Hamiltonian coupling the detector and the meter, H_{DM}, that is

$$[A, H_{DM}] = 0 \qquad\qquad (14.14)$$

This is known as the *back action evasion criterion*.

In this chapter we are primarily concerned with QND measurements on optical systems. This requires a slight change in nomenclature. We will refer to the field with respect to which the QND variable is defined as the 'signal' rather than the detector, and the field upon which measurements are ultimately made as the 'probe' rather than the meter.

14.3 Criteria for a QND Measurement

We need to clearly define the objectives of a quantum nondemolition measurement in an optical context. These objectives may differ depending on the situation of the measurement. For example, in a transmission with a series of receivers, the goal may be to tap information from the signal, without degrading the signal transmitted to the next receiver. In a system used to measure the magnitude of an external force the goal of the measurement may be state preparation. That is an initial measurement prepares the system in a known quantum state. The presence of the perturbing force will be detected by a subsequent measurement on the system. In order to evaluate the merits of a measurement scheme we shall define a set of criteria which we would like to be satisfied in a good measurement. We begin by considering the general measurement scheme depicted in Fig. 14.2 where an observable X_{in} of the input signal is determined by a measurement of an observable Y_{out} of the output probe. The measurement may be characterised by the following criteria [2]:

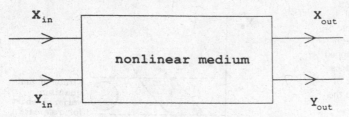

Fig. 14.2 General scheme for a QND measurement in an optical context

1. *How good is the measurement scheme?* This is determined by the level of correlation between the probe field measured by a detector and the signal field incident on the apparatus. The appropriate correlation function is

$$C^2_{X^{in}Y^{out}} = \frac{|\langle X^{in}\,Y^{out}\rangle_s - \langle X^{in}\rangle\langle Y^{out}\rangle|^2}{V_{X^{in}}V_{Y^{out}}} \tag{14.15}$$

where $V(A) = \langle A^2\rangle - \langle A\rangle^2$ is the variance in a measurement of A and $\langle AB\rangle_s = \langle AB + BA\rangle/2$. For a perfect measurement device the phase quadrature of the probe output is equal to the amplitude quadrature of the signal input multiplied by the QND gain, plus the input probe phase quadrature. In this case the correlation coefficient defined above is unity, for large gain.

2. *How much does the scheme degrade the signal?* The quantity of interest here is the correlation between the signal input field and the signal output field:

$$C^2_{X^{in}X^{out}} = \frac{|\langle X^{in}\,X^{out}\rangle_s - \langle X^{in}\rangle\langle X^{out}\rangle|^2}{V_{X^{in}}V_{X^{out}}} \tag{14.16}$$

This is a measure of the back action evasion, that is the ability of the scheme to isolate quantum noise introduced by the measurement process from the observable of interest. For an ideal QND scheme we require this correlation to be unity. Thus, for a perfect QND scheme we have $C^2_{X^{in}Y^{out}} + C^2_{X^{in}X^{out}} = 2$.

3. *How good is the scheme as a state preparation device?* If we have a perfect measurement device that does not degrade the signal at all, we satisfy the two previous criteria exactly, then we must be able to completely predict the state of the signal output. However, once we leave this ideal case the predictability of the signal output is no longer fully determined by correlations with the signal input. The extreme example is that of a destructive measurement: independently of how well the input is measured the output is always the vacuum. On the other hand, the correlation between the signal and probe output fields is not a good indicator of the quality of state preparation. Figure 14.3 shows a situation in which both output fields are well correlated, but a probe measurement does not allow inference of the signal output field to be better than the quantum limit. This situation arises when the interaction within the QND medium introduces significant correlated noise to both output fields.

Given that we have made a perfect measurement of some physical quantity X with the result x, what is the state of the system after such a measurement

Fig. 14.3 Illustration of a situation in which a value of the probe output has been measured, but when mapped onto the error ellipse, does not permit an inference of the signal to better than the quantum limit

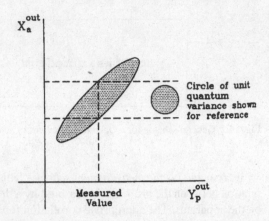

conditioned on the result x? In standard quantum mechanics the conditional state is generally assumed to be an eigenstate of X with an eigenvalue equal to the measured result, at least for perfect measurements. In the case of the QND measurement scheme above we then expect the state of the signal mode conditioned on the results of the probe measurements should in some limit be an eigenstate of X^{out}. Of course, the variance of X^{out} in such a state is zero. Thus as a measure of how well the scheme prepares eigenstates at the output we need to consider the conditional variance $V(X^{out}|Y^{out})$. This quantity is calculated as follows:

The probability to obtain the result Y^{out} for a probe measurement is given by

$$P(Y^{out}) = \text{Tr}\{\rho^{out}|Y^{out}\rangle\langle Y^{out}|\} \tag{14.17}$$

(assuming perfect readout of the probe state). The conditional state of the signal mode based on this result is

$$\rho^{out} = \frac{\text{Tr}_{probe}\{\rho^{out}|Y^{out}\rangle\langle Y^{out}|\}}{P(Y^{out})}. \tag{14.18}$$

Using this result we see that the conditional distribution for X^{out} is

$$P(X^{out}|Y^{out}) = \text{Tr}_{signal}\{\rho^{out}_{signal}|X^{out}\rangle\langle X^{out}|\}$$
$$= \frac{P(X^{out}, Y^{out})}{P(Y^{out})}$$

where

$$P(X^{out}, Y^{out}) = \text{Tr}\{\rho^{out}|X^{out}\rangle\langle X^{out}| \otimes |Y^{out}\rangle\langle Y^{out}|\}. \tag{14.19}$$

In many cases of interest $P(X^{out}, Y^{out})$ is a bivariate Gaussian. In that case one may show

$$V(X^{out}|Y^{out}) = V(X^{out})(1 - C^2_{X^{out}Y^{out}}). \tag{14.20}$$

Thus, the condition for a perfect state reduction in the conditional state is

$$C^2_{X_{out} Y_{out}} = 1 \,.$$ (14.21)

We shall now analyse some possible measurement schemes and see how well they approach the conditions for an ideal measurement.

14.4 The Beam Splitter

We consider first a beam splitter deflecting part of the incident signal field onto a homodyne detector, as shown in Fig. 14.4. This will serve as a standard of comparison for other measurement schemes. There is obviously little point in constructing complicated schemes involving cavities containing nonlinear media if they cannot improve on the performance of a beam splitter. We consider the case where the signal and probe fields are single mode with annihilation operators a and b, respectively. The amplitude and phase quadratures of the signal and probe fields are defined as

$$X_a = a + a^\dagger \,,$$ (14.22)

$$X_\phi = -\mathrm{i}(a - a^\dagger) \,,$$ (14.23)

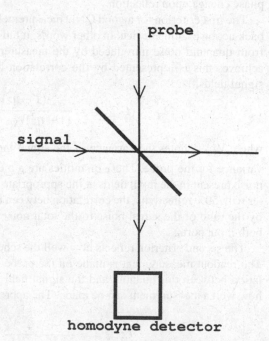

Fig. 14.4 An optical measurement scheme for the quadrature phase based on a beam splitter

$$Y_a = b + b^\dagger, \tag{14.24}$$

$$Y_\phi = -i(b + b^\dagger). \tag{14.25}$$

(This notation assumes that the coherent amplitude of each field is real). Note that according to the uncertainty principle

$$\Delta X_a \Delta X_\phi \geq 1 \tag{14.26}$$

where ΔX_a and ΔX_ϕ are the square root of the variances. A precise measurement of the amplitude quadrature must therefore be at the expense of uncertainty in the phase. A good back action evading scheme must be able to feed all the quantum noise induced by the act of measurement into the phase quadrature of the signal.

In the beam splitter the phase change on reflection gives a coupling between the amplitude quadrature of the signal and the phase quadrature of the probe. We consider making a measurement on the phase quadrature of the probe in order to determine the amplitude quadrature of the signal. The input quadrature fields can be related to the output quadrature fields using the transformation at the beam splitter:

$$\begin{pmatrix} X_a^{\text{out}} \\ Y_\phi^{\text{out}} \end{pmatrix} = \begin{pmatrix} \sqrt{1-\eta^2} & -\eta \\ \eta & \sqrt{1-\eta^2} \end{pmatrix} \begin{pmatrix} X_a^{\text{in}} \\ Y_\phi^{\text{in}} \end{pmatrix} \tag{14.27}$$

where η is real and represents the mirror amplitude reflectivity, and there is a $\pi/2$ phase change upon reflection.

The first criterion for a good QND measurement scheme is that it must be a good back action evading device. In other words, it must be able to isolate the signal field from quantum noise introduced by the measurement. How well the beam splitter achieves this is represented by the correlation between the input and the output signal fields,

$$C_{X_a^{\text{in}} X_a^{\text{out}}}^2 = \frac{(1-\eta^2) V_{X_a^{\text{in}}}}{(1-\eta^2) V_{X_a^{\text{in}}} + \eta^2 V_{Y_\phi^{\text{in}}}} \tag{14.28}$$

where $V_{X_a^{\text{in}}}$ denotes the variance of the signal input, and $V_{Y_\phi^{\text{in}}}$ is the corresponding variance for the probe. These quantities are a measure of the quantum or classical noise present in the input fields at the appropriate qudrature phase. For a beam splitter with 50% reflectivity, the correlation between the signal input and output is given by the ratio of the signal noise to the total noise introduced to the system through both input ports.

The second criterion reflects how well the scheme acts as a measurement device. The readout measurement is made on the probe output field, so the level of correlation between this quantity and the signal field incident on the device determines how well a measurement can be made. The appropriate correlation function is

$$C_{X_a^{\text{in}} Y_\phi^{\text{out}}}^2 = \frac{\eta^2 V_{X_a^{\text{in}}}}{\eta^2 V_{X_a^{\text{in}}} + (1-\eta^2) V_{Y_\phi^{\text{in}}}}. \tag{14.29}$$

Again, for a 50% beam splitter, the correlation is given by the ratio of the incident signal noise to the total noise introduced.

The third criteria is that the measurement must prepare the output observable in a well known state. This is given by the variance in the output state after the measurement has been performed. Using

$$C^2_{X_a^{out} Y_\phi^{out}} = \frac{\eta^2(1-\eta^2)(V_{X_a^{in}} - V_{Y_\phi^{in}})^2}{[(1-\eta^2)V_{X_a^{in}} + \eta^2 V_{Y_\phi^{in}}][\eta^2 V_{X_a^{in}} + (1-\eta^2)V_{Y_\phi^{in}}]} \tag{14.30}$$

and

$$V_{X_a^{out}} = (1-\eta^2)V_{X_a^{in}} + \eta^2 V_{Y_\phi^{in}}, \tag{14.31}$$

and the linearity predictor for the beam splitter, the conditional variance is given by

$$V(X_a^{out}|Y_\phi^{out}) = \frac{V_{X_a^{in}} V_{Y_\phi^{in}}}{\eta^2 V_{X_a^{in}} + (1-\eta^2)V_{Y_\phi^{in}}}. \tag{14.32}$$

We would like this variance to be zero. If both signal and probe inputs are in the vacuum or coherent states with unit quantum variance in both quadratures, then

$$C^2_{X_a^{in} Y_\phi^{out}} = \eta^2,$$
$$C^2_{X_a^{in} X_a^{out}} = 1 - \eta^2,$$
$$C^2_{X_a^{out} Y_\phi^{out}} = 0$$
$$V(X_a^{out}|Y_\phi^{out}) = 1.$$

As expected, the correlation between the signal input field and the signal output field is the intensity transmission coefficient of the mirror. To reduce the amount of noise added to the signal variable we would like to split off only a small portion of the light field. However, this reduces the correlation between the signal input field and the probe field upon which the readout is made, which is given by the intensity reflection coefficient. It is not possible therefore to simultaneously satisfy the first two criteria for a good QND scheme. Since the signal and probe fields are completely uncorrelated, a measurement of the probe does not reduce the signal output variable at all. The result is that you cannot use a beam splitter to prepare the state of the output signal with probe fluctuations at the vacuum level. Clearly the performance of the beam splitter improves if the input probe has squeezed fluctuations (Exercise 14.5).

Note that for the beam splitter $C^2_{X_a^{in} Y_\phi^{out}} + C^2_{X_a^{in} Y_a^{out}} = 1$, a typical result for a non-back-action evasion scheme, and significantly less than the maximum result of 2 for this quantity achieved in an ideal scheme. The quality of a QND scheme can thus be measured by the extent to which this quantity exceeds unity and approaches the upper limit of 2.

14.5 Ideal Quadrature QND Measurements

We now consider another scheme to make perfect QND measurements of the quadrature phase of a single mode field. We shall assume that the amplitude quadratures of each mode are coupled. That is, the interaction Hamiltonian has the form

$$\mathscr{H} = \hbar\chi \, X_a \, Y_a \tag{14.33}$$

where χ is the coupling strength and X_a, Y_a are defined by (14.22 and 14.24). Clearly X_a is a QND variable of the signal which satisfies the back action evading condition (14.12). The input and output quadratures are related by

$$X_a^{\text{out}} = X_a^{\text{in}} \, ,$$
$$Y_\phi^{\text{out}} = G \, X_a^{\text{in}} + Y_\phi^{\text{in}} \, ,$$

where $G = \chi t$ is known as the QND gain (t being the interaction time).

A measurement on the phase quadrature of the probe will be used to determine the amplitude quadrature of the input signal. To begin we calculate the correlation coefficients which define the measurement. Clearly $C_{X_a^{\text{in}} X_a^{\text{out}}} = 1$, the signal is completely unaffected by the measurement. The correlation between the input signal and the phase of the output probe is

$$C_{X_a^{\text{in}} Y_\phi^{\text{out}}} = \frac{G^2 V_{X_a^{\text{in}}}}{G^2 V_{X_a^{\text{in}}} + V_{Y_\phi^{\text{in}}}} \tag{14.34}$$

where we have taken $\langle Y_\phi^{\text{in}} \rangle = 0$. For a large QND gain $G^2 \gg 1$,

$$C_{X_a^{\text{in}} Y_\phi^{\text{out}}} \to 1 \, . \tag{14.35}$$

The conditional variance $V(X_a^{\text{out}} | Y_\phi^{\text{out}})$ which determines the value of the scheme as a state preparation device is given by

$$V(X_a^{\text{out}} | Y_\phi^{\text{out}}) = V_{X_a^{\text{in}}} \left(1 - \frac{G^2 V_{X_a^{\text{in}}}}{G^2 V_{X_a^{\text{in}}} + V_{Y_\phi^{\text{in}}}} \right)$$
$$\approx \frac{V_{Y_\phi^{\text{in}}}}{G^2}$$
$$\to 0 \text{ for } G^2 \gg 1 \, .$$

Again in the limit of high QND gain this device operates as a good state preparation device.

Another measure of the performance of the measurement is the signal-to-noise ratio of the probe output

$$\frac{\text{signal}}{\text{noise}} = \frac{\langle Y_\phi^{\text{out}} \rangle^2}{V_{Y_\phi^{\text{out}}}}$$

$$= \frac{G^2 \langle X_a^{\text{in}} \rangle^2}{G^2 V_{X_a^{\text{in}}} + V_{Y_\phi^{\text{in}}}}$$

$$\rightarrow \frac{\langle X_a^{\text{in}} \rangle^2}{V_{X_a^{\text{in}}}} \text{ for } G^2 \gg 1.$$

In the limit of large QND gain the signal-to-noise ratio of the output probe is equal to that of the signal input.

14.6 Experimental Realisation

It is possible to achieve a QND coupling of the form in (14.33) by considering two degenerate modes a and b with frequency ω and orthogonal polarisation, which undergo parametric amplification [3]. The two polarisation modes initially undergo a mixing interaction using polarisation rotators, after which a mixture of the signal and probe fields will propagate along each of the ordinary and orthogonal extrordinary axis of a KTP crystal pumped by a pulsed intense classical field. After this amplification step the fields then pass through a second polarisation rotator adjusted to give the same mixing angle as the first. In order to ensure that the device operates as an ideal QND scheme the mixing angle of the rotators must be carefully adjusted. The situation is depicted in Fig. 14.5.

The transformation performed by the polarisation rotators is given by

$$a(\theta) = a\cos\theta + ib\sin\theta, \tag{14.36}$$

$$b(\theta) = b\cos\theta + ia\sin\theta, \tag{14.37}$$

with θ being the mixing angle. The transformation in the parametric amplification process is

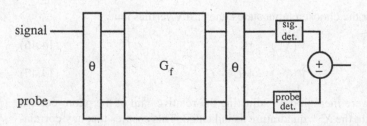

Fig. 14.5 Schematic representation of a perfect QND scheme based on a parametric interaction

$$a(r) = a\cosh r + ib^{\dagger}\sinh r \,, \tag{14.38}$$

$$b(r) = b\cosh r + ia^{\dagger}\sinh r \,, \tag{14.39}$$

where r is related to the parametric gain G_f by $G_f = e^r$. Using these transformations for the system in Fig. 14.5, we find that the transformations for the signal and probe quadratures are

$$X_a^{\text{out}} = X_a^{\text{in}} \,, \tag{14.40}$$

$$X_{\phi}^{\text{out}} = X_{\phi}^{\text{in}} + (G_f - G_f^{-1})Y_{\phi}^{\text{in}} \,, \tag{14.41}$$

$$Y_a^{\text{out}} = Y_a^{\text{in}} - (G_f - G_f^{-1})X_a^{\text{in}} \,, \tag{14.42}$$

$$Y_{\phi}^{\text{out}} = Y_{\phi}^{\text{in}} \,, \tag{14.43}$$

where we have taken the polarisation mixing angle to be

$$\theta = \arccos\left(\frac{G_f + 1}{\sqrt{2(G_f^2 + 1)}}\right) . \tag{14.44}$$

Clearly this represents an ideal QND scheme.

In the experiment of *La Porta* et al. [3], the incident signal was in a coherent state while the probe was in the vacuum state at input. The measured gain was $G_f = 1.33$ thus giving $\theta = 8°$. The output quadratures are measured by phase sensitive homodyne detection using polarisation beam splitters. To demonstrate that the experiment is operating as a back action evading measurement, three quantities were measured. Firstly the variances of each quadrature of the signal and probe were measured, with the gain both on and off. The probe shows a large noise in only one quadrature when the paramp was on. This is a reflection of the gain term appearing in the amplitude quadrature of the probe. The phase quadrature noise was at the shot-noise level. Secondly, the signal variances alone were measured showing a similar effect. Finally the variance of the quantities

$$X_{\pm} = (G_f - G_f^{-1})X_a^{\text{out}} \pm Y_a^{\text{out}} \tag{14.45}$$

was measured. For the choosen input states one easily verifies that

$$V(X_+) = 1 \,, \tag{14.46}$$

$$V(X_-) = 4(G_f - G_f^{-1})^2 \,. \tag{14.47}$$

These quantities were measured by adjusting the relative gain of the photo-current amplifiers to weight the X_a^{out} quadrature as indicated. This ensures that any correlation between X_a^{out} and Y_a^{out} will give maximum cancellation of the noise from each quadrature seperately in the variances for $X_{\pm}$. The results of this experiment are

Fig. 14.6 The combination of the amplified signal and probe quadrature at output versus the phase of the local oscillator. The variance of $X-$ occurs at integer multiples of π. The *dotted line* corresponds to the case of the paramp off, and is identical to randomly added shot-noise levels of the signal and probe. The noise reduction is 0.6 dB below the combined shot-noise level. From Arthur La Porta: Phys. Rev. Lett. **62**, 28 (1989)

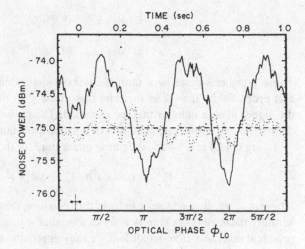

shown in Fig. 14.6. The results of all experiments taken together clearly indicate that the scheme is operating as a back action evasion device.

Recently, QND experiments have been performed [4] with two-photon transitions in three-level atoms, where the signal amplitude is strongly correlated with the probe phase. The measurement correlation between the signal in and the probe out is $C^2_{X^{in}_a Y^{out}_\phi} = 0.45$ and the back action evasion correlation between signal in and signal out is $C^2_{X^{in}_a Y^{out}_\phi} = 0.9$. The overall performance measure of this device as a QND optical tap is then determined by $C^2_{X^{in}_a Y^{out}_\phi} + C^2_{X^{in}_a X^{out}_\phi} = 1.35$, which exceeds the beam splitter limit of one, but is still well below the optimal value of 2.

14.7 A Photon Number QND Scheme

We turn now to a scheme to measure the photon number in the signal field. Conventional photon counting techniques absorb quanta. The scheme considered is a true nondemolition measurement of photons in that no photons are absorbed from the signal field [5].

Consider the coupled signal/probe system described by the interaction Hamiltonian

$$\mathscr{H}_1 = \hbar \chi a^\dagger a b^\dagger b \tag{14.48}$$

where a refers to the signal mode, and b to the probe. Such a coupling can occur in a four wave mixing process in which case χ is proportional to the third-order nonlinear susceptibility.

Clearly, $a^\dagger a$ is a constant of the motion and is thus a QND variable for the signal. The solution of the Heisenberg equations of motion gives

$$a(t) = e^{-i\chi b^\dagger b t} a(0) \, , \tag{14.49}$$

$$b(t) = e^{-i\chi a^\dagger a t} b(0) \, . \tag{14.50}$$

These equations describe a mutual intensity-dependent phase shift for the signal and probe fields. If we can measure this phase shift on the probe, information on the signal photon number may be obtained. The probe phase shift may, in fact, be determined by homodyne detection of a probe quadrature.

Using (14.24 and 14.25) the phase quadrature for the probe field becomes

$$Y_\phi^{\text{out}} = \cos(\kappa a^\dagger a) \, Y_\phi^{\text{in}} - \sin(\kappa a^\dagger a) \, Y_a^{\text{in}} \tag{14.51}$$

where $\kappa = \chi t$. It would appear from this equation that the signal operator that we actually measure is not simply $a^\dagger a$ but a nonlinear function of $a^\dagger a$. However in any practical scheme κ is so small that we may approximate the trignometric functions by the lowest order in κ. Thus, we use

$$Y_\phi^{\text{out}} = Y_\phi^{\text{in}} - \kappa a^\dagger a \, Y_a^{\text{in}} \, . \tag{14.52}$$

What quantity plays the role of the QND gain in this scheme? To answer this question we need to evaluate the correlation functions which provide criteria for the quality of the QND measurement. The first of these functions is

$$
\begin{aligned}
C^2_{a^\dagger a Y_\phi^{\text{out}}} &= \frac{|\langle a^\dagger a Y_\phi^{\text{out}}\rangle - \langle a^\dagger a\rangle\langle Y_\phi^{\text{out}}\rangle|^2}{V(a^\dagger a)V(Y_\phi^{\text{out}})} \\
&= \frac{\kappa^2 \langle Y_a^{\text{in}}\rangle^2 V(a^\dagger a)}{V(Y_\phi^{\text{in}}) + 2\kappa F_1 + \kappa^2 F_2}
\end{aligned}
$$

where

$$F_1 = \langle a^\dagger a\rangle\langle Y_a^{\text{in}}, Y_\phi^{\text{in}}\rangle_s \, ,$$
$$F_2 = V(a^\dagger a)\langle Y_\phi^{\text{in}}\rangle^2 + V(Y_a^{\text{in}})(V(a^\dagger a) + \langle a^\dagger a\rangle^2)$$

and the symmetrised correlation function is defined by

$$\langle A, B\rangle_s = \frac{1}{2}\langle AB + BA\rangle - \langle A\rangle\langle B\rangle \, . \tag{14.53}$$

If we now assume $\langle Y_a^{\text{in}}\rangle^2 \gg V(Y_a^{\text{in}})$, $V(Y_\phi^{\text{in}})$ (that is the coherent amplitude of the probe is much greater than the fluctuations in either quadrature), we find,

$$C^2_{a^\dagger a Y_\phi^{\text{out}}} \to 1 \tag{14.54}$$

when $\langle Y_a^{\text{in}}\rangle$ is large. It would thus appear that the coherent amplitude of the probe plays the role of the QND gain. This result is easily understood in terms of a complex amplitude diagram for the probe. If the vector representing the input state of the probe is very long a small rotation due to the signal makes a large change in the projection of the coherence vector onto the phase quadrature direction. In a similar way the signal-to-noise ratio of the output quadrature reduces to the signal-to-noise ratio for $a^\dagger a$ in the limit of $\langle Y_\phi^{in} \rangle \gg 1$. One easily verifies that the conditional variance of $a^\dagger a$ at the output approaches zero in the same limit. This last result indicates that the conditional state of the signal output will have sub-Poissonnian statistics.

If κ is not small, we cannot simply approximate the coupling between the signal and probe as being linear in the signal photon number. A measurement of the probe quadrature phase still provides information on the signal photon number, however, due to the multivalued nature of the trignometric functions, the signal is reduced to a superposition of number states in the case that the initial photon number distribution of the signal is sufficiently broad [5].

Exercises

14.1 Consider a signal beam and a probe beam coupled via a four wave mixing interaction;

$$\mathscr{H}_1 = \hbar \chi a^\dagger a b^\dagger b \tag{14.55}$$

Calculate the QND correlation coefficients between the amplitude quadrature of the signal $X_a = a^\dagger + a$ and the phase quadrature of the probe, $Y_\phi = -i(b - b^\dagger)$.

14.2 Consider a QND measurement in an optical cavity. Generalise the QND correlations to the frequency domain. For example, the stationary spectral covariance between signal X and probe Y is defined as

$$C_{XY}(\omega) = \int d\tau e^{-i\omega\tau} \frac{1}{2} \langle X(t)\, Y(t+\tau) + Y(t+\tau) X(t) \rangle \tag{14.56}$$

Using the input/output formulation developed in Chap. 7, calculate the spectral QND correlations between the signal amplitude and the probe phase in the input and output fields for the intracavity interaction

$$\mathscr{H}_1 = \hbar \frac{\chi}{2} X_a Y_\phi \,. \tag{14.57}$$

14.3 Consider the four wave mixing process with the Hamiltonian

$$\mathscr{H}_1 = \hbar \chi a^\dagger a b^\dagger b \tag{14.58}$$

taking place inside a cavity. Calculate the QND spectral correlations between the amplitude quadrature X_a of the signal and the phase quadrature Y_ϕ of the probe, in the input and output fields.

14.4 Consider a degenerate parametric amplifier inside a resonant two-sided optical cavity with mirrors with loss rates γ_1, γ_2. Treat the left-hand input as the signal and the right-hand input as the probe. Calculate the QND spectral correlations between the phase quadratures of the signal and the probe.

14.5 Calculate the QND correlations for a beam splitter with a squeezed input probe.

References

1. C.M. Caves, K.S. Thorne, R.W.P. Drever, V.P. Sandberg, M. Zimmerman: Rev. Mod. Phys. **52**, 341 (1980)
2. M.J. Holland, M.J. Collett, D.F. Walls, M.D. Levenson: Phys. Rev. A **42**, 2995 (1990)
3. A. La Porta, R.E. Slusher, B. Yurke: Phys. Rev. Lett. **62**, 28 (1989)
4. J. Ph. Poizat, P. Grangier: Phys. Rev. Lett. (1992)
5. G.J. Milburn, D.F. Walls: Phys. Rev A **28**, 2065 (1983)

Chapter 15
Quantum Coherence and Measurement Theory

Abstract The feature of quantum mechanics which most distinguishes it from classical mechanics is the coherent superposition of distinct physical states. Many of the less intuitive aspects of the quantum theory can be traced to this feature. Does the superposition principle operate on macroscopic scales? The famous Schroedinger cat argument highlights problems of interpretation were macroscopic superposition states allowed. In this chapter we discuss schemes to produce and detect superposition states in an optical context. We shall show that such states are very fragile in the presence of dissipation and rapidly collapse to a classical mixture exhibiting no unusual interference features.

The superposition principle is also the source of the "problem of measurement" in quantum mechanics. We do not attempt to present a solution to this problem here. Rather we show how the effect of the environment on superposition states enables a consistent description of the measurement process to be given and which avoids some of the problems inherent in previous approaches.

15.1 Quantum Coherence

The well known two slit experiment demonstrates the observational consequences of the coherent superposition of states, namely the possibility of interference patterns. In analogy with the classical theory of wave interference the visibility of the interference pattern in the probability distributions for various measurements can be used as a measure of quantum coherence. (There is a possibility of confusion here which should be cleared up. In the interference of waves we are usually concerned with the interference of two or more field modes. In this chapter however we are concerned with the superposition of different states of a single mode.)

The essential point in understanding quantum coherence is the physical distinction between the coherent superposition state

$$|\psi> = \sum_j c_j |\phi_j>$$

(15.1)

and the classical mixture

$$\rho_m = \sum_j |c_j|^2 \, |\phi_j><\phi_j| \quad . \tag{15.2}$$

The density operator corresponding to the pure state in (15.1) is

$$\rho_p = \rho_m + \sum_{i \neq j} c_i c_j^* |\phi_i><\phi_j| \tag{15.3}$$

How is one to distinguish these states in practice? Let X be the operator corresponding to some physical quantity with eigenvalues x. The probability distribution for X in the state $|\psi>$ is then given by

$$P_p(x) = P_m(x) + \sum_{i \neq j} c_i c_j^* < x|\phi_i>< \phi_j|x> , \tag{15.4}$$

where $P_m(x)$ is the probability distribution for the state ρ_m given by

$$P_m(x) = \sum_i |c_i|^2 | < x|\phi_i > |^2 = \sum_i |c_i|^2 P_{\phi_i}(x) \tag{15.5}$$

Measurements of X will distinguish the states ρ_m and ρ_p provided the second term in (15.3) is nonzero. We are thus led to define the quantum coherence with respect to the measurement of X by the coherence function

$$\mathscr{C}(x) = \sum_{i \neq j} c_i c_j^* < x|\phi_i>< \phi_j|x> \quad . \tag{15.6}$$

How does one choose an operator X such that the resulting probability distribution will exhibit interference fringes? Clearly one cannot choose operators which are diagonal in the basis $\{|\phi_i>\}$ as then the coherence function vanishes. The simplest example of an operator which distinguishes these states is the projector

$$P = |\psi><\psi| , \tag{15.7}$$

with eigenvalues $p \in \{0,1\}$. Then

$$P_p(p) = \delta_{1,p} \tag{15.8}$$

while

$$P_m(p) = \begin{cases} \sum_j |c_j|^4 & \text{if} \, p = 1 \\ 1 - \sum_j |c_j|^4 & \text{if} \, p = 0 \end{cases} \tag{15.9}$$

The coherence function is

$$\mathscr{C}(p) = (-1)^p (\sum_j |c_j|^4 - 1) \tag{15.10}$$

In practice however there may be no way to measure the operator P.

In quantum optics one either measures photon number (by photon counting) or quadrature phase (by balanced homodyne detection). As an example consider the superposition of two number states

$$|\psi> = \frac{1}{\sqrt{2}}(|n_1> + |n_2>) , \tag{15.11}$$

and the corresponding mixed state

$$\rho_m = \frac{1}{2}(|n_1><n_1| + |n_2><n_2|) . \tag{15.12}$$

A measurement of photon number will not distinguish these states, however as the quadrature phase does not commute with the number operator, quadrature phase measurements should distinguish the states.

Define the quadrature operator

$$X_\theta = (ae^{-i\theta} + a^\dagger e^{i\theta}) \tag{15.13}$$

with eigenstates $|x_\theta>$. Using the result

$$<x_\theta|n> = (2\pi)^{-1/4}(2^n n!)^{-1/2} H_n(\frac{x_\theta}{\sqrt{2}})e^{-x_\theta^2/4-in\theta} \tag{15.14}$$

one finds

$$P_m(x_\theta) = \frac{1}{2}(P^{(1)}(x_\theta) + P^{(2)}(x_\theta)) \tag{15.15}$$

and

$$\mathscr{C}(x_\theta) = (2\pi)^{-1/2}e^{-x_\theta^2/2}(2^{n_1+n_2}n_1!n_2!)^{-1/2} \tag{15.16}$$

$$\times H_{n_1}(\frac{x_\theta}{\sqrt{2}})H_{n_2}(\frac{x_\theta}{\sqrt{2}})\cos((n_1-n_2)\theta) \tag{15.17}$$

where

$$P^{(i)}(x_\theta) = (2\pi)^{-1/2}(2^{n_i}n_i!)^{-1}H_{n_i}(\frac{x_\theta}{\sqrt{2}})^2 e^{-x_\theta^2/2} \tag{15.18}$$

Thus a superposition of number states will exhibit interference fringes for some quadrature phase angle θ. In Fig. 15.1 we plot $P_p(x_\theta)$ versus x_θ for $\theta = 0$. It is surprising that, depending on whether $n_1 - n_2$ is even or odd, certain phase angles do not give interference. However it is quite clear that the superposition of two number states will exhibit phase dependent noise despite the fact that number states themselves have phase independent noise.

As a second example consider the superposition of two coherent states (so called *cat states*)

$$|\psi> = \mathscr{N}(|\alpha_1> + |\alpha_2>) , \tag{15.19}$$

where

$$\mathscr{N} = \left(2 + e^{-\frac{1}{2}(|\alpha_1|^2+|\alpha_2|^2)}(e^{\alpha_1^* \alpha_2} + e^{\alpha_1 \alpha_2^*})\right)^{-1/2} \tag{15.20}$$

Fig. 15.1 A plot of the probability density for the quadrature phase with phase angle
at zero for a superposition
of two number states, with
$n_1 = 0, n_2 = 5$

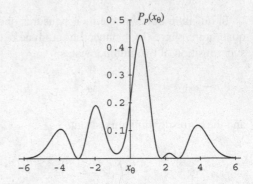

and

$$\rho_m = \frac{1}{2}(|\alpha_1 ><\alpha_1| + |\alpha_2 ><\alpha_2|) \tag{15.21}$$

In Chap. 17 we discuss how states of this kind have been produced for the centre of mass degree of freedom for a harmonically trapped ion. For cat states, measurements of either quadrature phase or photon number should exhibit interference fringes as neither of the corresponding operators commutes with a coherent state projector. However there is an optimal quadrature phase angle which leads to maximum interference.

In the case of photon number one finds

$$P_p(n) = \mathcal{N}^2(P^{(1)}(n) + P^{(2)}(n) + \mathscr{C}(n)) \tag{15.22}$$

where

$$\mathscr{C}(n) = (\frac{1}{n!}) \exp\left(\frac{-1}{2}(|\alpha_1|^2 + |\alpha_2|^2)\right) ((\alpha_1\alpha_2^*)^n + (\alpha_1^*\alpha_2)^n) \tag{15.23}$$

where

$$P^{(i)}(n) = (\frac{1}{n!})|\alpha_i|^{2n}e^{-|\alpha_i|^2} \tag{15.24}$$

If we write $\alpha_i = |\alpha_i|e^{i\phi}$ then

$$\mathscr{C}(n) = 2(P^{(1)}(n)P^{(2)}(n))^{1/2}\cos(n(\phi_1 - \phi_2)) \tag{15.25}$$

and we see that the degree of interference depends on the phase angle between the amplitudes of the superposed states. For simplicity let us take $|\alpha_1| = |\alpha_2| = |\alpha|$. Then

$$P_p(n) = 2\mathcal{N}^2(\frac{1}{n!})|\alpha|^{2n}e^{-|\alpha|^2}(1 + \cos(n(\phi_1 - \phi_2))). \tag{15.26}$$

When $\phi_1 - \phi_2 = \pi$, $P_p(n)$ is zero for n odd. Thus a superposition of coherent states π out of phase but of equal amplitude will contain only even photon number, a similar situation to that of a squeezed vacuum state.

In the case of quadrature phase measurements one finds

$$P^{(i)}(x_\theta) = (2\pi)^{-1/2} \exp\left(-|\alpha_i|^2 + \frac{x_\theta}{2} - \frac{(x_\theta - \alpha_i e^{-i\theta})^2}{2} - \frac{(x_\theta - \alpha_i^* e^{i\theta})^2}{2}\right)$$

$$(15.27)$$

and thus

$$\mathscr{C}(x_\theta) = \left(\frac{2}{\pi}\right)^{1/2} \mathfrak{R} \exp\left(\frac{-1}{2}(|\alpha_1|^2 + |\alpha_2|^2 - x_\theta^2 + (x_\theta - \alpha_1 e^{-i\theta})^2 + (x_\theta - \alpha_2^* e^{i\theta})^2)\right)$$

$$(15.28)$$

To gain some insight into these equations we take $|\alpha_1| = |\alpha_2| = |\alpha|$ and choose

$$\theta = \phi_+ = (\phi_1 + \phi_2)/2 \tag{15.29}$$

This phase angle bisects the angle between the two coherent states (see Fig. 15.2). We then have

$$P^{(1)}(x) = P^{(2)}(x) = (2\pi)^{-1/2} \exp\left(-(x - 2|\alpha|\cos\phi_-)^2/2\right) \tag{15.30}$$

and

$$\mathscr{C}(x) = 2P^{(1)}(x)\cos(2|\alpha|\sin\phi_-(x - |\alpha|\cos\phi_-)) \tag{15.31}$$

with $\phi_- = (\phi_1 - \phi_2)/2$ and we have put $x = x_{\phi_+}$. Thus

$$P_p(x) = 2\mathscr{N}^2 P^{(i)}(x)(1 + \cos(2|\alpha|\sin\phi_-(x - |\alpha|\cos\phi_-)) \tag{15.32}$$

This is a Gaussian centred at $2|\alpha|\cos\phi_-$ modulated by an interference envelope (see Fig. 15.3)

This result has a simple geometric interpretation. Referring to Fig. 15.2 we see that projecting the two coherent states onto the x_{ϕ_+} axis gives a maximum overlap centred on the mean value $< X_{\phi_+} >= 2|\alpha|\cos\phi_-$. We conclude that whenever the coherent states are projected onto a quadrature such that they overlap exactly, the interference will be maximal. Conversely we expect that if we project the coherent

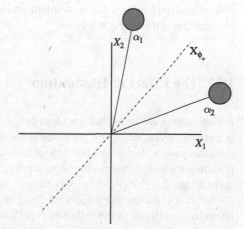

Fig. 15.2 Phase space representation of the superposition of two coherent states. The dashed line represents the direction of the quadrature phase angle which exhibits maximum interference

Fig. 15.3 The probability density for the quadrature phase at phase angle $\theta = \phi_+ = \pi/2$ for a superposition of two coherent states with $\phi_- = \pi/2$. This is the quadrature direction which bisects the angle between the two superposed coherent amplitudes, and which exhibits maximum interference

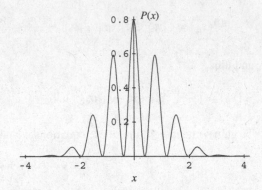

states onto a quadrature with $\theta = \phi_+ \pm \pi/2$ there will be minimum overlap and thus the least interference, (see Exercise 15.1).

We now need to consider the effect of dissipation on the interference features discussed above. With this in mind we write the coherence function in terms of the complex valued functions $C(x)$,

$$\mathscr{C}(x) = C(x) + C(x)^* \qquad (15.33)$$

where for a general superposition state

$$\mathscr{C}(x) = \sum_{i<j} c_i c_j^* <x|\phi_i><\phi_j|x> \qquad (15.34)$$

A convenient measure of the degree of quantum coherence is the quantum visibility defined by

$$\mathscr{V}(x) = \frac{|C(x)|}{(P^{(1)}(x)P^{(2)}(x))^{1/2}} \qquad (15.35)$$

In all the cases considered above $\mathscr{V}(x)$ is unity for all values of x and the states considered thus have maximum quantum coherence. However when dissipation is present this is no longer the case.

15.2 The Effect of Dissipation

In this section we show that quantum coherence associated with superposition states is extremely fragile in the presence of nonunitary effects such as damping. Such effects cause a decay of quantum coherence at a rate which is proportional to a parameter which measures the separation of the superposed states. For macroscopic separations this decay can be very rapid.

We first consider the effects of dissipation. The master equation for a damped harmonic oscillator in the interaction picture is

$$\frac{d\rho}{dt} = \frac{\gamma}{2}(2a\rho a^\dagger - a^\dagger a\rho - \rho a^\dagger a) \tag{15.36}$$

where γ is the damping constant and the reservoir is taken to be at zero temperature. We will solve this equation for an initial superposition of coherent states using the normally ordered characteristic function,

$$X(\lambda,t) = tr(\rho(t)e^{\lambda a^\dagger}e^{-\lambda^* a}) \tag{15.37}$$

Using (15.36) the equation of motion for $X(\lambda,t)$ is

$$\frac{\partial X}{\partial t} = -\frac{\gamma}{2}(\lambda^*\frac{\partial X}{\partial \lambda} + \text{c.c.}) \tag{15.38}$$

where c.c. means complex conjugate. The solution to this equation is

$$X_0(\lambda e^{-\gamma t/2}, \lambda^* e^{-\gamma t/2}) \tag{15.39}$$

where

$$X_0(\lambda,\lambda^*) = X(\lambda,0) \tag{15.40}$$

For the initial state given in (15.19), a superposition of coherent states, we find

$$X_0(\lambda,\lambda^*) = \mathcal{N}^2 \sum_{i,j=1}^{2} e^{(\lambda\alpha_i^* - \text{c.c.})} < \alpha_i|\alpha_j > \tag{15.41}$$

Thus

$$X(\lambda,t) = \mathcal{N}^2 \sum_{i,j=1}^{2} < \alpha_i|\alpha_j > \exp\left((\lambda\alpha_i^* - \text{c.c.})e^{-\gamma t/2}\right) \tag{15.42}$$

The corresponding solution for the density operator is

$$\rho(t) = \mathcal{N}^2 \sum_{i,j=1}^{2} < \alpha_i|\alpha_j >^{(1-e^{-\gamma t})} |\alpha_i e^{-\gamma t/2} >< \alpha_j e^{-\gamma t/2}| \tag{15.43}$$

Due to the coefficient with $i \neq j$ in this expansion the contribution of the off-diagonal terms to the coherence function will be small. Thus the superposition is reduced to a near mixture of coherent states.

The visibility for any measurement is easily found to be

$$\mathcal{V}(x) = | < \alpha_1|\alpha_2 > |^{(1-e^{-\gamma t})} \tag{15.44}$$

where $x = x_\theta$ for quadrature phase or $x = n$ for number measurements. For short times this is given approximately by

$$\mathcal{V}(x) \approx \exp(-\frac{\gamma t}{2}|\alpha_1 - \alpha_2|^2) \tag{15.45}$$

Thus the rate at which coherence decays is proportional to the square of the distance between the superposed coherent amplitudes. In the last section we considered the case $\alpha_1 = -\alpha_2$ which gave good interference fringes centered on the origin for the appropriate quadrature. However we now see that such fringes must decay rapidly causing the quadrature phase statistics to be indistinguishable from a classical mixture of coherent states.

This result is not confined to amplitude damping. For example the phase diffusion model discussed in Exercise 6.3 for which the master equation is

$$\frac{d\rho}{dt} = \frac{\gamma}{2}(2a^\dagger a\rho a^\dagger a - (a^\dagger a)^2 \rho - \rho(a^\dagger a)^2) \tag{15.46}$$

The solution for the matrix elements in the number basis is

$$< n|\rho(t)|m> = \exp(-\frac{\gamma}{2}t(n-m)^2) < n|\rho(0)|m> \tag{15.47}$$

If $\rho(0)$ is a superposition of number states $|n_1>$ and $|n_2>$ the visibility of the quadrature phase statistics is given by

$$\mathcal{V}(x) = \exp(-\frac{\gamma t}{2}(n_1 - n_2)^2) \tag{15.48}$$

which shows a rapid decay when $n_1 - n_2$ is large.

These examples are special; the visibility in each case decays in the same way for all measurements which give interference fringes. This is because the superposed states are eigenstates or near eigenstates of the operator appearing in the irreversible part of the evolution equation. This is an important point to which we shall return in the next section. In other cases the visibility is more complicated. For example consider the effect of phase diffusion on the quadrature phase statistics of a superposition of two coherent states with $\alpha_1 = -\alpha_2 = q_0$ where q_0 is real. If we choose $\theta = \pi/2$ (that is we project onto the imaginary axis in the complex amplitude diagram) we find that for short times

$$\mathcal{V}(x_{\pi/2}) \approx 1 - \gamma t(2q_0 x_{\pi/2})^2 \tag{15.49}$$

As expected the coherence decays from unity at a rate which is proportional to the square of the separation of the superposed states. In Fig. 15.4 we plot the probability distribution for quadrature phase at $\theta = \pi/2$ with a superposition of coherent states, with $q_0 = 2$, subject to phase diffusion with different damping rates.

A similar result may be derived for quadrature phase measurements on a superposition of number states undergoing damping. To show this we use the complex P-representation for the projector $|n_i><n_j|$

$$|n_i><n_j| = \oint_{c_1} d\alpha \oint_{c_2} d\beta \, P_{ij}(\alpha, \beta) \frac{|\alpha><\beta^*|}{<\beta^*|\alpha>} \tag{15.50}$$

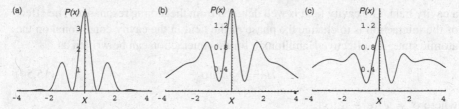

Fig. 15.4 A plot of the probability distribution for quadrature phase, $\theta = \pi/2$, on an oscillator prepared in a coherent superposition of coherent states, $|2\rangle + |-2\rangle$, which is subject to phase diffusion. **(a)** $\gamma t = 0$ **(b)** $\gamma t = 0.1$ **(c)** $\gamma t = 0.5$. The loss of visibility and spreading of the distribution is a conseqence of phase diffusion

where

$$P_{ij}(\alpha,\beta) = -\frac{1}{4\pi^2}e^{\alpha\beta}(n_i!n_j!)^{1/2}\alpha^{-(n_i+1)}\beta^{-(n_j+1)} \tag{15.51}$$

and c_1, c_2 are contours encircling the origin. Thus under time evolution $|n_i\rangle\langle n_j|$ evolves to

$$(|n_i\rangle\langle n_j|)_t = \oint_{c_1} d\alpha \oint_{c_2} d\beta\, P_{ij}(\alpha,\beta)\langle\beta^*|\alpha\rangle^{(1-e^{-\gamma t})}$$

$$\times \frac{|\alpha e^{-\gamma t/2}\rangle\langle\beta^* e^{-\gamma t/2}|}{\langle\beta^*|\alpha\rangle} \tag{15.52}$$

Using this result one may show

$$C(x_\theta, t) = (2\pi)^{-1/2}2^{-(n_1+n_2)/2}(n_1!n_2!)^{-1/2}(-1)^{n_1+n_2}$$

$$\exp\left(-\frac{x_\theta^2}{2} - i(n_1 - n_2)\theta\right)e^{-\gamma t(n_1+n_2)/2}$$

$$\sum_{p=0}^{\min(n_1,n_2)} 2^p e^{\gamma t p} p!\binom{n_1}{p}\binom{n_2}{p}H_{n_1+n_2-2p}(\frac{x_\theta}{2}) \tag{15.53}$$

The dominant term in the sum occurs for $p = \min(n_1, n_2)$. For example if $n_1 > n_2$ the time dependence of the dominant term is proportional to $\exp(-\gamma t(n_1 - n_2)/2)$. As expected the coherence decays at a rate which is proportional to the separation of the states.

15.2.1 Experimental Observation of Coherence Decay

The Haroche group in Paris demonstrated the rapid decay of coherence for a superposition of two coherent states. They used Rydberg atoms in microwave cavities [1]. Two Rydberg atomic levels with ground state $|g\rangle$ and excited state $|e\rangle$ interact with a cavity field. The cavity field is well detuned from the atomic resonance. The effect of the interaction is to change the phase of the field in the cavity conditional on the atomic state. An effective Hamiltonian for this interaction can be written as

$$H_C = \hbar\chi a^\dagger a \sigma_z \qquad (15.54)$$

where $\sigma_z = |e\rangle\langle e| - |g\rangle\langle g|$.

If the cavity C is initially prepared in a weak coherent state, $|\alpha\rangle$, (in the experiment $|\alpha| = 3.1$) and the atom is prepared in an equal superposition of the ground and excited states, the total system state evolves to

$$|\psi(\tau)\rangle = \frac{1}{2}\left(|g\rangle|\alpha e^{i\phi}\rangle + |e\rangle|\alpha e^{-i\phi}\rangle\right) \qquad (15.55)$$

where $\phi = \chi\tau$, for an interaction time τ. The state in (15.55) is an entangled state between a two level degree of freedom and an oscillator. To obtain a state which entangles the atomic degree of freedom with coherent superpositions of coherent states we use an independent laser to rotate the atomic states by $|g\rangle \rightarrow (|g\rangle + |e\rangle)/\sqrt{2}$, $|e\rangle \rightarrow (|g\rangle - |e\rangle)/\sqrt{2}$. The state at the end of this last pulse is

$$|\psi\rangle_{\text{out}} = \frac{1}{2}\left(|g\rangle(|\alpha e^{i\phi}\rangle + |\alpha e^{-i\phi}\rangle) + |e\rangle(|\alpha e^{i\phi}\rangle - |\alpha e^{-i\phi}\rangle)\right) \qquad (15.56)$$

If we now measure the atomic state $|g\rangle$ (by state selective ionisation in the experiment) the conditional state of the field is either

$$|\psi^{(g)}\rangle_{\text{out}} = \mathcal{N}_+(|\alpha e^{i\phi}\rangle + |\alpha e^{-i\phi}\rangle) \qquad (15.57)$$

or

$$|\psi^{(e)}\rangle_{\text{out}} = \mathcal{N}_-(|\alpha e^{i\phi}\rangle - |\alpha e^{-i\phi}\rangle) \qquad (15.58)$$

where $\mathcal{N}_\pm$ is the normalisation constant. These conditional states are superpositions of coherent states.

In this discussion we ignored the cavity decay as this is small on the time scale of the interaction time between a single atom and the cavity field. To observe decoherence we prepare the field in a coherent superposition of coherent states, as previously described, and then let it evolve for a time so that there is a significant probability that at least one photon is lost from the cavity. We then need to find a way to probe the field state at the end of that time. It was not possible to directly measure the state of a microwave cavity field directly. Instead the Haroche experiment used a second atom as a probe for the field state.

There are two possible results for the first atomic measurement (g, e) with the corresponding conditional states given by (15.57 and 15.58). Now consider sending in another atom after some delay time T and ask (for example) for the probability to find the second atom in the excited state given one or the other of the two conditional states, that is to say, we seek the conditional probabilities $p(e|g)$ and $p(e|e)$ (where the conditioning label refers to the result of the first atom measurement). As the respective conditional states are different, these probabilities should be different. The extent of the difference is given by

Fig. 15.5 A plot of the two-atom correlation η versus the delay time between successive atoms for two different values of the conditional phase shift. The theoretical results are shown as a dashed and solid line. From Haroche et al. Phys. Rev. Letts., **77**, 4887, (1996), Fig. 4b

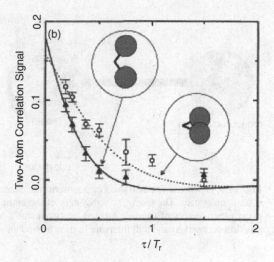

$$\eta = p(e|e) - p(e|g) \tag{15.59}$$

If the second atom is sent in a time T after the first the two conditional states, including decoherence, after this delay time can be written in the general form

$$|\psi^g_e\rangle = \mathcal{N}^2_\pm \left(|\alpha(t)e^{i\phi}\rangle\langle\alpha(t)e^{i\phi}| + |\alpha(t)e^{-i\phi}\rangle\langle\alpha(t)e^{-i\phi}| \right)$$
$$\pm \mathcal{D} \left(|\alpha(t)e^{i\phi}\rangle\langle\alpha(t)e^{-i\phi}| + |\alpha(t)e^{-i\phi}\rangle\langle\alpha(t)e^{i\phi}| \right)$$

where $\mathcal{D}$ is a measure of the decoherence. In the limit that $\mathcal{D} \to 0$, these two states are indistinguishable and the two conditional probabilities $p(e|g)$ and $p(e|e)$ are the same, so that $\eta \to 0$. Thus by repeating a sequence of double atom experiments the relevant conditional probabilities may be sampled and a value of η as a function of the delay time can be determined.

In Fig. 15.5 we reproduce the results of the experimental determination of η for two different values of the conditional phase shift, ϕ, as a function of the delay time in units of cavity lifetime. As expected the correlation signal decays to zero. Furthermore it decays to zero more rapidly for larger conditional phase shifts, that is to say it decays to zero more rapidly when the superposed states are further apart in phase-space. The agreement with the theoretical result is very good.

15.3 Quantum Measurement Theory

The object of any physical theory is to provide an explanation for the results of measurements. It is usually the case that measurements are made by coupling a macroscopic device to the system of interest which may be of any size, see Fig. 15.6. If the system is very small then some element of amplification is required. Can this process, considered purely as a physical interaction between systems, be described

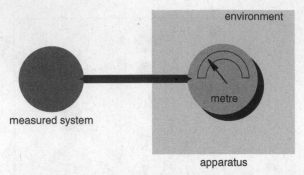

Fig. 15.6 A schematic illustration of a generic measurement illustrating that a measured system is an open system. The metre is a subsystem of the entire measurement apparatus which includes many other degrees of freedom labeled "environment". A particular metre variable, the "pointer" position is observed and will fluctuate in time around an average pointer position

entirely within the framework of quantum mechanics? Would such a description, if given, accord with our intuitive understanding of real measuring devices? How is the measured system effected by the measurement? These are questions which are generally included under the heading of *the measurement problem*.

We will primarily be concerned with measurements that are continuous in time, that is to say, a measurement for which the measurement results are a stochastic process. An example of such a measurement in quantum optics is balanced homodyne detection for which the measurement record is the difference photocurrent. For any measurement we can ask two generic questions:

1. What is the measured system state *averaged* over all measured results: *unconditional dynamics*.
2. What is the measured system state *given* a record of measured results: *conditional dynamics*.

We will take the first question to begin with.

15.3.1 General Measurement Theory

The unconditional state after the measurement is related to the initial state of the system via a *completely positive map* (CP map) which takes density operators to density operators.

$$\rho_0 \rightarrow \rho' = \Phi[\rho_0] \tag{15.60}$$

The Krauss representation theorem [2] enables us to write all CP maps as

$$\Phi[\rho_0] = \sum_\alpha \hat{E}_\alpha \rho \hat{E}_\alpha^\dagger \tag{15.61}$$

where

$$\sum_\alpha \hat{E}_\alpha^\dagger \hat{E}_\alpha = 1 \tag{15.62}$$

ensures normalisation. The sum in (15.61) could stand for an integral. The Krauss decomposition is not unique: there is a unitary equivalence class of possible Krauss decompositions.

Suppose we measure some metre quantity for which the result is a real number x. The statistics of measurement results is given in terms of a positive operator, $\hat{F}(x)$ as

$$P(x) = \mathrm{tr}(\rho \hat{F}(x)) \tag{15.63}$$

which is the most general way to construct probability distributions from density operators allowed by the quantum formalism. The conditional state *given* the measurement result is,

$$\rho^{(x)} = \frac{\hat{\Upsilon}(x)\rho\hat{\Upsilon}^\dagger(x)}{P(x)} \tag{15.64}$$

where $\hat{F}(x) = \hat{\Upsilon}^\dagger(x)\hat{\Upsilon}(x)$. The *unconditional* state is given by,

$$\rho' = \int dx P(x)\rho^{(x)} = \int dx \hat{\Upsilon}(x)\rho\hat{\Upsilon}^\dagger(x) \tag{15.65}$$

which relates the CP map to the measurement operators.

As an example of this formalism we consider the measurement of a generic two level system (a qubit) coupled to a measurement apparatus for which the pointer variable has real eigenvalues. To specify the nature of the measurement we need only specify the Krauss measurement operator $\hat{\Upsilon}(x)$,

$$\hat{\Upsilon}_\Delta(x) = (2\pi\Delta)^{-1/4} \exp\left[-\frac{(x - \kappa\hat{\sigma}_z)^2}{4\Delta} \right] \tag{15.66}$$

where $\sigma_z = |1\rangle\langle 1| - |0\rangle\langle 0|$ is the measured system observable and $\{|0\rangle, |1\rangle\}$ are an orthonormal basis for the two level system. (see Exercise 15.5). The probability density for measurement result, $x(t)$ at time t;

$$P(x,t) = \mathrm{tr}(\rho(t)\hat{\Upsilon}_\Delta^\dagger(x)\hat{\Upsilon}_\Delta(x)) \tag{15.67}$$

The moments of this distribution are related to the quantum moments of the measured system observable by

$$\mathcal{E}(x(t)) = \kappa\langle\hat{\sigma}_z(t)\rangle$$
$$\mathcal{E}((\Delta x(t)))^2 = \Delta + \kappa^2\langle(\Delta\hat{\sigma}_z(t))^2\rangle$$

Thus we regard Δ as the noise added to the signal by the apparatus. Using the fact that $\sigma_z^2 = 1$ the measurement operator may be written

$$\hat{\Upsilon}_\Delta(x) = \sqrt{P_1(x)}|1\rangle\langle 1| + \sqrt{P_0(x)}|0\rangle\langle 0| \tag{15.68}$$

where

$$P_\alpha(x) = (2\pi\Delta)^{-1/2} \exp\left[-\frac{(x+(-1)^\alpha \kappa)^2}{2\Delta}\right] \tag{15.69}$$

where $\alpha = 1,0$. The unconditional state, is then given by (15.65) as

$$\rho' = p_1|1\rangle\langle 1| + p_0|0\rangle\langle 0| + e^{-\kappa^2/2\Delta}(|1\rangle\langle 0| + |0\rangle\langle 1|) \tag{15.70}$$

where $p_\alpha = \langle\alpha|\rho|\alpha\rangle$, and $\alpha = 0,1$. We can now define the *good measurement* limit as $e^{-\kappa^2/2\Delta} << 1$, and the measurement result statistics easily enable the eigenstates of σ_z to be resolved. In this limit the unconditional post measurement state is diagonalised in the eigenbasis of σ_z.

15.3.2 The Pointer Basis

How does the construction of a physical measurement apparatus determine what system quantity is measured? This is one of the key components of the measurement problem. To understand why there might be a problem we will consider a simple model of the interaction between a measured system S and a measuring device M. Let us first ignore the macroscopic nature of M and simply treat it as a single quantum system with one degree of freedom. As we shall see such an assumption does not lead to an adequate description of the measurement process. The macroscopic nature of M is essential for a complete description.

Let $\{|m_i\rangle\}$ denote a set of eigenstates of some meter quantity and let $\{|s_i\rangle\}$ denote the eigenstates which diagonalise the system operator we are seeking to measure. Suppose the initial state of the system is $|m_0\rangle \otimes |s_i\rangle$, and further suppose that under unitary evolution this state evolves to

$$|\psi_i(t)\rangle = |m_i\rangle \otimes |s_i\rangle \tag{15.71}$$

Then for a general system state

$$|\psi\rangle = \sum_i c_i|s_i\rangle \otimes |m_0\rangle \tag{15.72}$$

the state at time t will be

$$|\psi(t)\rangle = \sum_i c_i|m_i\rangle \otimes |s_i\rangle \tag{15.73}$$

We regard m_i as the value read-out from the meter scale. If an observer finds the meter in a state $|m_j\rangle$ (with probability $|c_j|^2$) the system is subsequently (i.e. conditionally) described by the state $|s_j\rangle$. It is clear that in effect we have measured some physical quantity S which is diagonal in the basis $\{|s_i\rangle\}$,

$$S = \sum_i a_i |s_i> < s_i| \tag{15.74}$$

Suppose one only knows that a measurement has taken place but we do not select a particular result. The state of the system must then be described by the reduced density operator obtained by tracing out over the meter states,

$$\rho_s = \sum_i |c_i|^2 |s_i> < s_i| \tag{15.75}$$

Thus as a result of the measurement the density operator of the system is diagonal in the basis which diagonalises the measured operator. Alternatively the basis in which ρ_s is diagonal determines the system operator which has been measured. This is the standard description of the measurement process. Unfortunately it is inadequate as we now explain.

The description given above assumed we read out a meter quantity which is diagonal in the basis $\{|m_i>\}$. Suppose however we decide to read out another meter quantity diagonal in the basis $\{|\tilde{m}_i>\}$. We can then express the meter states $|m_i>$ in the alternative basis $\{|\tilde{m}_i>\}$,

$$|m_i> = \sum_j < \tilde{m}_j|m_i> |\tilde{m}_j> \tag{15.76}$$

The state of the combined system after the interaction is then

$$\begin{aligned} |\psi(t)> &= \sum_i c_i |m_i> \otimes |s_i> \\ &= \sum_j d_j |\tilde{m}_j> \otimes |r_j> \end{aligned} \tag{15.77}$$

where

$$d_j |r_j> = \sum_i c_i < \tilde{m}_j|m_i> |s_i> \tag{15.78}$$

Although the system meter coupling has not altered there is now some ambiguity as to what system operator has been measured; is it

$$S = \sum_i a_i |s_i> < s_i|$$

or

$$R = \sum_i b_i |r_i> < r_i| \ ? \tag{15.79}$$

This ambiguity can only be removed if we say that the meter is so constructed that the only physical property that we readout is the one which is diagonal in the basis $\{|m_i>\}$. What is the property of the meter which determines such a preferred or *pointer* basis?

We can answer this question by admitting that a true measuring device must be macroscopic and thus contain many degrees of freedom. Thus, in addition to the read-out variable, we must consider the meter as composed of many other degrees of freedom possibly coupled to the the read-out variable. These other degrees of freedom cannot be determined in any realistic scheme and thus may be treated as a environment. In order to model this situation we now divide the measurement scheme into system+meter+environment. The system and meter are directly coupled and the meter is coupled to the environment, see Fig. 15.6. As Zurek has shown it is the nature of the coupling between the meter and the reservoir which determines the pointer basis.

In order that the special correlation between the system and the meter given by (15.73) be preserved in the presence of the coupling to the reservoir, described by some Hamiltonian H_{ME}, we require that the pointer basis $\{|m_i>\}$ be a complete set of eigenstates for a pointer quantity M which commutes with H_{ME}. This ensures that fluctuations from the reservoir do not find there way back to the measured system quantity S. In many situations it may not be possible to find an operator which satisfies this condition exactly. However an approximate pointer basis may exist in as much as the diagonal elements of the density operator in such a basis relax on a very long time scale while the off-diagonal elements decay on a much shorter time scale.

Thus it is the meter-reservoir interaction which determines the pointer observable M and thus the corresponding pointer basis appropriate to the measurement. The to-be-measured quantity is only defined in the course of the meter-reservoir interaction; a situation consistent with Bohr's general description of measurement in quantum mechanics.

The meter cannot be observed in a superposition of pointer basis states as its state vector is being continually collapsed. It is the monitoring of the meter by the reservoir which results in the apparent state reduction of the system and the meter is so constructed to ensure that this occurs. The correlations between the corresponding system and pointer basis states are preserved in the final mixed state density operator

$$\sum_{i,j} c_i c_j^* |m_i><m_j| \otimes |s_i><s_j| \rightarrow$$

$$\sum_i |b_i|^2 |\tilde{m}_i><\tilde{m}_i| \otimes |r_i^p><r_i^p| \tag{15.80}$$

where $\{|r_i^p>\}$ are the system states determined by the pointer basis, referred to as the *relative states*.

There is a close connection between the concept of a quantum nondemolition measurement and the concept of a pointer basis. The condition that an operator $Q(t)$ be a QND variable is that (in the interaction picture)

$$[Q(t), Q(t')] = 0 \tag{15.81}$$

Applying this idea to the measurement description above it is clear that the pointer observable P, which determines the pointer basis, must be a QND variable of the meter. This ensures that an initial eigenstate of P evolves entirely within the pointer basis set.

In the theory of QND measurements we also require the variable Q to maintain its QND property in the presence of couplings to other systems, the reservoir in our example, which represent further stages of the meter device. This will be true provided the back action evasion criterion is satisfied

$$[Q(t), H_{\text{ME}}] = 0 \qquad (15.82)$$

where H_{ME} represents the interaction Hamiltonian between the system of the QND variable and other systems to which it is coupled (the reservoir in our case). This property is precisely the property that the pointer observable must satisfy. We may thus view the pointer observable as a QND variable of the meter which is coupled to the environment by a back action evasion coupling.

15.4 Examples of Pointer Observables

In Sect. 15.1 we considered a number of models which lead to the density operator becoming diagonal in a preferred basis. For example, if the meter is a harmonic oscillator coupled to the bath by

$$H_{\text{ME}} = a^{\dagger} a \Gamma_E \qquad (15.83)$$

so that it evolves according to the master equation (15.46) the state of the meter becomes diagonal in the number basis. The pointer observable is $a^{\dagger} a$ and (15.83) represents an ideal back action evasion coupling.

As another example suppose that the amplitude of an oscillator is coupled to the environment by

$$H_{\text{ME}} = a \Gamma^{\dagger} + a^{\dagger} \Gamma \qquad (15.84)$$

This is not a back action evasion coupling, however the meter state tends to become diagonal in the coherent state basis which are eigenstates of the operator a. Unfortunately the diagonal elements of the density operator in this basis are also changed. However as we saw in (15.43) for short times the diagonal elements do not change much, yet coherence between states separated by large coherent amplitudes decay quite rapidly. In this case we can say that the coherent states are an approximate pointer basis.

15.5 Model of a Measurement

We are now able to consider a full model of a quantum limited measurement including the interaction of the meter with the environment. The quantum system and meter are taken to be harmonic oscillators with annihilation operators a and b

respectively. The coupling between the system and the meter is taken to be quadratic in the system amplitude and the interaction Hamiltonian is

$$H_{\mathrm{SM}} = \frac{\hbar}{2} a^\dagger a (bE^* + b^\dagger E) . \qquad (15.85)$$

Such a system may represent a four wave mixing interaction in quantum optics, where one field is taken as classical of amplitude E. The measured system operator is, as we shall see, $a^\dagger a$ (or some function thereof).

We assume that mode b is coupled to the environment by amplitude coupling

$$H_{\mathrm{ME}} = b\Gamma^\dagger + b^\dagger \Gamma . \qquad (15.86)$$

This will determine a particular pointer basis. There are good physical reasons why this is a suitable choice for H_{ME}. First, if the oscillators are realised as field modes this coupling represents the usual system-bath interaction of a linear loss mechanism. In particular it could represent the interaction of a field mode with a photoelectron counter. Perhaps the most important reason for choosing H_{ME} in this form however is that it leads to the coherent state pointer basis. As coherent states have a well defined semiclassical limit this is a desirable basis for a real (i.e. classical) measuring device. We now solve for the complete dynamics of the system and meter coupled to the bath.

The density operator for the system and meter after tracing out the reservoir obeys the master equation

$$\frac{d\rho}{dt} = \frac{-i}{2}[(Eb^\dagger + E^* b)a^\dagger a, \rho] + \frac{\gamma}{2}(2b\rho b^\dagger - b^\dagger b\rho - \rho b^\dagger b) \qquad (15.87)$$

We have assumed the environment is at zero temperature. Initially the state of the system is arbitrary while the meter is in the ground state

$$\rho(0) = \sum_{n,m=0}^{\infty} (\rho_{nm}|n><m|)_S \otimes (|0><0|)_M \qquad (15.88)$$

where we have expanded the system state in energy eigenstates and $\rho_{nm} = < n|\rho_s(0)|m >$. Equation (15.87) may be solved by the characteristic function. The solution is,

$$\rho(t) = \sum_{n,m=0}^{\infty} \exp\left(\frac{|E|^2}{\gamma^2}(n-m)^2(1 - \gamma t/2 - e^{-\gamma t/2}) \right)$$
$$\times (|n><m|)_S \otimes \left(\frac{|\alpha_n(t)><\alpha_m(t)|}{<\alpha_m(t)|\alpha_n(t)>} \right)_M \qquad (15.89)$$

where $|\alpha_n(t) >$ is a meter coherent state with

$$\alpha_n(t) = \frac{E_n}{\gamma}(1 - e^{-\gamma t/2}) \qquad (15.90)$$

In the long time limit we have

$$\rho \to \sum_n P(n)\,(|n><n|)_S \otimes (|\alpha_n><\alpha_n|)_M \tag{15.91}$$

where $P(n)$ is the initial number distribution for the system. This is a mixture of number states in the system, perfectly correlated with a mixture of coherent states in the meter. It is thus of the general form of (15.80). The coherent states $|\alpha_n>$ are the pointer basis states and the number states the corresponding relative states. The amplitude of the coherent states can be made arbitrarily large by increasing the strength of the system meter coupling E. Hence this model includes amplification in a natural way. The large E limit is the appropriate limit for an accurate measurement [3]. In fact the states $|\alpha_n>$ for different values of n become approximately orthogonal as the coupling strength is increased.

We have assumed in this analysis that the environment is at zero temperature. For photoelectric detection this is a good assumption at optical frequencies. Were the environment not taken at zero temperature there would be an additional thermal spread in the diagonal part of the meter states.

This model contains all the features of the measurement process discussed in Sect. 15.3.2. The correlations between the system and the meter are created by unitary evoultion. The (almost complete) reduction of the meter states to the pointer basis (the coherent states) occurs as a result of nonunitary dissipative evolution, which causes the off-diagonal elements of the meter state in the pointer basis to decay. It is clear that the model represents the measurement of some function of the system number operator $a^\dagger a$. To determine what this function is we must reconsider the interpretation of the environment as a photoelectron counter. If we assume that every meter quanta lost to the environment is actually counted in the detector, a full analysis shows that in fact the model describes the measurement of the square of the number operator.

From (15.91) we can calculate the reduced state of the system by tracing out over the meter states. The resulting state has an exponential decay of off-diagonal elements in the number basis which goes as t^2 for short times. Such a dependence indicates that for short times a Markovian evolution equation does not describe the evolution of the system state. However if the rate of photon counting, γ, is very large then this short time behaviour is rapidly superceeded by an exponential linear dependence. In this case the effective master equation for the system state is

$$\frac{d\rho_S}{dt} = -\frac{|E|^2}{2\gamma}[a^\dagger a,[a^\dagger a,\rho_S]] \tag{15.92}$$

In this strong measurement limit we see that as far as the system is concerned the effect of the measurement of photon number is to induce a diffusion in the oscillator phase. In some sense the phase is the conjugate variable to the photon number so this result is consistent with the uncertainty principle.

15.6 Conditional States and Quantum Trajectories

We turn now to discuss the second question that may be asked of a measured system: what is the conditional state conditioned on a measurement result? In Chap. 6 we discussed an approach to solving master equations based on unravelling the solution as an average over a quantum trajectory: a solution to a stochastic master equation or stochastic Schrödinger equation. Here we will show that a quantum trajectory can describe the conditional evolution of the state of a continuously measured system, conditioned on the stochastic measurement record.

A simple model for continuous can be built out of the two-level mode discussed in Sect. 15.3.1. To get to a continuous measurement we let readouts occur at Poisson distributed times at rate γ. The system then obeys the master equation

$$\dot{\rho} = -i[H,\rho] + \gamma \left(\int_{-\infty}^{\infty} dx \hat{\Upsilon}_\Delta(x) \rho \hat{\Upsilon}_\Delta^\dagger(x) - \rho \right)$$

$$= -i[H,\rho] - \frac{\gamma}{4}\left(1 - e^{-\kappa^2/2\Delta}\right)[\hat{\sigma}_z,[\hat{\sigma}_z,\rho]]$$

Clearly this describes a QND measurement of σ_z. We now take the limit of weak, rapid measurements $\gamma \gg 1$, $\Delta \gg \kappa^2$ to get

$$\dot{\rho} = -i[H,\rho] - D[\hat{\sigma}_z,[\hat{\sigma}_z,\rho]] \qquad (15.93)$$

where $D = \frac{\gamma\kappa^2}{8\Delta}$ is the *decoherence* rate in the $\hat{\sigma}_z$ basis. When D is large decoherence is rapid, which should correspond to a good measurement.

The measurement record is a real valued classical stochastic variable, $x(t_i)$, conditioned on the state of the system. To get a continuous stochastic record we define a stochastic differential: $dy(t) = dN(t)x(t)$ where $dN(t)$ is a Poisson process:

$$dN(t)^2 = dN(t)$$
$$\mathcal{E}(dN(t)) = \gamma dt$$

We can now take the *diffusive limit*. Consider a time δt such that $\gamma\delta t \gg 1$ yet, $D\delta t \ll 1$.

$$\mathcal{E}(y(t+\delta t) - y(t)) \approx \gamma\delta t \kappa \langle \hat{\sigma}_z(t)\rangle_c \qquad (15.94)$$

$$\mathcal{E}((y(t+\delta t) - y(t))^2 \approx \gamma\delta t\Delta \qquad (15.95)$$

we can then approximate the observed process by the Ito stochastic d.e.

$$dy(t) = \gamma\kappa \left[\langle \hat{\sigma}_z(t)\rangle_c dt + (8D)^{-1/2} dW(t) \right] \qquad (15.96)$$

where $dW(t)$ is the Wiener process: $\mathcal{E}(dW(t)) = 0$, $\mathcal{E}(dW(t)^2) = dt$. If D is large, then we will have a good signal-to-noise ratio.

Under the same assumptions we can derive the conditional master equation and the stochastic Schrödinger equation for this measurement. Using (15.64 and 15.67) and taking the limit of weak rapid measurement as above we find that the conditional master equation is

$$d\rho_c(t) = -i[H, \rho_c(t)]dt - Ddt[\sigma_z, [\sigma_z, \rho_c(t)]] + \sqrt{2D}dW(t)\mathscr{H}[\sigma_z]\rho_c(t) \quad (15.97)$$

and the stochastic Schrödinger equation is

$$d|\psi_c(t)\rangle = -iHdt|\psi_c(t)\rangle - Ddt\left[1 - 2\langle\sigma_z\rangle_c\sigma_z + \langle\sigma_z\rangle^2\right]|\psi_c(t)\rangle$$

$$+ \sqrt{2D}(\sigma_z - \langle\sigma_z\rangle_c)dW(t)|\psi_c(t)\rangle \quad (15.98)$$

This leads us to define a *quantum limited measurement* as measurement for which the signal-to-noise is determined only by the intrinsic quantum noise of the measured system, and the minimum back-action noise consistent with the uncertainty principle. Under these conditions the decoherence rate is determined *only* by the back-action noise due to measurement. The example considered here is an example of a quantum limited measurement. Usually a measurement is not quantum-limited as the system dynamics has other irreversible channels (e.g. dissipation) and other sources of noise (e.g. thermal) are added to the measured signal.

The stochastic master equation has an interesting property. Using (15.97) we can derive equations of motion for the conditional mean value of σ_z. Define $z_c = \langle\sigma_z\rangle_c$

$$dz_c = (\ldots)dt + 2\sqrt{D}(1 - z_c^2)dW(t) \quad (15.99)$$

where ... refers to that part of the dynamics arising directly from the Hamiltonian term. If we assume that $[\sigma_z, H] = 0$, the stochastic dynamics is determined entirely by the measurement. It is then clear that there are two fixed points, $z_c = \pm 1$ which correspond to two eigenstates of σ_z. In a simulation the result is that the conditional state tends to localise on one or the other of the eigenstates of σ_z. This is the continuous measurement version of quantum state reduction.

15.6.1 Homodyne Measurement of a Cavity Field

Consider now a single mode cavity damped into a zero temperature heat bath (see Sect. 6.1). The environment is the multi-mode field external to the cavity. Some of these external modes are coupled into a balanced homodyne detection system. The change in the state of the system, over a time interval dt, can be described by a single jump operator a. In what follows we choose the units of time so that the cavity decay constant, γ is unity. We can define two Krauss operators. Firstly we define $\Omega_1 = a\sqrt{dt}$, corresponding to a conditional Poisson process with probability $\langle a^\dagger a\rangle dt$. Normalization requires a second Krauss operator, $\Omega_0 = I - a^\dagger adt/2 - iHdt$, where H is Hermitian. Then the unconditional master equation without feedback is just the familiar Lindblad form

$$
\begin{aligned}
\mathrm{d}\rho &= \Omega_0 \rho \Omega_0 + \Omega_1 \rho \Omega_1 - \rho \\
&= -\mathrm{i}[H,\rho]\mathrm{d}t + a\rho a^\dagger \mathrm{d}t - \frac{1}{2}(a^\dagger a \rho + \rho a^\dagger a)\mathrm{d}t \\
&\equiv -\mathrm{i}[H,\rho]\mathrm{d}t + \mathscr{D}[a]\rho\,\mathrm{d}t .
\end{aligned}
\tag{15.100}
$$

A fermionic example is given in [4].

In homodyne detection, the output field from the cavity is mixed with a strong coherent field, the local oscillator. This corresponds to a displacement of the cavity field a. Obviously this does not change the unconditional dynamics of the cavity field. With this in mind we first note that given some complex number $\alpha = |\alpha|e^{i\phi}$, we may make the transformation

$$
\begin{aligned}
a &\to a + \alpha \\
H &\to H - \frac{\mathrm{i}|\alpha|}{2}(e^{-\mathrm{i}\phi}a - e^{\mathrm{i}\phi}a^\dagger)
\end{aligned}
\tag{15.101}
$$

and obtain the same master equation. In the limit as $|\alpha|$ becomes very large, the rate of the Poisson process is dominated by the term $|\alpha|^2$. In this case it may become impossible to monitor every jump process, and a better strategy is to approximate the Poisson stochastic process by a Gaussian white-noise process.

For large α, we can consider the system for a time δt in which the system changes negligibly but the number of detections $\delta N(t) \approx |\alpha|^2 \delta t$ is very large; then we can approximate $\delta N(t)$ as

$$
\delta N(t) \approx |\alpha|^2 \delta t + |\alpha|\langle e^{-\mathrm{i}\phi}a + a^\dagger e^{\mathrm{i}\phi}\rangle_c \delta t + |\alpha|\delta W(t),
\tag{15.102}
$$

where $\delta W(t)$ is normally distributed with mean zero and variance δt (a *Wiener increment*).

We now define the stochastic measurement record as the current

$$
I_c^{\mathrm{hom}}(t) = \lim_{\alpha \to \infty} \frac{\delta N(t) - |\alpha|^2 \delta t}{|\alpha|\delta t}
\tag{15.103}
$$

$$
= \langle e^{-\mathrm{i}\phi}a + e^{\mathrm{i}\phi}a^\dagger\rangle + \mathrm{d}W(t)/\mathrm{d}t .
\tag{15.104}
$$

In balanced homodyne detection, this current corresponds to the photo-current. Given this stochastic measurement record, we can determine the conditional state of the quantum system by the stochastic master equation (SME)

$$
\begin{aligned}
\mathrm{d}\rho_a(t) &= -\mathrm{i}[H, \rho_c(t)]\,\mathrm{d}t + \mathscr{D}[e^{-\mathrm{i}\phi}a]\rho_c(t)\mathrm{d}t \\
&\quad + \mathscr{H}[e^{-\mathrm{i}\phi}a]\rho_c(t)\mathrm{d}W(t).
\end{aligned}
\tag{15.105}
$$

In the above equations, the expectation $\langle a \rangle_c$ denotes the average over the conditional state, $\mathrm{Tr}(\rho_c a)$ and $\mathscr{H}$ is a superoperator defined in Sect. 6.7

$$\mathcal{H}[c]\rho = c\rho + \rho c^\dagger - \rho \operatorname{Tr}[c\rho + \rho c^\dagger].$$ (15.106)

Let us define the quadrature phase operator

$$X_\theta = \mathrm{e}^{-\mathrm{i}\phi} a + \mathrm{e}^{\mathrm{i}\phi} a^\dagger$$ (15.107)

The corresponding stochastic Schrödinger equation is

$$\mathrm{d}|\psi_c(t)\rangle = \left\{ -\mathrm{i}H\mathrm{d}t - \frac{1}{2}\left[a^\dagger a - 2\langle X_\theta/2\rangle_c + \langle X_\theta/2\rangle_c^2\right]\mathrm{d}t \right.$$
$$\left. + [a - \langle X/2\rangle_c]\,\mathrm{d}W(t)\right\}|\psi_c(t)\rangle$$ (15.108)

Note that as $|\psi_c(t)\rangle$ approaches a coherent state for which $a|\psi_c(t)\rangle = \langle X/2\rangle_c|\psi_c(t)\rangle$ the noise term tends to zero. This leads to the stochastic localisation of the conditional state on the set of coherent states. If we ignore normalisation of the state, we get a very direct sense of how the measured photo current conditions the state,

$$\mathrm{d}|\bar{\psi}_c(t)\rangle = \mathrm{d}t\left[-\mathrm{i}H - \frac{1}{2}a^\dagger a + I_c^{\mathrm{hom}}(t)\right]|\bar{\psi}_c(t)\rangle$$ (15.109)

Exercises

15.1 Consider the superposition of two coherent states

$$|\psi\rangle = \mathcal{N}(|\alpha_0\rangle + |-\alpha_0\rangle)$$

where α_0 is real. Show that the probability distribution for $X_1 = a + a^\dagger$ is a double peaked Gaussian ($\alpha_0 > 1$) while the distribution for $X_2 = -\mathrm{i}(a - a^\dagger)$ shows interference fringes.

15.2 Consider the Hamiltonian

$$H = \hbar\omega a^\dagger a + \hbar\chi(a^\dagger a)^2$$ (15.110)

Show that this Hamiltonian can generate a superposition of two coherent states or two squeezed states of the type discussed in the preceeding questions.

15.3 The spin coherent states are defined by

$$|j;\gamma\rangle = \exp\left(-\frac{\theta}{2}(J_+ \mathrm{e}^{-\mathrm{i}\phi} - J_- \mathrm{e}^{\mathrm{i}\phi})\right)|j,j\rangle$$ (15.111)

where $\gamma = \mathrm{e}^{\mathrm{i}\phi}\tan\frac{\theta}{2}$ and $|j,j\rangle$ is a J_z eigenstate with eigenvalue j. Consider the superposition state

$$|\psi\rangle = \mathcal{N}(|j;\gamma_0\rangle + |j;-\gamma_0\rangle)$$ (15.112)

with γ_0 real. Calculate the probability distributions for J_x, J_y, J_z and show that the J_y distribution exhibits interference effects. Give a geometrical interpretation of this result.

15.4 Consider the two mode squeezed vacuum state

$$|\psi> = (\cosh r)^{-1} \sum_{n=0}^{\infty} (\tanh r)^n |n>_1 \otimes |n>_2 \qquad (15.113)$$

and the mixed state with the same diagonal distribution

$$\rho = (\cosh r)^{-2} \sum_{n=0}^{\infty} (\tanh r)^{2n} |n>_1 <n| \otimes |n>_2 <n| \qquad (15.114)$$

Show that these states may be distinguished by the probability distributions for the two mode quadrature phase operators $X_\pm = a_\pm + a_\pm^\dagger$ where

$$a_\pm = \frac{1}{\sqrt{2}}(a_1 \pm a_2) \ . \qquad (15.115)$$

15.5 Consider a two-level system with basis states $\{|0\rangle, |1\rangle\}$ coupled impulsively to the momentum of a free particle via the Hamiltonian

$$H_I(t) = \kappa \hat{p} \sigma_z \delta(t - t_r) \qquad (15.116)$$

with $\sigma_z = |1\rangle\langle 1| - |0\rangle\langle 0|$. Just prior to the readout at time t_r the state of the particle is a Gaussian with the wave function $\psi(p) = (2\pi\Delta)^{-1/4} \exp[-x^2/(4\Delta)]$. Immediately after the coupling at time t_r the position of the particle is measured with perfect accuracy and projected onto the position eigenstate $|x_r\rangle$. Show that the conditional state of the two level system immediately after the readout is given by (15.70) with $x = x_r$.

15.6 Show that the measurement operator

$$\hat{\Upsilon}_\Delta(x) = (2\pi\Delta)^{-1/4} \exp\left[-\frac{(x - \kappa\hat{\sigma}_z)^2}{4\Delta} \right] \qquad (15.117)$$

reduces to a projection operator in the limit $\sigma \to 0$.

References

1. M. Brune, E. Hagley, J. Dryer, X. Maitre, A. Maali, C. Wunderlich, J.M. Raimond S. Haroche: *Observing the Progressive Decoherence of the 'Meter' in a Quantum Measurement*, Phys. Rev. Letts. **77**, 4887 (1996)
2. K. Krauss: *States, Effects and Operations: Fundamental Notions of Quantum Theory* (Springer-Verlag, Berlin 1983)
3. D.F. Walls, M.J. Collett and G.J. Milburn: Phys. Rev. D **32**, 3208 (1985)
4. G.J. Milburn, H.B. Sun: Phys. Rev. B **59**, 10748 (1999) ·

Chapter 16
Quantum Information

16.1 Introduction

Quantum information theory is the study of communication and information processing tasks using physical systems that obey the rules of quantum theory. Information theory was largely the creation of Claude Shannon working at Bell laboratories over 50 years ago. Shannon produced an elegant mathematical theory for information encoding, transmission and decoding in the presence of noise. The work was grounded in a deep intuitive knowledge of the nature of noise in classical electronics and electromagnetism although it made little reference to the physical carrier of the information or the physical source of the noise. In the early 1980s a number of pioneers, including Feynman, Fredkin, Bennett, Landauer and Deutsch, began to re-consider these issues in the light of quantum noise. We now know that quantum mechanics provides powerful new ways to communicate and process information that are impossible, or difficult, in a classical world. Many of these new ideas have had an impact on quantum optics and some of the first experiments in this burgeoning field involve quantum optical systems. In this chapter we will consider some of these developments including quantum cryptography, quantum teleportation and quantum computation.

In classical information theory, the elementary unit of communication and information processing is the binary digit, or bit, which can take the mutually exclusive values 0 or 1. All communication and information processing can be reduced to operations on strings of binary digits. In 1946 Shannon [1] established a number of theorems for such operations and founded the subject of information theory [2]. Somewhat paradoxically, the key for this development lay in asking how much information is gained when the result of a random binary choice is known. Consider, for example, a fair coin toss. If we code a head as 1 and a tail as 0, it is clear that to record the result of a single coin toss we require one binary digit. When the result is known we have gained one bit of information. If we toss N coins there are 2^N possible outcomes, yet to record a single outcome requires only N bits. It would appear from this that an intuitive definition for a numerical measure of information is the logarithm of the number of possible alternative ways a given outcome can be realised. If all outcomes are equally likely, as in the case of a fair coin toss of N coins, the probability of each outcome is 2^{-N}. The information content of a the ith outcome is then $H = -\log_2 p_i$ where $p_i = 2^{-N}$ is the probability of the outcome. The dependence of the information measure on the logarithm of the probability ensures that information is additive as our intuition with coin tosses would suggest. In general all outcomes are not equally likely. In that case we are led to define the average information of an outcome as $H = -\sum_i p_i \log_2 p_i$. We choose to define our logarithms base two as this leads to a measure of information in bits, which appears more natural in this context.

16.1.1 The Qubit

Quantum mechanics indicates that, at its most fundamental level, the physical world is irreducibly random. Given complete knowledge of the state of a physical system (that is a pure state) there is at least one measurement the results of which are completely random. The simplest example is provided by a two-state system such as a spin-half particle, a polarised photon, or a two-level atom. An elementary optical two-state system is a beam splitter, shown in Fig. 16.1. A single photon directed towards a 50/50 beam splitter will be reflected or transmitted with equal probability (we assume an ideal device that does not absorb the photon). If we place a perfect photon detector in both output ports of the beam splitter we will get a count at one or the other detector with equal probability. At first sight it would appear that a single two-state system such as this is a perfect quantum coin toss, but the reality is more subtle.

To understand why this is so consider the example depicted in Fig. 16.2 in which we try to toss the quantum coin twice in succession by redirecting the photon towards another identical beam splitter. In a real coin toss the outcome is no less

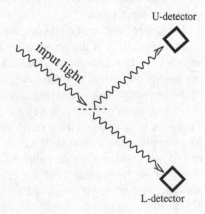

Fig. 16.1 A single photon at a 50/50 beam splitter can be reflected or transmitted with equal probability. A perfect photon detector in both output ports of the beam splitter, labeled U (*upper*) or L (*lower*), will register a count at one or the other detector with equal probability

Fig. 16.2 Tossing a quantum
coin twice. After the first
beam splitter in Fig. 16.1, a
single photon is redirected,
using perfectly reflecting
mirrors, towards an identical beam splitter. The device
is now a Mach–Zehnder
interferometer and can be
adjusted, by moving a mirror
as indicated, so that the photon emerges with certainty
in the upper output mode.
The adjustment can be made
by small displacements on
one of the perfectly reflecting
mirrors

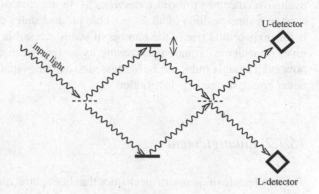

uncertain than the first coin toss. Such is not the case for this "quantum coin toss".
In Fig. 16.2 we illustrate a possible way to make the photon choose twice in succession whether to be reflected or transmitted, and immediately recognise the form
of a Mach–Zehnder interferometer. Clearly we can set up this device so that the
photon will be detected with certainty in say the upper photon detector. This is very
different from tossing two coins in succession.

The explanation of this phenomenon takes us to the heart of why quantum information theory will necessarily be different from classical information theory. Immediately after the first beam splitter the photon is in a quantum superposition of two
distinct spatial modes of the field;

$$|\psi\rangle = \frac{1}{\sqrt{2}}(|1\rangle_U \otimes |0\rangle_L + |0\rangle_U \otimes |1\rangle_L) \qquad (16.1)$$

where we have labelled the two output modes as up (U) or lower (L). If we place a
photon detector in both output modes it is easy to see that we will count a photon in
each mode with equal probability. However the state is not a truly random state. In
fact it is a pure state, the entropy of which is zero. The beam-splitter has unitarily
transformed the initial pure state $|0\rangle_U \otimes |1\rangle_L$. If the system is caused to pass through
another beam splitter a further unitary transformation takes place which, for appro-
priate path lengths, will produce the state $|1\rangle_U \otimes |0\rangle_L$ at the output and the photon
will be detected with certainty in the upper detector.

We need to distinguish a true coin toss from a "quantum" coin toss. The key distinguishing feature is the ability of the quantum system to be prepared in a coherent
superposition of the two mutually exclusive alternatives. This is not possible for a
classical coin toss which is either heads or tails but not both. While it is true that the
result of an arbitrary measurement on a single two state system will give one bit of
information in general, the system state is not like a classical coin toss, as the single photon example illustrates. To distinguish a true one bit classical system from a
quantum one bit system we will refer to the quantum case as a *qubit*. A qubit is then

a quantum system which can yield at most one bit of information upon measurement, but which can be in a coherent quantum superposition of the two mutually exclusive outcomes prior to measurement. In the case of N qubits the system can exist in a superposition of all 2^N possible product states of each individual qubit. It is this exponential rise in the number of states accessible to an N qubit system that gives quantum information processing its power. We discuss below an example of how even a single qubit can be harnessed to do things that a classical one bit system never could; secure key distribution.

16.1.2 Entanglement

The key feature of quantum mechanics that lies behind quantum information theory is quantum entanglement. Quantum entanglement refers to correlations between the results of measurements made on distinct subsystems of a composite system that cannot be explained in terms of standard statistical correlations between classical properties inherent in each subsystem. An example is provided by the violation of the Bell inequality for two distinct two-state quantum systems (see Chap. 13). If the subsystems are time like separated, quantum entanglement implies non-locality. Non-locality means that measurements on distinct subsystems, local measurements, are incapable of determining the joint state of the composite system. While quantum entanglement and non-locality are related they are not the same. It is possible to have non-locality without entanglement [3].

In quantum optics the simplest source of entanglement is provided by the non-degenerate squeezed vacuum state produced by spontaneous parametric down conversion (see Sect. 5.2.1),

$$|\mathscr{E}\rangle = (1 - \lambda^2)^{1/2} \sum_{n=0}^{\infty} \lambda^n |n\rangle_a \otimes |n\rangle_b \qquad (16.2)$$

where $\lambda = \tanh r$ with r the squeezing parameter. Note that this state is a zero eigenstate of the photon number difference operator, $\hat{n}_a - \hat{n}_b$, between the two modes. The entanglement here results from a superposition of the infinite number of indistinguishable ways we can distribute equal numbers of photons in each mode. The reduced state of each subsystem (modes a and b) is in fact a thermal state (see Sect. 5.2.5). This is the maximum entropy state for a mode with a fixed average energy. Thus while the total state is a pure state with zero entropy, the state of each subsystem is as uncertain as it can be given the constraint on the average energy.

Measurements on the component sub-systems of entangled states are insufficient to completely determine the joint state of the system. In some cases local measurements may give no information at all about the joint state and the entropy of the subsystem reduced states are maximal. Such states are called maximally entangled.

An example is provided by the following eigenstates of total photon number,

$$|\psi_N\rangle = \frac{1}{\sqrt{N+1}} \sum_{n=0}^{N} |n\rangle_a \otimes |N-n\rangle_b \tag{16.3}$$

Local measurements on either mode, say a, are described by the reduced density operator

$$\rho_a = \text{Tr}_b(\rho) \tag{16.4}$$

where Tr_b refers to the partial trace over mode b. In this case the resulting reduced density operator for each mode is the identity matrix in $N+1$ dimensions. The entropy of such a state is

$$S_{a,b} = -\text{Tr}\rho_{a,b} \ln\rho_{a,b} = \ln(N+1) \tag{16.5}$$

which, given the constraint on total photon number, is maximal. In general the entropy of each subsystem satisfy an important inequality, the Araki–Lieb inequality,

$$|S_a - S_b| \le S \le S_a + S_b \tag{16.6}$$

where S is the entropy of the state of the joint system. In the case of a pure entangled state this implies that $S_a = S_b$.

Entangled states do not necessarily need to be pure states. Furthermore there can be non-entangled states that still exhibit classical correlations between the subsystems. If an entangled state interacts with an environment entanglement can be reduced to zero while classical correlations remain. An example is provided by a two-mode squeezed vacuum state undergoing phase diffusion in each mode. The steady state density operator describing such a system is

$$\rho = (1-\lambda^2) \sum_{n=0}^{\infty} \lambda^{2n} |n\rangle_a\langle n| \otimes |n\rangle_b\langle n| \tag{16.7}$$

which still retains a perfect classical correlation between the photon numbers in each mode. However as the state is a convex sum of states which factorise, the state in (16.7) is not entangled and in fact is defined as separable.

It seems reasonable to suggest that between pure entangled states and totally separable mixed states there is a gradation of entanglement. To quantify this we require a measure of entanglement, and a number of such measures for finite dimensional Hilbert spaces have been proposed [4]. The situation for infinite dimensional Hilbert spaces, which is the case for much of quantum optics, is complicated except for a special class of states known as Gaussian states. Such states have a Gaussian Wigner function. The two mode squeezed state is an example (see Sect. 5.2.4). Further discussion on mixed Gaussian entangled states is given in [5]

When a pure state, $|\psi\rangle$ interacts with an environment it undergoes decoherence (see Chap. 15) and generally becomes a mixed state, ρ. We can then ask for the probability of finding the initial state, ψ in the ensemble represented by ρ.

This probability is given by

$$F = \mathrm{Tr}(\rho|\psi\rangle\langle\psi|) \tag{16.8}$$

which is called the fidelity. Fidelity has a deeper significance in terms of the statistical distinguishability of quantum states [6].

16.2 Quantum Key Distribution

For millennia, communicating parties have devised schemes whereby messages can be authenticated (the signature) and secured from unauthorised access (cryptography). Modern methods (symmetrical crypto-systems) for secure electronic communication involve the prior exchange of a random number which is called the key. If the communicating parties share this number with each other and no one else, messages can be securely encrypted and decoded. The method however is vulnerable to a third party acquiring access to the key. In this section we will describe how quantum mechanics enables two communicating parties to arrive at a shared secure key via Quantum Key Distribution (QKD).

The idea that quantum mechanics might enable more secure communication was hinted at in the work of Wiesner [7] and made explicit in the pioneering work of Bennett and Brassard [8], in which the first QKD protocol, BB84, was presented. It uses a set of four qubit states to encode one bit. The first experimental demonstration of QKD was made by Bennett, Brassard and co-workers in 1989 [9]. The first practical implementation over a kilometer of optical fibre was achieved by Gisin's group in Geneva [10]. The idea has since been elaborated by a number of authors including Ekert [11] who in 1991 showed that EPR entangled states of pairs of photons could also be used for QKD. However here we will describe the minimal QKD scheme of Bennettt, B92 [12], as this scheme is simpler than BB84 (it uses only two non orthogonal states) and has been successfully implemented in optical fibres over long distances and in free space communication. In practice however a two state QKD scheme is not desirable as it is possible to distinguish two non orthogonal states provided we accept inconclusive outcomes in some trials.

The key idea behind QKD is the Heisenberg uncertainty principle which ensures that any attempt to measure a quantum state will change it, and thus eavesdropping can in principle be detected. This is related to a powerful theorem in quantum information theory, the "no-cloning" theorem: an unknown quantum state cannot be duplicated [13]. Thus experimental QKD offers important new insights into the nature of quantum physics. Let us now follow a classical protocol to establish a shared random key between two communicating parties, called Alice (A) and Bob (B).

Alice and Bob are assumed to have a means to generate completely random binary numbers. Alice generates a random binary number and sends it to Bob, and Bob generates a random binary number and compares it to the binary number received from Alice. If it is the same he tells Alice publicly that this is the case, but does *not* reveal what the value actually was. If it was the same, Alice and Bob keep this binary number, otherwise they discard it. Alice and Bob then repeat the procedure for another binary number and continue in this way until they share a binary string that is a subset of the total binary string that Alice sent to Bob. If for some reason Alice's binary number fails to get to Bob in a particular run, it makes no difference to the final shared binary string (although it does reduce the rate of communication for the shared binary string). The big problem with this method is that classically it is possible for an eavesdropper, Eve, to copy Alice's transmitted binary number without disturbing it. Then Eve can listen to the public channel and hear Bob telling Alice that this number was the same as his binary number. QKD avoids this problem by making it impossible for Eve to measure (or copy) an unknown quantum state without also disturbing it in general. If Alice and Bob chose carefully the quantum state encoding their binary numbers an eavesdropper can be detected by Alice and Bob.

Alice and Bob will communicate with polarised single photon pulses (see Sect. 16.4.2 for further discussion). They first need to agree on how to physically implement the encoding. Suppose Alice decides to transmit only vertically (V) and $+45^\circ$ $(+D)$ photons. She will send a V-photon when a previously generated random binary number is a 0 and a +D-photon when the random binary number is a 1.

$$A: \quad V \leftrightarrow 0; \quad +D \leftrightarrow 1. \tag{16.9}$$

Bob and Alice also agree that Bob can make a polarisation *measurements* of Alice's photon in only two directions; horizontally (H) or at -45° $(-D)$. These measurements project onto non-orthogonal polarisation states. Bob randomly decides which of his two allowed measurements he will make on any photon he receives from Alice. The choice of measurement is made by referring to a previously generated random bit according to the code,

$$B: \quad 0 \leftrightarrow -D; \quad 1 \leftrightarrow H. \tag{16.10}$$

When Bob measures the polarisation he records the result as a yes (Y) or a no (N) depending on whether the photon was indeed found to have that particular polarisation. Bob will never record a Y if his bit is different from Alice's (crossed polarisers), and he records a Y on 50% of runs in which their bits are the same. Thus Bob can only get a Y if his bit is the same as Alice's (although he may get a N in that case as well). Finally Bob sends a copy of his results to Alice, over a public channel, but he does not tell Alice what measurement he made on each bit. Now Alice and Bob retain only those bits for which Bob's result was "Y". These bits are the shared key.

If Eve, an eavesdropper, makes a QND polarisation measurement of Alice's transmitted photons in an attempt to learn what was sent, she will introduce a 25% error rate between Alice and Bob's shared key. This occurs because her measurement will project the transmitted state into the eigenstates corresponding to her measurement result and this state may be different from that sent by Alice. Alice and Bob can test for eavesdropping by agreeing to sacrifice part of their shared key to check the error rate. If the error rate is 25% or higher they will suspect an eavesdropper and discard the entire shared key. In reality errors are inevitable, and Alice and Bob will need to agree on an acceptable error threshold less than 25%.

This protocol has been demonstrated experimentally by the group of Hughes at Los Alamos National Laboratory. The requirement that we use single photon states to code the bits of information places considerable demands on the physical resources required to implement B92. A considerable effort is being expended to realise single photon pulsed sources. Given such a source we also need to be able to reliably detect single photons, with a small dark count rate, and we need to propagate single photon pulses over possibly large distances with as little loss as possible. If the loss rate is too high very few counts will be available to Alice and Bob to construct their shared key and thus the data transmission rate could be unacceptably low. Finally, if we are to use standard optical fibres to transmit the photons, polarisation encoding is difficult owing to the birefringence of optical fibres.

To overcome this last problem the Los Alamos experiment used an interferometric implementation of B92 [14]. We can use any two state system to represent a qubit and thus any two state system can in principle be used for QKD. An example is provided by a single photon Mach-Zhender (M-Z) interferometer, see Fig. 16.3. The M-Z interferometer couples two input modes to two output modes, labelled U (for upper) and L (for lower) in Fig. 16.3. The device provides two possible paths for a single photon input at say U to be transmitted to the output. If path lengths are equal we can set up the device so that a photon input at U will be counted with certainty at the L-detector. However we also have the freedom to insert phase shift devices into either path and thus change the interference conditions. In particular Alice can insert a phase shift ϕ_A into one end of the device co-located with her transmission while

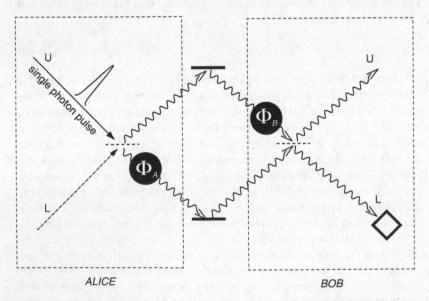

Fig. 16.3 A method to use phase shift coding for coherent pulses in a QKD protocol

Bob can insert another phase shift ϕ_B co-located with his reception. Let us assume Alice injects photons into the U input mode and that Bob counts photons only from the L detector mode. The input state is $|1\rangle_U|0\rangle_L$. The beam splitters implement the state transformations,

$$|1\rangle_U|0\rangle_L \rightarrow \frac{1}{\sqrt{2}}(|1\rangle_U|0\rangle_L + |0\rangle_U|1\rangle_L) \qquad (16.11)$$

$$|0\rangle_U|1\rangle_L \rightarrow \frac{1}{\sqrt{2}}(|0\rangle_U|1\rangle_L - |1\rangle_U|0\rangle_L) \qquad (16.12)$$

The output state is thus given by

$$|\Psi\rangle_{\text{out}} = \frac{1}{2}\left(e^{i\phi_B} - e^{i\phi_A}\right)|1\rangle_U|0\rangle_L + \frac{1}{2}\left(e^{i\phi_B} + e^{i\phi_A}\right)|0\rangle_U|1\rangle_L \qquad (16.13)$$

The probability that a photon injected by Alice is detected by Bob is then

$$P_D = \cos^2\left(\frac{\phi_A - \phi_B}{2}\right) \qquad (16.14)$$

Now if Alice and Bob use phase angles $(\phi_A, \phi_B) = (0, 3\pi/2)$ to encode 0 and $(\phi_A, \phi_B) = (\pi/2, \pi)$ to encode 1, they have an exact realisation of B92, where polariser angels are replaced by path length differences.

To realise a M-Z interferometer using optical fibres for each of the paths is difficult if the arms extend over large distances. Small fluctuations in phase shifts along each path would lead to a very unstable interferometer. In the Los Alamos experiment this problem was overcome using a single optical fibre for both arms with a unbalanced M-Z interferometer (a Franson-type interferometer) at either end, see Fig. 16.4.

In this scheme there are two paths for a single photon at each end, a "long" path and a "short" path. Thus the four possible histories of a photon can be conveniently described as; short-short (SS), long-long (LL), short-long (SL) and long-short (LS). As the last two histories are indistinguishable we expect to see interference between these two processes. When a single photon pulse passes though Alice's M-Z interferometer the output state is a superposition of two pulses delayed by a time T

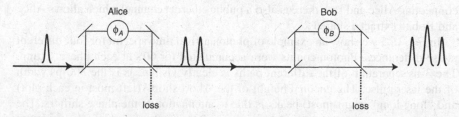

Fig. 16.4 A method to implement time multiplexed codes for QKD. From "Secure communications using quantum cryptography" [14]

equal to the delay between the short and the long path. At the output of Bob's M-Z interferometer there will be three pulses, one leading 'prompt' pulse corresponding to the history SS, one delayed pulse corresponding to LL and one central pulse corresponding to the two interfering possibilities SL and LS. Of course not every photon makes it to Bob's output port: each M-Z interferometer has two output ports and some photons exit in the "loss" ports in Fig. 16.4. As there is no interference in LL and SS the photons arrive in these time windows with a probability of 1/16 for 50/50 beam splitters. However the probability of detecting a photon in the central pulse is

$$P_D = \frac{1}{4} \cos^2 \left(\frac{\phi_A - \phi_B}{2} \right) \tag{16.15}$$

Note that this is the same as the previous single M-Z scheme apart from the additional factor of 1/4. Thus we can implement the M-Z version of B92 provided we are prepared to sacrifice detection events so that the data rate is at least reduced by a factor of 4.

In the Los Alamos experiment the two unbalanced M-Z interferometers were constructed using 50/50 fibre couplers. The long arm of the device corresponded to a standard underground optical fibre link 24 km long. The total travel time over the underground link is about 80 μs, with 10 DBE of attenuation due to the fibre's 0.3-dB/km attenuation and four connections along the path. Photons emerge from one of the output legs of Bob's interferometer into a cooled InGaAs APD detector. The photons at Alice's end are generated by a 1.3 μm pulsed semiconductor laser. A 300-ps electrical pulse is applied to the laser, with a 10-kHz repetition rate. A laser of course does *not* produce pulses with one and only one photon per pulse but rather generates a coherent state with a Poisson distribution of photons per pulse. However if we attenuate the output pulse so that on average there is only a singe photon we can get a very close realisation of the B92 protocol. The possibility that a pulse contains 2 or more photons is a potential loop hole for an eavesdropper to exploit, and thus there is some motivation to consider developing a true single photon source for this implementation. Each "single-photon" pulse is preceded by a bright reference pulse, introduced to the lower input port of Alice's interferometer, to provide arrival time information to Bob. This bright pulse triggers a room-temperature detector in the upper output port of Bob's interferometer, which provides the "start" signal for a time-interval analyser. In addition to the quantum channel (24 km of optical fibre) connecting Alice and Bob there is also a public ethernet channel which allows Alice and Bob to extract a shared key.

In Fig. 16.5 we show an example of photon arrival time spectra for four different phase differences. Photon counts were accumulated for 60 s at each phase setting. The 3-ns separation of the different paths is clearly visible, as is the 300-ps width of the laser pulse. The unequal height of the "short-short" (left-most in each plot) and "long-long" (right-most) peaks is due to attenuation at the phase shifters. The average number of photons per laser pulse arriving in the central peak maximum was $n = 0.4$. After accounting for background noise an underlying interferometric visibility of $98.4 \pm 0.6\%$ was determined for the central peak. This visibility is not as

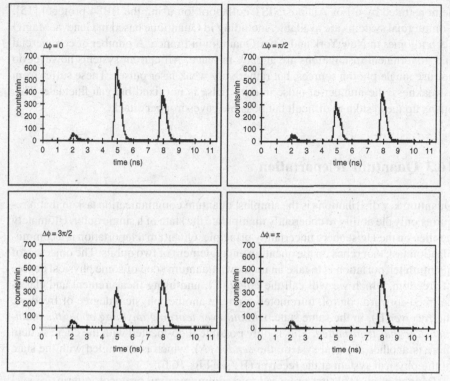

Fig. 16.5 Photon arrival time spectra for a QKD protocol discussed in the text. From "Secure communications using quantum cryptography", [14]

useful as the total probability of a count in the central time-window when the phase difference is π, because this quantity determines the error rate of the B92 protocol.

The performance of an experimental QKD system is stated in terms of the number of bits per second of a shared secret key (the distilled bit rate R_{dis}) and the distance between the communicating parties. It is usually easier to determine the raw bit rate (R_{raw}). This is determined by actual losses in the quantum channel, sources and detectors as well as possible intervention of an eavesdropper. From this the error rate in the sifted key (obtained after Alice and Bob perform a round of classical communication to reconcile their bases) is called the quantum bit error rate (QBER). The QBER for the Los Alamos experiment was 1.6%. The rate of key generation is necessarily lower than the laser pulse rate if attenuated coherent states are used instead of single photon pulses as most pulses contain no photons at all. Obviously a single photon source is desirable. We will return to this issue in Sect. 16.4.2. Furthermore there is the intrinsic inefficiency in the protocol due to a factor of 4 reduction (16.15). Along the way fibre losses and detector inefficiencies diminish the rate still further.

Quantum key distribution systems are now functioning in many laboratories around the world. A QKD system using optical fibre over 148.7 km was demonstrated by a Los Alamos/NIST collaboration using the BB84 protocol [15]. Commercial systems are available, including id Quantique based in Geneva, MagiQ Technologies in New York and SmartQuantum in France. A number of commercial and government installations are already in place. All current systems however do not use single photon sources, but rather very weak laser pulses. These suffer from a weakness: the number of pulse in each pulse is not fixed but can fluctuate. This opens up the system to difficult but possible eavesdropper attack.

16.3 Quantum Teleportation

Quantum key distributions is the simplest quantum communication task in that it requires only the ability to coherently manipulate the state of a single qubit. Ultimately it relies on the Heisenberg uncertainty principle. Quantum teleportation is a communication task that relies on the quantum entanglement of two qubits. The objective of quantum teleportation is to take an unknown quantum state of some physical degree of freedom, which we will call the *client* (C), and using measurement and classical feed-forward control, to remotely prepare another physical degree of freedom, the *receiver*(B), in the same state, *without ever learning anything about the quantum state thus transmitted*. This is only possible if co-located with the client system there is another physical system, the *sender* (A), which is entangled with the state of the physical system at the receiver (B)(see Fig. 16.6).

Bennett et al. [16] first proposed this communication protocol in terms of systems with a two dimensional Hilbert space (qubits[17]). Inspired by a proposal of Braunstein and Kimble[18], Furasawa et al. [19] demonstrated that the method can also be applied to entangled systems with an infinite dimensional Hilbert space, specifically for harmonic oscillator states. This is known as continuous variable teleportation as it requires the ability to make measurements of observables with a continuous spectrum.

The scheme of Braunstein and Kimble was itself based on a simpler, though less practical, scheme proposed by Vaidman[20] . Vaidman showed that continuous variable teleportation is possible using the EPR entangled state (see Sect. 13.1) of two degrees of freedom. This state is the result of making a perfect quadrature phase QND (quantum nondemolition, see Chap. 14) measurement between two optical modes, A and B, to create the entanglement resource. To take this example further see Exercise 16.1. The EPR state is not a physical state because quadrature phase eigenstates are infinite energy states. However we can use arbitrary close approximations to these states in terms of a squeezed vacuum state, (16.16). This is essential feature exploited in the scheme of Furasawa et al.

Suppose that at some prior time a two mode squeezed vacuum state is generated and that one mode is available for local operations and measurements at the sender's location A by observer Alice, while the other mode is open to local operations and measurements in the receiver's location B, by observer Bob. Alice and Bob can communicate via a classical communication channel. Thus Alice and Bob each have

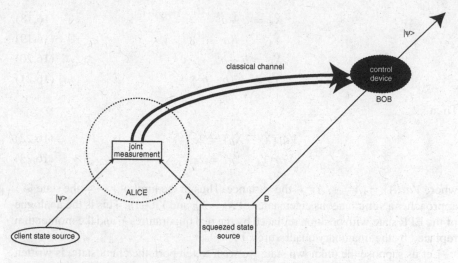

Fig. 16.6 A teleportation protocol. The sender, Alice, shares one mode A of a two-mode squeezed state, and another mode, the client, the state of which is unknown to her. Alice makes a measurement of the sum of the quadrature phase amplitudes of the client mode and mode A. The results of the measurement are sent via a classical channel to the receiver, Bob, who conditional on the information received, applies a unitary control to his share of the two-mode entangled state, mode B. The output of Bob's action is a mode now prepared in the same state as the client mode, but neither Alice or Bob learn what this state is

access to one of the two entangled subsystems described by

$$|\mathscr{E}\rangle_{AB} = \sqrt{(1-\lambda^2)} \sum_{n=0}^{\infty} \lambda^n |n\rangle_A \otimes |n\rangle_B \qquad (16.16)$$

This state is generated from the vacuum state by the Unitary transformation

$$U(r) = e^{r(a^\dagger b^\dagger - ab)} \qquad (16.17)$$

where $\lambda = \tanh r$ and where a, b refer to the mode accessible to Alice and the mode accessible to Bob respectively (see figure 16.6).

The entanglement of this state can be viewed in two ways. Firstly as an entanglement between quadrature phases in the two modes (EPR entanglement) and secondly as an entanglement between number and phase in the two modes (see exercise 2). We can easily show that this state approximates the entanglement of an EPR state in the limit $\lambda \to 1$ or $r \to \infty$. The quadrature phase entanglement is easily seen by calculating the effect of the squeezing transformation Eq(16.17) in the Heisenberg picture. We first define the quadrature phase operators for the two modes

$$\hat{X}_A = a + a^\dagger \tag{16.18}$$

$$\hat{Y}_A = -i(a - a^\dagger) \tag{16.19}$$

$$\hat{X}_B = b + b^\dagger \tag{16.20}$$

$$\hat{Y}_B = -i(b - b^\dagger) \tag{16.21}$$

Then

$$Var(\hat{X}_A - \hat{X}_B) = 2e^{-2r} \tag{16.22}$$

$$Var(\hat{Y}_A + \hat{Y}_B) = 2e^{-2r} \tag{16.23}$$

where $Var(A) = \langle A^2 \rangle - \langle A \rangle^2$ is the variance. Thus in the limit of $r \to \infty$ the state $|\mathscr{E}\rangle$ approaches a simultaneous eigenstates of $\hat{X}_A - \hat{X}_B$ and $\hat{Y}_A + \hat{Y}_B$. This is the analogue of the EPR state with position replaced by the real quadratures $\hat{X}$ and the momentum replaced by the imaginary quadratures, $\hat{Y}$.

Let us suppose the unknown state we wish to teleport, the client state, is written as $|\psi\rangle_C$. By this we mean that some party has prepared this mode in state $|\psi\rangle_C$, but this preparation procedure remains unknown to A and B. Perfect (projective) measurements are made of the joint quadrature phase quantities, $\hat{X}_C - \hat{X}_A$ and $\hat{Y}_C + \hat{Y}_A$ on the client mode and the Alice's part of the entangled mode, A, with the results X, Y respectively. The conditional state resulting from this joint quadrature measurement (see Exercise 16.2) is described by the projection onto the state $|X, Y\rangle_{CA}$ where

$$|X, Y\rangle_{CA} = e^{-\frac{i}{2}\hat{X}_A \hat{Y}_C} |X\rangle_C \otimes |Y\rangle_A \tag{16.24}$$

The (unnormalised) conditional state of total system after the measurement is then seen to be given by

$$|\tilde{\Psi}^{(X,Y)}\rangle_{out} = {}_{CA}\langle X, Y | \psi \rangle_C |\mathscr{E}\rangle_{AB} \otimes |X, Y\rangle_{CA} \tag{16.25}$$

The state of mode B at the receiver, denoted as Bob, is the pure state

$$|\phi^{(X,Y)}(r)\rangle_B = [P(X,Y)]^{-1/2} {}_{CA}\langle X, Y | \psi \rangle_C \otimes |\mathscr{E}\rangle_{AB} \tag{16.26}$$

with the wave function (in the $\hat{X}_B$ representation),

$$\phi_B^{(X,Y)}(x) = \int_{-\infty}^{\infty} dx' e^{-\frac{i}{2}x'Y} \mathscr{E}(x, x') \psi(X + x') \tag{16.27}$$

where $\psi(x) = {}_C\langle x | \psi \rangle_C$ is the wavefunction for the client state we seek to teleport. The kernel is simply the wave function for the two mode squeezed state resource.

The state in (16.27) is clearly not the same as the state we sought to teleport. However in the limit of infinite squeezing, $r \to \infty$, we find that $\mathscr{G}(x_1, x_2; r) \to \delta(x_1 + x_2)$ and the state of mode B approaches

$$|\phi_{XY}(r)\rangle_B \to e^{-\frac{i}{2}Y\hat{X}_B} e^{\frac{i}{2}X\hat{Y}_B} |\psi\rangle_B \tag{16.28}$$

which, up to the expected unitary translations in phase-space, is the required teleported state.

For finite value of the squeeze parameter, r, the state after Bob's conditional control is not an exact replica of the client state. We can quantify how the state differs by computing the probability that the state in Bob's mode, after displacement, is the same as the state of the client mode. This probability is called the *fidelity* and is given by

$$F = |\langle\psi|e^{\frac{i}{2}\mu\hat{X}_B}e^{-\frac{i}{2}\nu\hat{Y}_B}|\phi^{(X,Y)}\rangle|^2 \qquad (16.29)$$

with $\mu = gY, \nu = gX$ which allows for some flexibility in the choice of displacements in the non ideal case. The quantity g is called the *gain*. In the limit of infinite squeezing we expect $g \to 1$.

Quite apart from the limitations on the fidelity that arise from finite squeezing, other limitations arise in the real world. Noise and uncertainty can enter through imperfect measurements, though the classical communication channel, through degradation of the entanglement due to uncontrollable interactions with the environment and though imperfections in the local unitary transformations in the feed forward correction stage. These problems all limit the extent to which mean that Bob's state matches the state of the client degree of freedom. For these reasons it is necessary to validate the teleportation channel explicitly by using known client states. This will require running a number of trials with different client states and repeated measurements upon the output state at mode B. From trial to trial the state that leaves the channel at mod B will fluctuate, which means we must describe the teleported state as a mixed state, ρ_B, in general. For a fixed input client state, the probability of reproducing it at the output is given by the fidelity

$$F = \langle\psi|\rho_B|\psi\rangle \qquad (16.30)$$

If we use an ensemble of client states, an overall measure of performance in terms of the average fidelity $\bar{F}$ obtained by averaging the fidelity over the ensemble of client states, $|\psi\rangle$, with some appropriate measure on the set of pure states. If the client states are drawn from an ensemble of coherent states we can obtain an explicit result. In the extreme case that A and B share no entanglement, $\bar{F} = \frac{1}{2}$., which gives a classical boundary for teleportation of a coherent state. A demonstrable quantum teleportation channel would need to give an average fidelity greater than this.

The group of Kimble at Caltech[19] were the first to demonstrate a teleportation channel using squeezed states. Similar experiments have been reported by the group of di Martini in Rome [21] and Zeilinger in Innsbruck[22]. We will take a closer look at the Caltech experiment to explain how some of the formal steps in the preceding analysis are done in the laboratory. This will also enable us to identify the sources of imperfections, such as photon loss, and noise. A schematic diagram of the experiment is given in Fig. 16.7.

In order to effect a joint measurement of the combined quadratures $\hat{X}_C - \hat{X}_A, \hat{Y}_C + \hat{Y}_A$, the experiment first combined the client and sender field amplitudes on a 50/50 beam splitter, followed by direct homodyne measurements of the output fields after the beam splitter. After the beam splitter we then make a homodyne measurement of

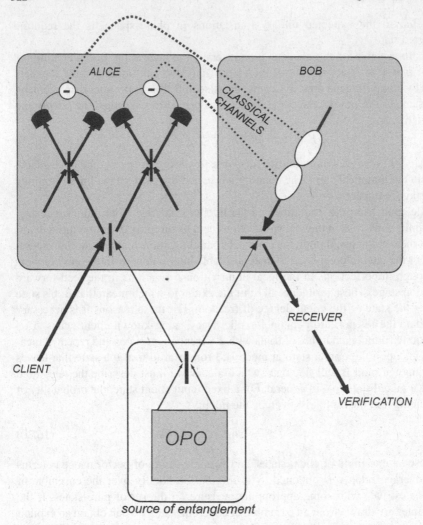

Fig. 16.7 A schematic of the Caltech teleportation experiment. An unknown quantum state is received from the client and mixed with one mode of a two mod entangled state at the sending party, Alice (A). A joint quadrature phase measurement is made by A and the results sent to a receiving party, Bob (B) though a classical channel. Given this information B then transforms the component of the shared entangled state held at B, by conditional displacements, to complete the protocol. In a checking step a state verification is undertaken by the client to determine the success of the teleportation

X-quadrature on mode C and the Y-quadrature on mode A. In the case of homodyne detection, the actual measurement records are two photo-currents (I_X, I_P). For unit efficiency detectors, this is an optimal measurement of the corresponding quadratures $\hat{X}_C, \hat{Y}_A$. In reality however efficiency is not unity and some noise is added to the measurement results. We shall return to this point below.

The measured photo-currents are a classical stochastic processes and may be sent to the receiver, B, over a standard communication channel. On receipt of this information the receiver must apply the appropriate unitary operator, a displacement, to complete the protocol. Displacement operators are quite easy to apply in quantum optics. To displace a mode, say B, we first combine it with another mode, prepared in a coherent state with large amplitude, $\alpha \to \infty$, on a beam splitter with very high reflectivity, $R \to 1$, for mode B. If $|\phi\rangle_B$ is the state of B, then after the combination at the beam splitter the state of B is transformed by

$$|\phi\rangle_B \to D(\beta)|\phi\rangle_B \tag{16.31}$$

where $D(\beta) = \exp(\beta b^\dagger - \beta^* b)$ is the unitary displacement operator, and

$$\beta = \lim_{R \to 1} \lim_{\alpha \to \infty} \alpha \sqrt{1-R} \tag{16.32}$$

In terms of the quadrature operators for B the displacement operator can be written

$$D(x,y) = e^{iy\hat{X}_B + ix\hat{Y}_B} \tag{16.33}$$

with $\beta = x + iy$. A suitable choice of β will produce the required displacements to complete the teleportation protocol. This was achieved by using the measured photocurrents to control the real and imaginary components of the displacement field using electrically controlled modulators. As the measurement records, the photocurrents, are classical stochastic processes they can be scaled by a gain factor, g, to produce the required β.

The experiment included an additional step to verify to what extent the state received by Bob faithfully reproduced the stat of the client field. In this experiment the state of the client was a coherent state. In essence another party, Victor, is verifying the fidelity of the teleportation using homodyne detection to monitor the quadrature variances of the teleported state.

The key feature that indicates success of the teleportation is a drop in the quadrature noise seen by Victor when Bob applies the appropriate unitary operator to his state. This is done by varying the gain g. If Bob simply does nothing to his state ($g = 0$), then Victor simply gets one half of a squeezed state. Such a state has a quadrature noise level well above the vacuum level of the coherent state. As Bob varies his gain, Victor finds the quadrature noise level fall until, at optimal gain, the teleportation is effected and the variance falls to the vacuum level of a coherent state. In reality of course extra sources of noise introduced in the detectors and control circuits limit the extent to which this can be achieved.

In a perfect system the fidelity should be peaked at unit gain. However photon loss in the shared entanglement resource and detector inefficiencies reduce this. In the experiment, the average fidelity at unit gain was found to be $F = 0.58 \pm 0.002$. As discussed previously, this indicates that entanglement is an essential part of the protocol.

16.4 Quantum Computation*

In 1982 *Richard Feynman* [23] suggested there were certain problems that would be difficult to perform on a computer running according to classical mechanics but which would be easy to do on a computer running according to quantum principles. The reason why this is so is easy to see. A quantum system consisting of say N interacting spins requires a simulation using vectors of 2^N dimensions in general. This exponential growth of the basis size is what makes classical simulations of complex quantum problems so difficult. On the other hand if we built a system with N interacting spins and allowed it to evolve unitarily, no such difficulty would be encountered. It would appear that a computer executing unitary evolution on a system of two level systems could significantly outperform a classical computer set to solve the equivalent problem.

In 1985 David Deutsch [24] showed in more detail what would be required for a quantum computer and gave examples of problems that might be solved more efficiently on such a machine when compared to a classical machine. The promise of quantum computation suggested by Feynman and elaborated by Deutsch was made very apparent in the factoring algorithm of Shor in 1994 [25]. Shor gave a quantum algorithm by which a large integer could be factored into its prime components with high probability, more efficiently than any known algorithm for a classical computer. As the supposed difficulty of factoring large integers is used in modern encryption schemes, Shor's algorithm indicated that such schemes would be open to attack by anyone with a quantum computer.

Quantum computers are as constrained as classical computers in the kinds of functions they can evaluate (so called computable functions) however a quantum computer can potentially solve a problem more efficiently than a classical computer. The efficiency of an algorithm is related to how many computational steps are required to solve the problem as the "size" of the problem increase. The size of a problem can often be expressed by the number of bits in a single number, for example in the case of the factoring problem, the size of the problem is just the number of bits required to store the number to be factored. If the number of steps required to implement an algorithm grows exponentially with the size of the problem, the algorithm is not efficient. If however the number of steps grows only as a polynomial power of the size of the problem, the algorithm is efficient. Shor's algorithm is an efficient factoring algorithm for a quantum computer, while all known algorithms for factoring on a classical computer require an exponentially increasing number of steps as the size of the integer to be factored increases.

How does a quantum computer achieve this enormous increase in efficiency? The answer lies in the quantum superposition principle. Suppose we wish to evaluate a function f on some binary input string x to produce a binary output string, $f(x)$. We can code the input and output binary string as the product state of N qubits. The output qubits however are preset to zero. Now we set up a machine so that under

---

* This section first appeared in "Springer Handbook of Lasers and Optics" ed. Träger, (Springer, New York, 2007)

unitary quantum evolution the state transforms as

$$|x\rangle|0\rangle \rightarrow |x\rangle|f(x)\rangle \tag{16.34}$$

Why do we demand that the transformation be unitary? Consider what happens when we prepare the input qubits in a uniform superposition of all possible input states;

$$\sum_x |x\rangle|0\rangle \rightarrow \sum_x |x\rangle|f(x)\rangle \tag{16.35}$$

If the dynamics is unitary the linearity of quantum mechanics ensures that (16.34) implies (16.35). It would appear that in a single run of the machine we have evaluated all possible values of the function.

This is not quite as interesting as it seems. If we measure the output qubits we will get one value at random. That does not seem very useful. To see why it is useful to do this let us ask; when would we ever want to evaluate every value of a particular function? The answer, is when we are not so much interested in a particular value of the function as a *property* of the function. The power of quantum computation arises in what we do next, after the transformation in (16.35). In the next step we continue to unitarily process the output register to extract, in one go, a property of the function, while simultaneously giving up information on the output of any particular evaluation. In all of this we emphasise the need to perform perfect unitary transformations of the qubits. Moreover the unitary transformations necessarily entangle many qubit degrees of freedom. A quantum computer must produce highly entangled states of many qubits without suffering any decoherence. It is this requirement that makes a physical realisation of a quantum computer so difficult to achieve as we shall see below.

How can we use the superposition state in (16.35) to determine properties of functions? To see this consider a function f which maps the binary numbers $\{0, 1\}$ to $\{0, 1\}$. There must be four such functions, two of which are constant functions with $f(0) = f(1)$, and two have $f(0) \neq f(1)$, so called balanced functions. Suppose now the problem involves determining if a function is balanced or constant. On a classical computer to answer this we need to make two evaluations of the function, $f(0), f(1)$. We would then need to run the computer twice. However a quantum computer can determine this property in only a single run.

Suppose we have two qubits. One qubit will be used to encode the input data and the other qubit, the output qubit, will contain the value of the function after the machine is run. The output qubit is initially set to 0. The machine might then run according to (16.34). However there is a problem with this expression. If f is a constant function we have two distinct input states unitarily transformed to the same output state. Clearly this is not a reversible transformation and thus cannot be implemented unitarily. The problem is easily fixed however by setting up the machine to evolve the states according to

$$|x\rangle|y\rangle \rightarrow |x\rangle|y \oplus f(x)\rangle \tag{16.36}$$

where the addition is defined modulo two and we have allowed all possible settings of both qubits. The unitary transformation which realises this operation has been called the f-controlled NOT gate [26]. The input qubit x is the control qubit while the output qubit y is the target. If the value of f on the control qubit is one, the bit on the target is flipped; thus the name. Every unitary transformation on qubits can be realised as suitable networks of simple one and two qubit gates using primitive gate operations.

The quantum algorithm that solves this problem is a version of a quantum algorithm first proposed by Deutsch. It proceeds as follows. In the first step we prepare the output qubit in the state $|0\rangle - |1\rangle$ (we ignore normalisation in what follows for simplicity). This can be done using a single qubit rotation $|1\rangle \rightarrow |0\rangle - |1\rangle$. Such a rotation is called a Hadamard transformation. In the second step the input qubit is prepared in the 0 state and is then subjected to a Hadamard gate as well, which immediately produces a superposition of the two possible inputs for the function f. In the third step we couple the input and output qubit via the f-controlled NOT gate. The transformation is

$$(|0\rangle + |1\rangle)(|0\rangle - |1\rangle) \rightarrow ((-1)^{f(0)}|0\rangle + (-1)^{f(1)}|1\rangle)(|0\rangle - |1\rangle) \qquad (16.37)$$

In the last step we apply a Hadamrd gate to the input qubit so that

$$((-1)^{f(0)}|0\rangle + (-1)^{f(1)}|1\rangle)(|0\rangle - |1\rangle) \rightarrow (-1)^{f(0)}|f(0) \oplus f(1)\rangle(|0\rangle - |1\rangle) \quad (16.38)$$

Thus the input qubit is in state 0 if f is constant and is in state 1 if f is balanced and measurement of the qubit will determine if the function is balanced or constant with certainty in a single run of the machine.

There is a simple quantum optical realisation of this algorithm based on a Mach-Zehnder interferometer, see Fig. 16.8. The interferometer couples two modes of the field, labeled upper (U) and lower (L). A single photon in the mode-U encodes logical 1 while a single photon in mode-L encodes logical 0. At the input a single photon in mode-U is transformed by the first beam splitter into a superposition state in which it is in either mode-1 or mode-0. If we encode our qubits so that a $|1\rangle$ corresponds to the photon in mode-1 and a $|0\rangle$ corresponds to a photon in mode-0, the first beam splitter performs a Hadamard transformation. Now we insert into

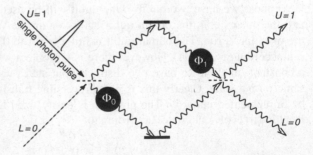

Fig. 16.8 An optical realisation of the Deutsch algorithm in terms of a Mach-Zehnder interferometer. The phase shifts are chosen according to the values of a binary function f as $\phi_0 = f(0)\pi$, $\phi_1 = f(1)\pi$

each arm a phase shift ϕ_i which can only be set at 0 or π phase shift. We encode the value of the functions as $\phi_0 = f(0)\pi$, $\phi_1 = f(1)\pi$. Set the interferometer so that in the absence of the phase shift the photon emerges with certainty at the upper detector, which encodes a 1. The lower detector encodes a zero. It is then clear that if $f(0) = f(1)$ a single photon will emerge at the upper detector, while if $f(0) \neq f(1)$ the photon will be detected at the lower detector, that is the result is a 0.

The previous example illustrates the key features of a quantum algorithm. Firstly it involves unitary transformations of pure quantum states. Secondly we need both single qubit and two qubit interactions to produce entangled states. These were the Hadamard transformation (H-gate) and a controlled NOT transformation (CNOT-gate). It turns out that suitable networks of an arbitrary single qubit rotations, together with a controlled NOT gate, can perform any computation involving arbitrarily many qubits. These features guide us in the search for a suitable physical implementation of a quantum computer. The requirement of unitary is most severe. In general small imperfections in an actual machine will not enable perfect unitary evolution. The pure states are necessarily degraded by unwanted interactions with extraneous degrees of freedom, the environment. The necessity for at least two qubit interactions means we must necessarily seek interactions that entangle at least two quantum systems. Fortunately even in the presence of nonunitary transformations we can use quantum error correction methods to mitigate the deleterious effects of environment induced errors.

16.4.1 Linear Optical Quantum Gates

In the interferometric implementation of Deutsch algorithm we used a simple physical qubit based on a single photon excitation of one of a pair of spatial modes. This is know as a "dual rail" logic. The relationship between logical states and the physical photon number state is

$$|0\rangle_L = |1\rangle_1 \otimes |0\rangle_2 \qquad (16.39)$$
$$|1\rangle_L = |0\rangle_1 \otimes |1\rangle_2 . \qquad (16.40)$$

The modes could be two input modes to a beam splitter distinguished by the different directions of the wave vector, or they could be distinguished by polarisation. In the case of a beam splitter a single qubit gate is easily implemented by the linear transformation

$$a_i(\theta) = U(\theta)^\dagger a_i U(\theta) \qquad (16.41)$$

with $U(\theta) = \exp\left[\theta(a_1 a_2^\dagger - a_1^\dagger a_2)\right]$. Thus

$$a_1(\theta) = \cos\theta a_1 - \sin\theta a_2 \qquad (16.42)$$
$$a_2(\theta) = \cos\theta a_2 + \sin\theta a_1 \qquad (16.43)$$

The description in the logical basis becomes,

$$|0\rangle_L \rightarrow \cos\theta_1 |0\rangle_L - \sin\theta_1 |1\rangle_L \qquad (16.44)$$

$$|1\rangle_L \rightarrow \cos\theta_1 |1\rangle_L + \sin\theta_1 |0\rangle_L \qquad (16.45)$$

While single qubit gates are readily implemented by linear optical devices such as beam splitters, quarter wave plates, phase shifters etc., two qubit gates are difficult. In order to implement the controlled phase gate (CSIGN) defined by

$$|x\rangle_L |y\rangle_L \rightarrow U_{CP} |x\rangle_L |y\rangle_L = (-1)^{x.y} |x\rangle_L |y\rangle_L \qquad (16.46)$$

In a dual rail, single photon code, this can be implemented using a two mode Kerr nonlinearity. A simple nonlinear optical model of a Kerr nonlinearity was discussed in Chap. 5 in relation to optical bistability. The two mode generalisation is described by the Hamiltonian

$$H = \hbar\chi a_1^\dagger a_1 a_2^\dagger a_2 \qquad (16.47)$$

At the level of single photons this Hamiltonian produces the transformation, $|x\rangle|y\rangle \rightarrow e^{-ixy\chi t}|x\rangle|y\rangle$ and it is a simple matter to implement the CSIGN gate in the logical basis for the dual rail single photon code.

There are at least two problems in pursuing this approach; (a) the difficulty of realising number states in the laboratory, (b) the difficulty of producing one photon phase shifts of the order of π. We will say more about the fist of these problems below. The second difficulty is very considerable. Third order optical nonlinearities are very small for a field with such a low intensity as a single photon. However experimental advances may eventually overcome this.

A quite different approach to achieve large single photon conditional phase shifts is based on the non-unitary transformation of a state that results when a measurement is made. Consider the situation shown in Fig. 16.9. Two modes of an optical field are coupled via a beam splitter. One mode is assumed to be in the vacuum state (a) or a one photon state (b), while the other mode is arbitrary. A single photon counter is placed in the output port of mode-2. What is the conditional state of mode-1 given a count of n photons?

Consider two modes, a_1, a_2, coupled with a beam splitter interaction, described by the one parameter unitary transformation, given in (16.42 and 16.43) We now assume that photons are counted on mode a_2 and calculate the conditional state

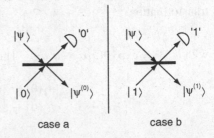

Fig. 16.9 A conditional state transformation conditioned on photon counting measurements

case a   case b

for mode a_1 for two cases: no count and also for a single count at mode a_2. The conditional state of mode a_1 is given by (unnormalised),

$$|\tilde{\psi}^{(i)}\rangle_1 = \hat{\Upsilon}(i)|\psi\rangle_1 \tag{16.48}$$

where

$$\hat{\Upsilon}(i) = {}_2\langle i|U(\theta)|i\rangle_2 \tag{16.49}$$

with $i = 1, 0$. The probability to observe each event is given by

$$P(i) = \langle\psi|\hat{\Upsilon}^{\dagger}(i)\hat{\Upsilon}(i)|\psi\rangle_1 \tag{16.50}$$

which fixes the normalisation of the state,

$$|\psi^{(i)}\rangle_1 = \frac{1}{\sqrt{P(i)}}|\tilde{\psi}^{(i)}\rangle_1 \tag{16.51}$$

In Exercise 16.5 we find that

$$\hat{\Upsilon}(0) = \sum_{n=0}^{\infty} \frac{(\cos\theta - 1)^n}{n!}(a_1^{\dagger})^n a_1^n$$

$$\hat{\Upsilon}(1) = \cos\hat{\Upsilon}^{(0)} - \sin^2\theta a_1^{\dagger}\hat{\Upsilon}^{(0)}a_1$$

This can be written more succinctly using normal ordering,

$$\hat{\Upsilon}(0) =: e^{\ln(\cos\theta)} : \tag{16.52}$$

In order to see how we can use these kind of transformations to effect a CSIGN gate, consider the situation shown in Fig. 16.10. Three optical modes are mixed on a sequence of three beam splitters with beam splitter parameters θ_i. The *ancilla* modes, a_1, a_2 are restricted to be in the single photon states $|1\rangle_2, |0\rangle_3$ respectively. We will assume that the *signal* mode, a_0, is restricted to have *at most* two photons, thus

$$|\psi\rangle = \alpha|0\rangle_0 + \beta|1\rangle_0 + \gamma|2\rangle_0 \tag{16.53}$$

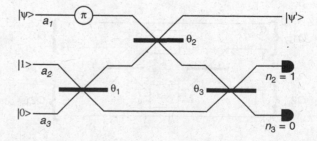

Fig. 16.10 A conditional state transformation on three optical modes, conditioned on photon counting measurements on the ancilla modes a_2, a_3. The signal mode, a_1 is subjected to a π phase shift

This captures the fact that in the dual rail encoding a general two qubit state can have at most two photons. The objective is to chose the beam splitter parameters so that when the two detectors at the output of modes $2, 3$ detect $1, 0$ photons respectively (that is detect no change in their occupation), the signal state is transformed as

$$|\psi\rangle \rightarrow |\psi'\rangle = \alpha|0\rangle + \beta|1\rangle - \gamma|2\rangle \tag{16.54}$$

with a probability that is *independent* of the input state $\psi\rangle$. This last condition is essential as in a quantum computation, the input state to a general two qubit gate is completely unknown. We will call this transformation the NS (for nonlinear sign shift) gate. In Exercise 16.7 we find that this can be achieved using: $\theta_1 = -\theta_3 = 22.5°$ and $\theta_2 = 65.53°$. The probability of the conditioning event ($n_2 = 1, n_3 = 0$) is $1/4$. Note that we can't be sure in a given trial if the correct transformation will be implemented. Such a gate is called a *nondeterministic* gate. However the key point is that success is heralded by the results on the photon counters (assuming ideal operation).

We can now proceed to a CSIGN gate in the dual rail basis. Consider the situation depicted in Fig. 16.11. We first take two dual rail qubits encoding for $|1\rangle_L|1\rangle_L$. The single photon components of each qubit are directed towards a 50/50 beam splitter where they overlap perfectly in space and time. This is precisely the case of the Hong-Ou-Mandel interference discussed in Exercise 3.4(c), and produces a state of the form $|0\rangle_2|2\rangle_3 + |2\rangle_2|0\rangle_3$. We then insert an NS gate into each output arm of the HOM interference. When the conditional gates in each arm work, which occurs with probability $1/16$, the state is multiplied by an overall minus sign. Finally we direct these modes towards another HOM interference. The output state is thus seen to be $-|1\rangle_L|1\rangle_L$. One easily checks the three other cases for the input logical states to see that this device implements the CSIGN gate with a probability of $1/16$ and successful operation is heralded.

Clearly a sequence of nondeterministic gates is not going to be much use: the probability of success after a few steps will be exponentially small. The key idea in using nondeterministic gates for quantum computation is based on the idea of gate teleportation of Gottesmann and Chuang [27]. We saw in Sect. 16.3 that in quantum teleportation an unknown quantum state can be transferred from A to B provided A and B first share an entangled state. Gottesmann and Chuang realised that it is

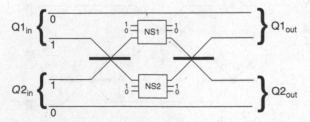

Fig. 16.11 A conditional state transformation conditioned on photon counting measurements. A CSIGN gate that works with probability of 1/16. It uses HOM interference and two NS gates

possible to simultaneously teleport a two qubit quantum state and implement a two qubit gate in the process by first applying the gate to the entangled state that A and B share prior to teleportation.

We use a non deterministic NS gate to prepare the required entangled state, and only complete the teleportation when the this stage is known to work. The teleportation step itself is non deterministic but, as we see below, by using the appropriate entangled resource the teleportation step can be made near deterministic. The near deterministic teleportation protocol requires only photon counting and fast feed-forward. We do not need to make measurements in a Bell basis.

A nondeterministic teleportation measurement is shown in Fig. 16.12. The client state is a one photon state in mode-0 $\alpha|0\rangle_0 + \beta|1\rangle_0$ and we prepare the entangled ancilla state

$$|t_1\rangle_{12} = |01\rangle_{12} + |10\rangle_{12} \tag{16.55}$$

where mode-1 is held by the sender, A, and mode-2 is held by the receiver, B. For simplicity we omit normalisation constants wherever possible. This an ancilla state is easily generated from $|01\rangle_{12}$ by means of a beam splitter.

If the total count is $n_0 + n_1 = 0$ or $n_0 + n_1 = 2$, an affective measurement has been made on the client state and the teleportation has failed. However if $n_0 + n_1 = 1$, which occurs with probability 0.5, teleportation succeeds with the two possible conditional states being

$$\alpha|0\rangle_2 + \beta|1\rangle_2 \text{ if } n_0 = 1, n_1 = 0 \tag{16.56}$$

$$\alpha|0\rangle_2 - \beta|1\rangle_2 \text{ if } n_0 = 0, n_1 = 1 \tag{16.57}$$

This procedure implements a partial Bell measurement and we will refer to it as a nondeterministic teleportation protocol, $\mathbf{T}_{1/2}$. Note that teleportation failure is detected and corresponds to a photon number measurement of the state of the client qubit. Detected number measurements are a very special kind of error and can be easily corrected by a suitable error correction protocol. For further details see [28]

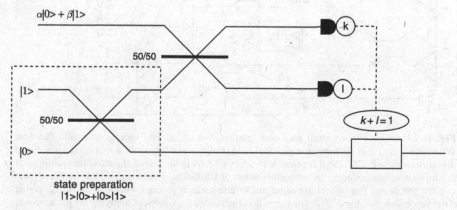

Fig. 16.12 A partial teleportation system for single photons states using a linear optics

The next step is to use $\mathbf{T}_{1/2}$ to effect a conditional sign flip $\text{csign}_{1/4}$ which succeeds with probability $1/4$. Note that to implement csign on two bosonic qubits in modes $1,2$ and $3,4$ respectively, we can first teleport the first modes of each qubit to two new modes (labelled 6 and 8) and then apply csign to the new modes. When using $\mathbf{T}_{1/2}$, we may need to apply a sign correction. Since this commutes with csign, there is nothing preventing us from applying csign to the prepared state before performing the measurements. The implementation is shown in Fig. 16.13 and now consists of first trying to prepare two copies of $|t_1\rangle$ with csign already applied, and then performing two partial Bell measurements. Given the prepared state, the probability of success is $(1/2)^2$. The state can be prepared using $\text{csign}_{1/16}$, which means that the preparation has to be retried an average of 16 times before it is possible to proceed.

To improve the probability of successful teleportation to $1 - 1/(n+1)$, we generalise the initial entanglement by defining

$$|t_n\rangle_{1\ldots 2n} = \sum_{j=0}^{n} |1\rangle^j |0\rangle^{n-j} |0\rangle^j |1\rangle^{n-j} . \tag{16.58}$$

The notation $|a\rangle^j$ means $|a\rangle|a\rangle\ldots$, j times. The modes are labelled from 1 to $2n$, left to right. Note that the state exists in the space of n bosonic qubits, where the kth qubit is encoded in modes $n+k$ and k (in this order).

To teleport the state $\alpha|0\rangle_0 + \alpha|1\rangle_0$ using $|t_n\rangle_{1\ldots 2n}$ we first couple the client mode to half of the ancilla modes by applying an $n+1$ point Fourier transform on modes 0 to n. This is defined by the mode transformation

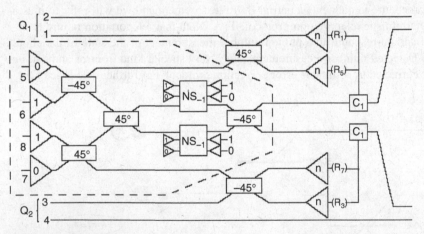

Fig. 16.13 A CSIGN two qubit gate with teleportation to increase success probability to $1/4$. When using the basic teleportation protocol (T1), we may need to apply a sign correction. Since this commutes with CSIGN, it is possible to apply CSIGN to the prepared state before performing the measurements, reducing the implementation of CSIGN to a state-preparation (outlined) and two teleportations. The two teleportation measurements each succeed with probability $1/2$, giving a net success probability of $1/4$. The correction operations C1 consist of applying the phase shifter when required by the measurement outcomes

$$a_k \to \frac{1}{\sqrt{n+1}} \sum_{l=0}^{n} \omega^{kl} a_l \qquad (16.59)$$

where $\omega = e^{i2\pi/(n+1)}$ This transformation does not change the total photon number and is implementable with passive linear optics. After applying the Fourier transform, we measure the number of photons in each of the modes 0 to n. If the measurement detects k bosons altogether, it is possible to show [28] that if $0 < k < n+1$, then the teleported state appears in mode $n + k$ and only needs to be corrected by applying a phase shift. The modes $2n - l$ are in state 1 for $0 \le l < (n-k)$ and can be reused in future preparations requiring single bosons. The modes are in state 0 for $n - k < l < n$. If $k = 0$ or $k = n + 1$ an effective measurement of the client is made, and the teleportation fails. The probability of these two events is $1/(n+1)$, regardless of the input. Note that again failure is detected and corresponds to measurements in the basis $|0\rangle, |1\rangle$ with the outcome known. Note that both the necessary correction and the receiving mode are unknown until after the measurement.

• The linear optics quantum computing (LOQC) model described above can be drastically simplified by adopting the cluster state method of quantum computation [29]. The cluster state model was developed by Raussendorf and Breigel [30] and is quite different from the circuit models that we have been using. In cluster state QC, an array of qubits is initially prepared in a special entangled state. The computation then proceeds by making a sequence of single qubit measurements. Each measurement is made in a basis that depends on prior measurement outcomes. Nielsen realised that the LOQC mode of [28] could be used to efficiently assemble the cluster using the nondeterministic teleportation t_n. As we saw the failure mode of this gate constituted an accidental measurement of the qubit in the computational basis. The key point is that such an error does not destroy the entire assembled cluster but merely detaches one qubit from the cluster. This enables a protocol to be devised that produces a cluster that grows on average. The LOQC cluster state method dramatically reduces the number of optical elements required to implement the original LOQC scheme. Of course if large single photon Kerr nonlinearities were available, the optical cluster state method could be made deterministic [31].

A number of LOQC protocols have been implemented in the laboratory. The first experiment was performed by Pittmann and Franson [32]. This used entangled ancillas that are readily produced as photon pairs in a spontaneous parametric down conversion process. A simplified version of the LOQC model was implemented by O'Brien et al. [33], based on a proposal of Ralph et al. [34] for a CNOT gate shown in Fig. 16.14. The simplification results by firstly setting the beam splitter parameters θ_1, θ_3 to zero in the NS gate implementation and secondly only detecting photon coincidences at the output. This gate performs all the operations of a CNOT gate but requires only a two photon input. Detecting only coincidences means that the device must be configured so that correct operation leads to a coincidence detection of both photons at the output. The gate is non deterministic but gate failures are simply not detected at all. In essence, the control (C) and target (T) qubits act as their own ancilla. When the control is in the logical one state, the control and target photons interfere non-classically at the central $1/3$ beam splitter which causes

Fig. 16.14 A simplified CNOT gate that gives correct operation only when both input photons are detected coincidentally at the output

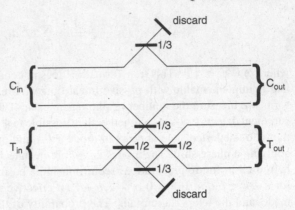

a π phase-shift in the upper arm of the central interferometer and the target state qubit is flipped. The qubit value of the control is unchanged. Successful operation is heralded by coincidence detection of both photons and success will occur with probability $1/9$.

In the UQ experiment the two modes of each qubit are distinguished by orthogonal polarisations. This may be converted to spatial mode encoding by using polarising beam splitters and a half wave plate, as shown in Fig. 16.15. The key advantage in using a gate based on two photon coincidence detection is that spontaneous parametric down conversion (SPDC) may be used in place of true single photon sources. An SPDC produces a photon pair in two distinct spatio-temporal modes at random times. There is a small probability of producing more than two photons, but this can be neglected.

The truth table for a CNOT operation was experimentally determined by preparing each of the four possible input states to the gate, $CT\rangle = |00\rangle, |11\rangle, |10\rangle, |01\rangle$. A comparison of the experimental results and the ideal CNOT gate are shown in Fig. 16.14. A single classical interference event is required when the control is in

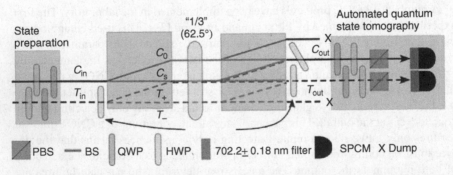

Fig. 16.15 The optical design schematic of the polarisation encoding implementation of the experiment of O'Brien et al., PBS: polarising beam splitter, QWP: quarter wave plate, HWP: half wave plate, SPCM: single photon counting module and X: beam dump. Reproduced, with permission, from Nature, **426**, 264 (2003)

state $|0\rangle_L$ and experimentally the correct output is obtained in roughly 95% of cases. In the case of the control in state $|1\rangle_L$, correct operation is obtained in only 75% of cases as this situation requires a non classical interference event for which very careful mode matching is required.

While this provides good evidence the gate is working the key test of a quantum CNOT gate is that it produce maximally entangled states when the control is in a superposition of the two logical states. For example if the control is in state $|0\rangle_L - |1\rangle_L$, while the target is in $1\rangle_L$, the two-qubit output state is $|\psi^-\rangle = |01\rangle_L - |10\rangle_L$ (where $|xy\rangle_L = |x\rangle_L \otimes |y\rangle_L$, with the first factor the control qubit and the second factor the target). In testing the truth table, the output logical states were measured in the logical basis. Such a "computational basis" measurement of course will not reveal the classical correlation imposed by the truth table but not quantum coherence. To see the quantum coherence implicit in the entangled state we need to measure in a basis other than the computational basis. In the experiment this was done by measuring the coincidence count rate while using a half wave plate set to pass control photons in either of the states $|0\rangle_L$ or state $|0\rangle_L + |1\rangle_L$. The experimental results are show in Fig. 16.16. The visibilities in the two curves are greater than 90% which is the signature of entanglement in the output state.

An even better diagnostic of the gate operation is provided by state tomography, a reconstruction of the full density matrix of the output state [35], when the output is entangled. State tomography requires sampling the statistics for the measurement outcomes of 16 different two qubit projections. Given these statistics data inversion can be devised to reconstruct the density matrix for the output state. Given the density matrix, we can then compute its overlap, or fidelity, with respect to the pure ideal entangled state $|\psi^-\rangle$ that the ideal gate would produce. In the case of $|\psi^-\rangle$ the fidelity obtained in the experiment was 0.87 ± 0.08. This is sufficiently high that such a state were it not destroyed in the detection process, would violate a Bell inequality test.

More recent experiments have improved on these early experiments. An NS gate close to the original proposal, was implemented in the Zeilinger group, using a polarisation encoding and the four photon state emitted by spontaneous parametric down conversion [36]. As in the UQ experiment, a coincidence detection configuration was used to signal correct operation of the gate. The experimentally observed

Fig. 16.16 Conditional coincidence rtes for non-orthogonal measurement bases. The control analyser was set to pass $|0\rangle_L + |1\rangle_L$ (*circles*) and $|0\rangle_L$ (*triangles*) when the input to the control is ($|0\rangle_L - |1\rangle_L$) and the input to the target is $|1\rangle_L$. Reproduced, with permission, from Nature, **426**, 264 (1003)

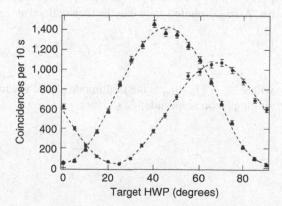

conditional phase shift was $1.05 \pm 0.06\pi$. Future progress on linear optical quantum computing schemes will most likely be based on cluster state implementations. A four photon cluster state implementation was recently implemented by the Zeilinger group [37].

16.4.2 Single Photon Sources

As we have seen both QKD and LOQC motivate the development of single photon sources. Single photon sources enable QKD to escape beam splitter attacks that are possible with weak coherent pulses. In order to progress to scaleable architectures, LOQC will certainly require the development of good single photons sources and highly efficient single photon detectors that can in fact discriminate between $0, 1$ and 2 photons. Fortunately steady progress is being made on both technologies. The requirements on single photon sources are much more demanding than those for single photon sources in QKD and we now briefly discuss some of these.

What we need is an optical pulse source in which each pulse contains one and only one photon. Clearly such a source is going to produce photon antibunching and the $g^{(2)}(\tau)$ (see Sect. 3.6) is clearly a key diagnostic for such a source. However a more stringent requirement is the ability of such a source to produce a strong Hong-Ou-Mandle interference dip. (See Exercise 3.4(c)).

In order to define single photon states, let us begin by defining the positive frequency field component as,

$$a(t) = \sum_{n=1}^{\infty} a_n e^{-int} \tag{16.60}$$

The allowed wave vectors for plane wave modes in a box of length L, form a denumerable set given by $k_n = \frac{n\pi}{L}$ with corresponding frequencies $\omega_n = ck_n$. If we measure time in units of $\pi L/c$, the allowed frequencies may simply be denoted by an integer $\omega_n = n = 1, 2, \ldots$. The Bose annihilation and creation operators obey the usual commutation relations. Following the standard theory of photo-detection (see Sect. 3.10) the probability per unit time for detecting a single photon is given by $p_1(t) = \gamma n(t)$ where $n(t) = \langle a^\dagger(t)a(t)\rangle$ and the parameter γ characterises the detector. A single-photon state may be then defined as

$$|1; f\rangle = \sum_{m=1}^{\infty} f_m a_m^\dagger |0\rangle \tag{16.61}$$

where $|0\rangle = \prod_m |0\rangle_m$ is the multimode global vacuum state, and we require that the single photon amplitude, f_m satisfies

$$\sum_{m=0}^{\infty} |f_m|^2 = 1 \tag{16.62}$$

The counting probability is then determined by

$$n(t) = \left| \sum_{k=1}^{\infty} f_k e^{-ikt} \right|^2 \tag{16.63}$$

This function is clearly periodic with a period 2π. As the spectrum is bounded from below by $n = 1$, it is not possible to choose the amplitudes f_n so that the functions $n(t)$ have arbitrarily narrow support on $t \in [0, 2\pi)$.

While a field for which exactly one photon is counted in one counting interval, and zero in all others, is no doubt possible, it does not correspond to a more physical situation in which a source is *periodically* producing pulses with exactly one photons per pulse. To define such a field state we now introduce time-bin operators. For simplicity we assume that only field modes $n \leq N$ are excited and all others are in the vacuum state. It would be more physical to assume only field modes are excited in some band, $\Omega - B \leq n \leq \Omega + B$. Here Ω is the carrier frequency and $2B$ is the bandwidth. However this adds very little to the discussion.
Define the operators

$$\tilde{a}_v = \frac{1}{\sqrt{N}} \sum_{m=1}^{N} e^{-i\tau m v} a_m \tag{16.64}$$

where $\tau = \frac{2\pi}{N}$ This can be inverted to give

$$a_m = \frac{1}{\sqrt{N}} \sum_{v=1}^{N} e^{i\tau m v} \tilde{a}_v \tag{16.65}$$

The temporal evolution of the positive frequency components of the field modes then follows from (16.60)

$$a(t) = \sum_{\mu=1}^{N} g_\mu(t) \tilde{a}_\mu \tag{16.66}$$

where

$$g_\mu(t) = \frac{1}{\sqrt{N}} \left[1 - e^{i(v\tau - t)} \right]^{-1} \tag{16.67}$$

The time-bin expansion functions, $g_\mu(t)$ are a function of $v\tau - t$ alone and are thus simple translations of the functions at $t = 0$. The form of (16.66) is a special case of a more sophisticated way to define time-bin modes. If we were to regard $a(t)$ as a classical signal then the decomposition in (16.66) could be generalised as a wavelet transform where the integer μ labels the translation index for the wavelet functions. In that case the functions $g_\mu(t)$ could be made rather less singular. In an experimental context however the form of the functions $g_\mu(t)$ depends upon the details of the generation process.

The linear relationship between the plane wave modes a_m and the time bin modes $\tilde{a}_v$ is realised by a unitary transformation that does not change particle number, so the vacuum state for the time-bin modes is the same as the vacuum state for the global plane wave modes. We can then define a one-photon time-bin state as

$$|\tilde{1}\rangle_v = \tilde{a}_v^\dagger |0\rangle \tag{16.68}$$

The mean photon number for this state is,

$$n(t) = |g_v(t)|^2 \tag{16.69}$$

This function is periodic on $t \in [0, 2\pi)$ and corresponds to a pulse localised in time at $t = v\tau$. Thus the integer v labels the temporal coordinate of the single photon pulse.

We are now in a position to define an N-photon state with one photon per pulse. In addition to the mean photon number, $n(t)$ we can now compute two-time correlation functions such as the second order correlation function, $G^{(2)}(\tau)$ defined by

$$G^{(2)}(T) = \langle a^\dagger(t) a^\dagger(t+T) a(t+T) a(t) \rangle \tag{16.70}$$

The simplest example for $N = 2$ is

$$|1_\mu, 1_v\rangle = \tilde{a}_\mu^\dagger \tilde{a}_v^\dagger |0\rangle \quad \mu \neq v \tag{16.71}$$

The corresponding mean photon number is

$$n(t) = |g_\mu(t)|^2 + |g_v(t)|^2 \tag{16.72}$$

as would be expected. The two-time correlation function is,

$$G^{(2)}(\tau) = |g_\mu(t)g_v(t+T) + g_v(t)g_\mu(t+T)|^2 \tag{16.73}$$

Clearly this has a zero at $T = 0$ reflecting the fact that the probability to detect a single photon immediately after a single photon detection is zero, as the two pulses are separated in time by $|\mu - v|$. This is known as *anti-bunching* and is the first essential diagnostic for a sequence of single photon pulses with one and only one photon per pulse. When $T = |\mu - v|\tau$ however there is a peak in the two-time correlation function as expected.

An example of such a single photon source producing a second order two-time correlation function of this kind was implemented by the group of Yamamoto [38]. The source was based on spontaneous emission from exciton recombination from a single InAs quantum dot in a micropillar cavity using distributed Bragg reflecting mirrors. The devices operate at low temperature $(3 - 7 \text{ K})$ and are pumped by a pulsed TiSi laser with 3 ps pulses every 13 ns. Three quantum dots were reported producing light with wavelengths $931, 932$ and 937 nm. In Fig. 16.17 we show the experimental results for the second order two time correlation function using a Hanbury-Brown and Twiss configuration.

We now consider an interferometer with single photon input states. The most relevant example for LOQC protocols is the Hong-Ou-Mandel (HOM) interferometer. This example has been considered by Rohde and Ralph [39]. In this case two fields, distinguished by momentum or polarisation are coupled by a linear optical device (referred to for simplicity as a beam-splitter). After the interaction, each field is

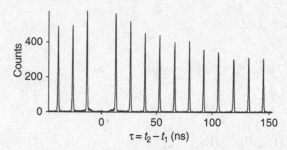

Fig. 16.17 The photon correlation histogram for emission form a single exciton quantum dot. The results were obtained using a Hanbury-Brown and Twiss experiment. The emission from the quantum dot was split into two paths via a beam splitter and each path directed towards a photon detector. The number of events in which a photon was detected on one detector at time t_1 and on the other detector at time $t_2 = t_1 + \tau$. The suppression of the peak at zero delay is characteristic of a single photon pulse source. (From [38]) reproduced with permission from Nature

directed onto a photon counter, and the probability for a coincidence count is determined. We label the two sets of modes by the latin symbols a, b so for example the positive frequency parts of each field are simply $a(t), b(t)$. The coupling between the modes is described by a scattering matrix connecting the input and output plane waves

$$a_n^{\text{out}} = \sqrt{\eta} a_n + \sqrt{1-\eta} b_n \qquad (16.74)$$

$$b_n^{\text{out}} = \sqrt{\eta} b_n - \sqrt{1-\eta} a_n \qquad (16.75)$$

where $0 \leq \eta \leq 1$. This is realised by a unitary transformation, U, for example, $a_n^{\text{out}} = U^\dagger a_n U$. The total photon number at the input is $N(t) = \langle a^\dagger(t)a(t)\rangle + \langle b^\dagger(t)b(t)\rangle$. It is easy to see that this is unchanged by the beam-splitter transformation. The probability, per unit time, for there to be a coincident detection of a single photon at each output beam is easily seen to be proportional to

$$C = \overline{\langle a^\dagger(t)b^\dagger(t)b(t)a(t)\rangle} \qquad (16.76)$$

The overline represents a time average over a detector response time that is long compared to the period of the field carrier frequencies. In this example we only need consider the case of one photon in each of the two distinguished modes, so we take the input state to be

$$|1\rangle_a \otimes |1\rangle_b = \sum_{m,n=1}^{\infty} \alpha_n \beta_m a_n^\dagger b_m^\dagger |0\rangle \qquad (16.77)$$

where α_n, β_n refer to the excitation probability amplitudes for modes a_n, b_n respectively. This state is transformed by the unitary transformation, U, to give $|\psi\rangle_{\text{out}} = U|1\rangle_a \otimes |1\rangle_b$. In the case of a 50/50 beam splitter, for which $\eta = 0.5$, this is given as

$$|\psi\rangle_{out} = \sum_{n,m=1}^{\infty} \alpha_n \beta_m U a_n^\dagger b_m^\dagger$$

$$= \frac{1}{2} \sum_{n,m=1}^{\infty} \alpha_n \beta_m (a_n^\dagger + b_n^\dagger)(b_m^\dagger - a_m^\dagger)|0\rangle$$

$$= \frac{1}{2} \sum_{n,m=1}^{\infty} \alpha_n \beta_m [|1\rangle_{a_n}|1\rangle_{b_m} - |1\rangle_{a_n}|1\rangle_{a_m}|0\rangle_b$$

$$+ |1\rangle_{b_n}|1\rangle_{b_m}|0\rangle_a - |1\rangle_{b_n}|1\rangle_{a_m}]$$

Note that the second and third terms in this sum have no photons in modes b and a respectively. We then have that

$$C = \frac{1}{2} - \frac{1}{2} \sum_{n,m=1}^{\infty} \alpha_n \alpha_m^* \beta_m \beta_n^* \tag{16.78}$$

If the excitation probability amplitudes at each frequency are identical, $\alpha_n = \beta_n$ this quantity is zero. In other words only if the two-single photon wave packets are identical do we see a complete cancellation of the coincidence probability. This is the second essential diagnostic for a single photon source. Of course in an experiment complete cancellation is unlikely. The extent to which the coincidence rate approaches zero is a measure of the quality of a single photon source as far as LOQC is concerned. Whether or not this is the case depends on the nature of the single photon source.

In Fig. 16.18 we show the results of a HOM interference experiment using the exciton quantum dot source of Yamamoto [38].

Currently the two schemes used to realise single photon sources are: I conditional spontaneous parametric down conversion, II cavity-QED Raman schemes. As discussed by Rohde and Ralph, type-I corresponds to a Gaussian distribution of α_n as a function of n. The second scheme, type-II, leads to a temporal pulse structure that

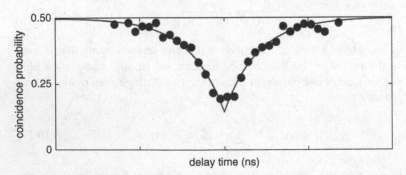

Fig. 16.18 The coincidence count probability for two photons incident on a beam splitter as a function of the time delay in arrival times of the photons at the beam splitter for the exciton quantum dot source of Santori et al. [38]. Reproduced with permission from Nature

is the convolution of the excitation pulse shape and the Lorentzian line shape of a cavity. If the cavity decay time is the longest time in the dynamics, the distribution α_n takes a Lorentzian form.

As an example of the experimental constraints on the generation of single photon states we now review an example of a cavity-QED Raman scheme implemented by Keller et al. [40]. Photon anti-bunching from resonance fluorescence was discussed in Sect. 10.3. If an atom decays spontaneously from an excited to a ground state, a single photon is emitted and a second photon cannot be emitted until the atom is re-excited. Unfortunately the photon is emitted into a dipole radiation pattern over a complete solid angle. Clearly we need to engineer the electromagnetic environment with mirrors, dielectrics, etc, to be sure a preferred mode for emission. However single photon sources based on spontaneous emission are necessarily compromised by the random nature of spontaneous emission. The decay process is a conditional Poison process. This means that after a fast excitation pulse there is a small random time delay in the emission of the photon. This leads to time jitter in the single photon pulse period. A similar situation prevails in the case of single exciton sources [38], where spontaneous recombination leads to time jitter for the same reason. Clearly what we need is a *stimulated* emission process not a spontaneous emission process. A number of schemes based on stimulated Raman emission into a cavity mode have been proposed to this end [40, 41, 42].

Consider a three-level atomic system in Fig. 16.19. The ground state is coupled to the excited state via a two-photon Raman process mediated by a well detuned third level. In this experiment a calcium ion ($^{40}Ca_{+}$) was trapped in a cavity via an rf ion trap. The cavity field is nearly resonant with the $4^2P_{1/2} \rightarrow 3^2D_{3/2}$ transition. Initially there is no photon in the cavity. An external laser is directed onto the ion and is nearly resonant with the $4^2P_{1/2} \rightarrow 4^2P_{1/2}$ transition. When this laser is on, the atom can be excited to the $3^2D_{3/2}$ level by absorbing one pump photon and *emitting* one photon into the cavity. This is a stimulated Raman process and thus time of emission of the photon into the cavity is completely controlled by the temporal structure of the pump pulse. The photon in the cavity then decays through the end mirror, again as a Poisson process this time at a rate given by the cavity decay rate. This can be made very fast.

In principle one can now calculate the probability per unit time to detect a single photon emitted from the cavity. If we assume every photon emitted is detected, this

Fig. 16.19 A possible three-level atomic system for a two-photon Raman single photon source. The pump beam is a strong classical coherent pulse. The cavity field is an intracavity field mode initially prepared in a vacuum state

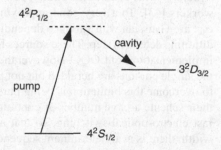

probability is simply $p_1(t) = \kappa \langle a^\dagger(t)a(t) \rangle$ where κ is the cavity decay rate and $a, a^\dagger$ are the annihilation and creation operators for the *intracavity* field and

$$\langle a^\dagger(t)a(t) \rangle = \mathrm{tr}(\rho(t)a^\dagger a) \tag{16.79}$$

where $\rho(t)$ is the total density operator for ion-plus-cavity-field system. This may be obtained by solving a master equation describing the interaction of the electronic states of the ion and the two fields, one of which is the time dependent pump. Of course for a general time-dependent pump pulse-shape this can only be done numerically. Indeed by carefully controlling the pump pulse shape considerable control over the temporal structure of the single photon detection probability may be achieved. In the experiment of *Keller* et al. [40] the length of the pump pulse was controlled to optimise the single photon output rate. The efficiency of emission was found to be about 8%, that is to say, 92% of pump pulses did not lead to a single-photon detection event. This was in accordance with the theoretical simulations. These photons are probably lost through the sides of the cavity. It is important to note that this kind of loss does not effect the temporal mode structure of the emitted (and detected) photons.

In a similar way we can compute the second order correlation function via

$$G^{(2)}(T) = \kappa^2 \mathrm{tr}(a^\dagger a e^{\mathscr{L}T}(a\rho(t)a^\dagger) \tag{16.80}$$

where $e^{\mathscr{L}T}$ is a formal specification of the solution to the master equation for a time T after the "initial" conditional state $a\rho(t)a^\dagger$. Once again, due to the non stationary nature of the problem, this must be computed numerically. However if the pump pulse duration is very short compared to the cavity decay time and further the cavity decay time is the fastest decay constant in the system, the probability amplitude to excite a single photon in a frequency at frequency ω is very close to Lorentzian. The experiment of Keller et al. [40] revealed a clear suppression of the peak at $T = 0$ in the (normalised) correlation function $g^{(2)}(T)$, thus passing the first test of a good single photon source.

A very different approach to single photon sources is based on the spontaneous parametric down conversion using a crystal with a significant second order optical non linearity. In these systems, a pair of photons is produced simultaneously, but at random times. However if we detect one photon of the pair in a given time window, we know the temporal coordinates of the other photon. A detailed study of the mode structure of the conditional photon pulse has been undertaken by Walmsley and co-workers [43]. To a very good approximation the probability amplitude functions, α_ω, are Gaussian with variance depending ultimately on the filters used in the conditioning detection step. These sources have been the sources of choice for the early implementations of LOQC. However the random time of pair production means that the single photons are heralded but not produced on-demand. An ingenious scheme to overcome this limitation is being pursued by the NIST group of Migdall [44]. In their scheme a large number of conditional sources are multiplexed, together with fast electro-optical switching, so that at some repetition rate and detection bandwidth, there is near determinant sequence of singe photon detection events.

Exercises

16.1 Consider the following EPR entangled state of two modes A and B,

$$|X,Y\rangle_{AB} = e^{-\frac{i}{2}\hat{Y}_A\hat{X}_B}|X\rangle_A \otimes |Y\rangle_B \qquad (16.81)$$

where the states appearing on the left hand side of this equation are the quadrature phase eigenstates. Verify that this state is a simultaneous eigenstate of $\hat{X}_A - \hat{X}_B$ and $\hat{Y}_A + \hat{Y}_B$ with respective eigenvalues, X, Y.

16.2 The two-mode squeezed vacuum state, (16.16), is also entangled with respect to the correlation specified by the statement: *an equal number of photons in each mode*. However it is not a perfectly entangled state, which would require the (unphysical) case of a uniform distribution over correlated states. Show that the state is an eigenstate of the photon number difference and the phase sum. To show this compute the canonical joint phase distribution $P(\phi_A, \phi_B)$, for the two modes using the projection operator valued measure (see Sect. 2.8). Show that as $\lambda \to 1$ this distribution becomes very sharply peaked at $\phi_A = -\phi_B$. Thus the photon number in each mode are perfectly correlated while the phase in each mode is highly anti correlated.

16.3 Joint quadrature phase measurement of $\hat{X}_C - \hat{X}_A$ and $\hat{Y}_C + \hat{Y}_A$ are made on two modes, A and C, with the results X, Y respectively. Show that the conditional state resulting from this joint quadrature measurement is described by the projection onto the state $|X,Y\rangle_{CA}$ where

$$|X,Y\rangle_{CA} = e^{-\frac{i}{2}\hat{X}_A\hat{Y}_C}|X\rangle_C \otimes |Y\rangle_A \qquad (16.82)$$

16.4 In the protocol for teleportation based on the state in Exercise 16.1, let the total input state for the teleportation protocol be

$$|\psi\rangle_{in} = |\psi\rangle_C \otimes |X_0, Y_0\rangle_{AB} \qquad (16.83)$$

Joint quadrature phase measurements of $\hat{X}_C - \hat{X}_A$ and $\hat{Y}_C + \hat{Y}_A$ are made on the client and sender modes C and A, yielding two real numbers, X, Y respectively. Show that the conditional state of mode B after the measurement on sender and the client, in the special case of $X_0 = Y_0 = 0$, is then given by

$$|\phi^{(X,Y)}\rangle_B = e^{\frac{i}{2}XY}e^{\frac{i}{2}X\hat{Y}_B}e^{-\frac{i}{2}Y\hat{X}_B}|\psi\rangle_B \qquad (16.84)$$

16.5 Consider the beam splitter unitary transformation $U = e^{\theta(a_2^\dagger a_1 - a_2 a_1^\dagger)}$. Show that

$$_2\langle 0|U(\theta)|0\rangle_2 = \sum_{n=0}^{\infty} \frac{(\cos\theta - 1)^n}{n!}(a_1^\dagger)^n a_1^n$$

$$_2\langle 1|U(\theta)|1\rangle_2 = \cos\hat{\Upsilon}^{(0)} - \sin^2\theta a_1^\dagger \hat{\Upsilon}^{(0)} a_1$$

16.6 A linear optical device acting on N modes may be described by a unitary transformation of the form

$$U(H) = \exp[-i\vec{a}^\dagger H \vec{a}] \tag{16.85}$$

where

$$\vec{a} = \begin{pmatrix} a_1 \\ a_2 \\ \vdots \\ a_N \\ a_N \\ \vdots \\ a_N \end{pmatrix} \tag{16.86}$$

and H is a hermitian matrix. Show that this transformation leaves the total photon number invariant,

$$U^\dagger(H)\vec{a}^\dagger \cdot \vec{a} U(H) = \vec{a}^\dagger \cdot \vec{a} \tag{16.87}$$

and induces a linear unitary transformation on the vector $\vec{a}$ as

$$U^\dagger(H)\vec{a} U(H) = S(H)\vec{a} \tag{16.88}$$

16.7 Consider the three mode optical device shown in Fig. 16.10. Mode a_1 is the signal mode prepared in an arbitrary two photon state $|\psi\rangle$. Modes a_2, a_3 are ancilla modes prepared in the photon number states $|1\rangle_2$ and $|0\rangle_3$, respectively. Using the notation of Exercise 16.6, let the $S(H)$ be the orthogonal matrix with matrix elements s_{ij}. Show that the (unnormalised) conditional state of the signal mode, conditioned on counting one photon on mode a_2 and no photons in mode a_3 is given by $|upsi'\rangle = \hat{E}|\psi\rangle$ with $\hat{E} = s_{22}\hat{A} + s_{12}s_{21}a_1^\dagger \hat{A} a_1$ where $\hat{A} = \sum_{n=0}^\infty \frac{(s_{11}-1)^n}{n!}(a_1^\dagger)^n a_1^n$. Verify that, for the choice given in Fig. 16.10, this implements a conditional sign shift gate with probability of 0.25 .

References

1. C.E. Shannon: Bell System. Tech. J. **27**, 379 (1948); Reprinted in, C.E. Shannon, W. Weaver: *The Mathematical Theory fo Communication* (The University of Illinois Press, Urbana 1949)
2. R.B. Ash: *Information Theory* (Dover, New York 1965)
3. Charles H. Bennett 1, David P. DiVincenzo 1, Christopher A. Fuchs, Tal Mor, Eric Rains, Peter W. Shor, John A. Smolin, William K. Wootters: *Quantum Nonlocality Without Entanglement*, quant- ph/ 9804053 v4 2 Nov 1998

4. A. Peres: Phys. Rev. Lett **77**, 1413 (1996)
5. G. Adesso, F. Illuminati: Phys. Rev. A **72**, 032334 (2005)
6. I. Bengtsson, K. Życzkowski, "Geometry of Quantum States", pp. 333, Cambridge University press (Cambridge, 2006)
7. S. Wiesner: "Conjugate Coding," SIGACT News **15**, 78 (1983)
8. C.H. Bennett, G. Brassard: "Quantum Cryptography: Public Key Distribution and Coin Tossing," Proceedings of IEEE International Conference on Computers, Systems and Signal Processing, Bangalore (New York, IEEE 1984)
9. C.H. Bennett, G. Brassard: "The Dawn of a New Era for Quantum Cryptography: The Experimental Prototype is Working," SIGACT NEWS **20** (4), 78 (1989); C.H. Bennett et al.: Experimental Quantum Cryptography, J. Crypto. **5**, 3 (1992)
10. A. Muller, J. Breguet, N. Gisin: Europhys. Lett. **23**, 383 (1993)
11. A.K. Ekert: Quantum Cryptography Based on Bell's Theorem, Phys. Rev. Lett. **67**, 661 (1991)
12. C.H. Bennett: Quantum Cryptography Using Any Two Non-Orthogonal States, Phys. Rev. Lett. **68**, 3121 (1992)
13. W.K. Wootters, W.H. Zurek: Nature **299**, 802 (1982); D. Dieks: Phys. Let. A, **92**, 271 (1982)
14. R.J. Hughes, W.T. Buttler, P.G. Kwiat, G.G. Luther, G.L. Morgan, J.E. Nordholt, C.G. Peterson, C.M. Simmons: Proc. SPIE **3076**, 2 (1997)
15. P.A. Hiskett et al., New Journal of Physics, **8**, 193 (2006)
16. C.H. Bennett, G. Brassard, C. Crepeau, R. Jozsa, A. Peres, W.K. Wooters: Phys. Rev. Lett. **70**, 1895 (1993)
17. B. Schumaker: Phys. Rev. A **51**, 2783 (1995)
18. A. Furasawa, J.L. Sørensen, S.L. Braunstein, C.A. Fuchs, H.J. Kimble, E.S. Polzik: Science, **282**, 23 October, 706, (1998)
19. S.L. Braunstein, H.J. Kimble: Phys. Rev. Letts. **80**, 869 (1998)
20. L. Vaidman: Phys. Rev. A **49** 1473 (1994)
21. refer to di Martini teleportation experiment
22. refer to Zeilinger teleportation experiment
23. Richard P. Feynman: *Simulating Physics with Computers*, Int. J. Theoretical Phys., **21**, 467 (1982)
24. D. Deutsch: Quantum-Theory, the Church-Turing Principle and the Universal Quantum Computer. Proc. R. Soc. Lond. A **400**, 97–117 (1985)
25. P. Shor: Algorithms for Quantum Computation: Discrete Logarithms and Factoring. Proc. 35th Annual Symposium on Foundations of Computer Science (1994). See also LANL preprint quant-ph/9508027
26. Richard Cleve, Artur Ekert, Leah Henderson, Chiara Macchiavello and Michele Mosca: On quantum algorithms, Complexity, **4**, 33 (1999)
27. D. Gottesman, I.L. Chuang: Nature, **402**, 390–393 (1999)
28. E. Knill, R. Laflamme, G.J. Milburn: Nature, **409**, 46, (2001)
29. Michael A. Nielsen: Phys. Rev. Lett. **93**, 040503 (2004)
30. R. Raussendorf, H.J. Briegel: Phys. Rev. Lett. **86**, 5188 (2001)
31. G.D. Hutchinson, G.J. Milburn: J. Mod. Opt. **51**, 1211–1222 (2004)
32. T.B. Pittman, B.C. Jacobs, J.D. Franson: Phys. Rev. A **64**, 062311 (2001)
33. J.L. OBrien, G.J. Pryde, A.G. White, T.C. Ralph, D. Branning: Nature, **426**, 264 (2003)
34. T.C. Ralph, N.K. Langford, T.B. Bell, A.G. White: Phys. Rev. A **65**, 062324 (2002)
35. D.F.W. James, P.G. Kwiat, W.G. Munro, A.G. White: Phys. Rev. A**64**, 052312 (2001)
36. K. Sanaka,T. Jennewein, Jian-Wei Pan, K. Resch, A. Zeilinger: Phys. Rev. Lett. **92**, 017902–1 (2004)
37. P. Walther, K.J. Resch, T. Rudolph, E. Schenck, H. Weinfurter, V. Vedral, M. Aspelmeyer, A. Zeilinger: Nature, **434**, 169 (2005)
38. C. Santori, D. Fattal, J. Vuckovic, G.S. Solomon, Y. Yamamoto: Nature, **419**, 594, (2002)
39. P.P. Rohde, T.C. Ralph: PRA **71**, 032320 (2005)
40. M. Keller, B. Lange, K. Hayasaka, W. Lange, H. Walther: Nature, **431**, 1075 (2003)
41. M. Hennrich, T. Legero, A. Kuhn, G. Rempe: New J. Phys. **6**, 86 (2004)

42. Christian Maurer, C. Becher, C.s Russo, J. Eschner, R. Blatt: New J. Phys. **6**, 94(2004)
43. W.P. Grice et al.: Physical Review A, **64**, 063815 (2001)
44. A.L. Migdall, D. Branning, S. Castelletto, M. Ware: SPIE Free-Space Laser Communication and Laser Imaging II. To Appear in the Proceeding of SPIE Free-Space Laser Communication and Laser Imaging II, Proc. of the SPIE, **4821**, 455–465, 2002

Further Reading

Nielsen M., I. Chuang: *Quantum Computation and Quantum Information* (Cambridge University press, Cambridge 2000)

Bouuwmester D., A. Ekert, A. Zeilinger: (eds) *The Physics of Quantum Information* (Springer, Berlin 2000)

Hoi-Kwong Lo, Tim Spiller, Sandu Popescu: *Introduction to Quantum Computation and Information* (World Scientific, Singapore 1998)

Loepp S., W.K. Wootters: *Protecting Information: From Classical Error Correction to Quantum Cryptography* (Cambridge University Press, New York 2006)

Gisin N., G. Ribordy, W. Tittel H. Zbinden: *Quantum Cryptography* Rev. Mod. Phys. **74**, 145 (2002)

Chapter 17
Ion Traps

17.1 Introduction

Ion trap technology currently leads the way in the effort to gain complete control of quantum coherence in isolated systems. The seminal paper of Cirac and Zoller thrust the technology to the forefront of the quantum processing agenda in a seminal paper in 1995 [1]. Beginning over 30 years ago, experimentalists began trapping clouds of atomic ions in order to achieve higher spectroscopic resolution [2]. The heritage of this effort is the current ability to define time standards using ion trap clocks. Future developments will depend on the ability to make smaller trap arrays for quantum computing applications.

Ion trap technology also enables quantum limited measurements of the electronic and vibrational states of a single trapped ion using the method of cycling fluorescent transitions. This ability led to a complete reappraisal of how quantum mechanics of single systems, subject to continuous observation, should be interpreted [3] enabling an explicit physical demonstration of the concept of a quantum trajectory discussed in Chap. 6. It is possible to trap and cool a single ion close to its vibrational ground state [4], carefully prepare complex superpositions of energy eigenstates through optical excitation and then monitor the subsequent dynamics with extraordinary sensitivity.

17.2 Trapping and Cooling

Laplace's equation indicates that it is not possible to trap a charged particle in three dimensions with a static potential: there is always one unstable (untrapped) direction in an electrostatic potential. We must resort to time periodic potentials. In Fig. 17.1 we show a possible configuration of electrodes.

The time dependent potential can be written

$$V(x,y,z,t) = \frac{\bar{V}}{2}(k_x x^2 + k_y y^2 + k_z z^2) + \frac{1}{2}V\cos(\omega_{rf}t)(k_x' x^2 + k_y' y^2) \qquad (17.1)$$

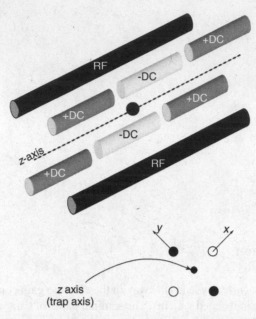

Fig. 17.1 A schematic representation of a linear radio frequency ion trap (after [5])

where ω_{rf} is the frequency of the time dependent potential. Laplace's equation implies, $k_x + k_y + k_z = 0$, $k'_x + k'_y = 0$. If we assume [5]

$$a_x = \frac{4Z|e|\bar{V}k_x}{m\omega_{rf}^2} \ll 1$$

$$q_x = \frac{2Z|e|Vk'_x}{m\omega_{rf}^2} \ll 1$$

then the motion in the x-direction is approximately harmonic, as is motion in the y-direction. Given isotropy in the $x-y$ plane, so that $k_x = k_y$, $k'_x = k'_y$ and the motion is harmonic with the secular frequency

$$v = \left(a_x + q_x^2/2\right)^{1/2} \omega_{rf}/2 \qquad (17.2)$$

A small amplitude oscillatory motion at frequency ω_{rf} is superimposed on the secular motion, called the micromotion, which we neglect. In an experiment with $^9Be^+$, the axial frequency was about 3 MHz while the transverse frequency was about 8 MHz. The static potential due to end caps gives harmonic confinement along the trap axis (z-direction). If this is kept weak, multiple ions can be trapped in a line along the z-direction. Typically the transverse frequencies are three to four times more than the axial.

The centre-of-mass quantum dynamics of the ion is determined by the eigenstates of the Hamiltonian

$$H = \hbar v a_z^\dagger a_z + \hbar v_t (a_x^\dagger a_x + a_y^\dagger a_y) \qquad (17.3)$$

The motion is thus separable into axial and transverse motion and, to be specific, we now concentrate on the axial motion alone. As we neglect the transverse motion, we will drop the subscript on $a_z, a_z^\dagger$.

The availability of lasers at appropriate atomic transition frequencies determine which ions can be successfully trapped and cooled. The Wineland group at NIST Colorado uses $^9\text{Be}^+$ while the Blatt group in Innsbruck uses $^{40}\text{Ca}^+$. After trapping it is necessary to remove vibrational energy from the ion, that is to say, it must be cooled. The initial temperature is of the order of 10^4 K. The first stage of cooling is based on Doppler cooling and is very efficient, the second stage is based on resolved sideband cooling (see below).

The extraordinary degree of control over quantum coherence that can be achieved in an ion trap is due to a number of reasons. Firstly, it is possible to coherently couple the vibrational motion and the internal electronic state using an external laser. Secondly, resolved sideband cooling enables the vibrational motion to be prepared in its ground state with probability approaching unity. External lasers induce Raman transitions between the ground and excited internal electronic state in which one phonon of vibrational energy is absorbed per excitation cycle. Finally, the internal electronic state of a single trapped ion can be determined with efficiency approaching unity by the method of fluorescence shelving enables.

We will assume that external lasers drive a two level transition from the ground state $|g\rangle$ to the excited state $|e\rangle$. This could be a direct dipole transition, but for quantum computing it typically involves a Raman two-photon transition connecting the ground state to an excited meta-stable state. In either case the Hamiltonian describing the system is (see 10.1),

$$H = \hbar v a^\dagger a + \hbar \omega_A \sigma_z + \frac{\hbar \Omega}{2} \left(\sigma_- e^{i(\omega_L t - k_L \hat{q})} + \sigma_+ e^{-i(\omega_L t - k_L \hat{q})} \right) \tag{17.4}$$

where $\hat{q}$ is the operator for the displacement of the ion from its equilibrium position in the trap, v is the trap (secular) frequency, Ω is the Rabi frequency for the two level transition, ω_A is the atomic transition frequency, and ω_L, k_L are the laser frequency and wave number. The sigma matrices are defined in Sect. 10.1. There are three frequencies in the problem: v, ω_A and ω_L. By carefully choosing relationships between these three frequencies various quantum interactions between electron and vibrational degrees of freedom can be driven. The key point is that the phase of the laser field as seen by the ion depends on the position of the ion. As the ion vibrates this phase is modulated at the trap frequency, which leads to sidebands in the absorption spectrum for the two level system.

The ion position operator may be written,

$$\hat{q} = \left(\frac{\hbar}{2mv} \right)^{1/2} (a + a^\dagger) \tag{17.5}$$

We now define the Lamb-Dicke parameter, η

$$\eta = k_L \left(\frac{\hbar}{2mv} \right)^{1/2} = 2\pi \Delta x_{rms} / \lambda_L \tag{17.6}$$

where the r.m.s position fluctuations in the oscillator ground state is Δx_{rms}. Then moving to an interaction picture via the unitary transformation

$$U_0(t) = \exp[-iva^\dagger at - i\omega_A \sigma_z t] \tag{17.7}$$

the interaction Hamiltonian can be written as

$$H_I(t) = \frac{\hbar\Omega}{2}\left(\sigma_- \exp[-i\eta(ae^{-ivt} + a^\dagger e^{ivt})]\exp[-i(\omega_A - \omega_L)t] + \text{h.c}\right) \tag{17.8}$$

The exponential of exponentials make this a complicated Hamiltonian system. However in most ion trap experiments the ion is confined to a spatial region that is significantly smaller than the wavelength of the exciting laser so that we may assume that the Lamb-Dicke parameter is small $\eta < 1$ (typically $\eta \approx 0.01 - 0.1$). Expanding the interaction to second order in the Lamb-Dicke parameter gives

$$H_I(t) = \frac{\hbar\Omega}{2}[1 - \eta^2 a^\dagger a]\left(\sigma_- e^{-i\delta t} + \sigma_+ e^{i\delta t}\right)$$
$$-i\frac{\hbar\Omega\eta}{2}\left(ae^{-ivt} + a^\dagger e^{ivt}\right)e^{-i\delta t}\sigma_- + i\frac{\hbar\Omega\eta}{2}\left(ae^{-ivt} + a^\dagger e^{ivt}\right)e^{i\delta t}\sigma_+$$
$$-\frac{\hbar\Omega\eta^2}{4}\left(a^2 e^{-2ivt} + (a^\dagger)^2 e^{2ivt}\right)\left(e^{-i\delta t}\sigma_- + e^{i\delta t}\sigma_+\right)$$

where the detuning of the laser from the atomic frequency is $\delta = \omega - \omega_L$.

Tuning the frequencies so that δ is a positive or negative integer multiple of the trap frequency leads to resonant terms, and all time dependent terms are neglected. For carrier excitation, $\delta = 0$, the resonant terms are

$$H_c = \hbar\Omega(1 - \eta^2 a^\dagger a)\sigma_x \quad \text{carrier excitation} \tag{17.9}$$

where $\sigma_x = (\sigma_- + \sigma_+)/2$. If we take $\delta = v$ so that the laser frequency is detuned below (to the red of) the carrier frequency by one unit of trap frequency, $\omega_L = \omega_A - v$, the resonant terms are

$$H_r = i\frac{\hbar\eta\Omega}{2}\left(a\sigma_+ - a^\dagger\sigma_-\right) \quad \text{first red sideband excitation} \tag{17.10}$$

This is the Jaynes-Cummings Hamiltonian except that it involves the absorption of a trap *phonon* as well as one laser photon. If we instead choose $\delta = -v$ so that $\omega_L = \omega_A + v$, the laser is detuned one unit of vibrational frequency above the carrier (a blue detuning), the resonant interaction Hamiltonian is

$$H_b = i\frac{\hbar\eta\Omega}{2}\left(a^\dagger\sigma_+ - a\sigma_-\right) \quad \text{first blue sideband excitation} \tag{17.11}$$

This corresponds to an excitation process in which one photon is absorbed from the laser and one trap phonon is *emited* . We can continue to define the second red sideband excitation $\delta = 2v$ and second blue sideband excitation $\delta = -2v$, and so

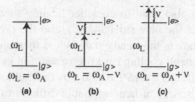

Fig. 17.2 Energy level diagram for (**a**) carrier (**b**) first red sideband and (**c**) first blue sideband excitation

on. In fig. 17.2 we give an energy level diagram that represents the carrier, red and blue sideband excitations.

An ion that is excited to $|e\rangle$ can spontaneously decay to the ground state, enabling another excitation. If we are tuned to the first red side band these cycles of excitation and emission remove one phonon per excitation cycle, thus cooling the vibrational degree of freedom. The external laser has coupled the vibrational motion to a very low temperature heat bath: the vacuum radiation field at frequency ω_A. For obvious reasons this is know as *side band cooling*. Needless to say this is only possible if we can spectroscopically resolve the red sideband. The width of each resonance is due to the spontaneous emission rate, γ, so we require that $\nu > \gamma$. The spectrum of resonance fluorescence for a single trapped ion follows from the methods given in Chap. 11. A detailed calculation in the low intensity limit ($\Omega < \gamma$) for a traveling wave field, by Cirac et al. [6] shows that the spectrum of the motional sidebands exhibits the following features:

- The first red side band is centred on $\omega_L = \omega_A - \nu$ and the first blue sideband is centred on $\omega_L = \omega_A + \nu$ with linewidths determined by

$$\gamma_s = \eta^2 \left(\frac{\Omega}{2}\right)^2 [P(\nu + \delta) - P(\nu - \delta)] \tag{17.12}$$

where $P(\delta) = \gamma/(\gamma^2 + \delta^2)$ and $\delta = \omega_L - \omega_A$ and γ is the spontaneous emission rate.
- The ratio of the peak height of the red sideband to the blue side band is $(\bar{n}+1)/\bar{n}$ where $\bar{n}$ is the steady state mean photon number of vibrational excitation.

Note that the heights of the peaks are different reflecting the fact that the red transition involves the absorption of a phonon while the blue involves the emission of a phonon.

In the Lamb-Dicke limit relaxation is dominated by spontaneous emission into the spectral peak at the carrier frequency ($\omega = \omega_A$) so that one unit of vibrational energy is removed on average per excitation cycle. We can understand this via a simple rate equation approach. In Exercise 17.1, we find that the rate of change of the average phonon number is given by

$$\frac{d\bar{n}}{dt} = -\gamma \left(\frac{\eta^2 \Omega^2 \bar{n}}{2\eta^2 \Omega^2 \bar{n} + \gamma^2}\right) \tag{17.13}$$

To be more realistic we also need to consider heating mechanisms, for example off-resonant excitation of the blue sideband [5], and the probability of populating the vibrational ground state in the steady state is less than unity. Despite the effects of heating, resolved sideband cooling care prepare an ion in the vibrational ground state with a probability greater than 99%, and was first achieved by the NIST group in Boulder [7]. Another source of heating that is difficult to control is due to fluctuating charge distributions on the trap electrodes. As these potentials change randomly in time, they produce a stochastic displacement of the centre of the trap. In Exercise 17.2 we consider this example in more detail.

We now turn to the problem of reading out the state of the ion. This is done by the technique of a cycling fluorescent transition which requires an additional auxiliary level, coupled by a strong probe laser to one or the other of the ground or excited states. We will consider the readout of the ground state (see Fig. 17.3). If the ion is in the ground state when the probe laser is turned on, fluorescent photons are scattered in all directions and can easily be detected. On the other hand if the ion is in the excited state, it is not resonant with the probe laser and no photons are scattered: the ion remains dark.

The interesting phenomenon of fluorescent shelving will now arise if a second weak laser induces incoherent transitions on $|g\rangle \leftrightarrow |e\rangle$. These transitions are incoherent as the strong coupling to the $|a\rangle$ state destroys coherence between the ground and excited states, see [8]. The fluorescent signal due to the strong probe laser switches on and off in the fashion of a random telegraph process. A typical signal is shown in 17.3. In so far as fluorescence indicates that the ion is in the ground state, the random switching of the fluorescence is a direct indicator of *quantum jumps* between the ground and excited states [3].

The quality of the readout can be reduced to a single number, called the efficiency, which is the conditional probability for a fluorescent photon to be detected *given* the ion is in the ground state. This is a function of the integration time of the fluorescent signal and the overall detection efficiency of the detection system.

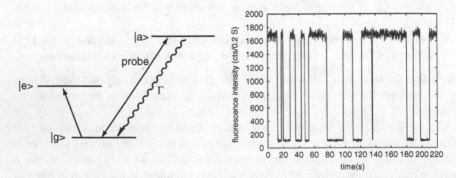

Fig. 17.3 Energy level diagram showing fluorescence readout of the ground atomic state. A strong probe laser drives a dipole allowed transition between the ground state $|g\rangle$ and an auxilary state $|a\rangle$ which decays back to the ground state at a rate Γ scattering many many photons. Also show is fluorescent signal on the probe transition when a weak laser couples the ground and excited state (reproduced, with permission, from Leibfried et al. [5]

The integration time should be at least of the order of the average time between photon emission events. In practice other sources of error must be considered,such as dark counts in the detector. Typically the minimum time to distinguish ground and excited states is of the order of 2 ms.

How efficient is the process of sideband cooling? We can only answer this if we have an independent way to determine the vibrational state of the ion at the end of a cooling phase. This may be done by coupling the vibrational motion to the internal state of the ion and then using the fluorescent readout technique described above. Suppose the electronic state of the ion can be coupled to its vibrational motion for a time T using either the first red and blue sideband transitions. If we write the probability for the atom to found in the excited state after time T as $P_e^R(T)$ and $P_e^B(T)$ for red and blue sideband excitation respectively it can be shown (see Exercise 17.3) that the mean phonon number $\bar{n}$ is given by $\bar{n}/(1+\bar{n}) = P_e^R(T)/P_e^B(T)$. Thus measurement of the ratio of excitation probability on the first red and blue sideband yields $\bar{n}$ directly.

17.3 Novel Quantum States

The ability to carefully control the coupling between internal electronic state of the ion and its vibrational motion in the trap enables us to carefully engineer novel quantum states of the vibrational degree of freedom. As an example we will here consider the preparation of a "cat state": a pure quantum state in which the two internal electronic states are correlated with different coherent states of the oscillator.

There are a number of ways to prepare the vibrational motion in a coherent state, $|\alpha\rangle$. The ion is first cooled to the vibrational ground state. A classical uniform driving force oscillating at the secular frequency, ν, can then be applied by changing the bias conditions on the trap electrode. Alternatively a non adiabatic displacement of the trap centre can be made again by changing the bias conditions. Finally a spatially varying Stark shift can be applied by using the moving standing wave that results from two laser beams with frequency difference $\Delta\omega = \nu$ to resonantly drive the motion of the ion in the trap. If the laser polarisation is carefully chosen this will result in a force that depends on the internal electronic state. From the point of view of the electronic and vibrational states, this is a two photon Raman process depicted in Fig. 17.4 that Stark shifts the excited state $|e\rangle$. We will refer to this choice of Raman pulses as the Raman displacement pulse. If we detune the Raman lasers by a frequency $\Delta\omega = \omega_A$ we can drive a carrier transition that coherently superposes the ground and excited states. We will refer to this choice of Raman pulses as the carrier pulse.

The state-dependent displacement is described by the interaction picture Hamiltonian

$$H_d = \hbar\chi(t)(a+a^\dagger)|e\rangle\langle e| + \hbar\Omega(t)\sigma_x \qquad (17.14)$$

the coupling constants χ and Ω are shown as time dependent as they can be turned on and off with the external Raman displacement pulse (χ) or the external carrier pulse (Ω).

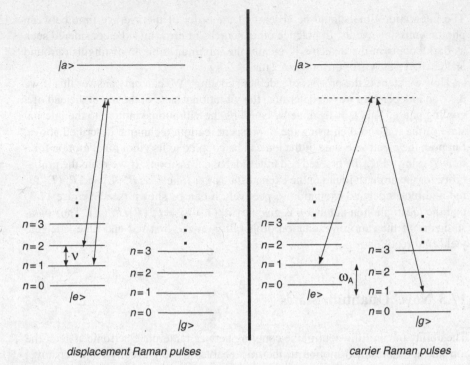

Fig. 17.4 A schematic indication of the optical transitions required to prepare a single ion in a linear superposition of displaced ground states (coherent states). On the left is the Raman pulse excitation scheme for applying a force to the ion conditional on it being prepared in the excited electronic state. On the right is the carrier pulse excitation scheme for producing coherent excitations of the internal electronic state leaving the vibrational motion unaffected. The vibrational frequency is ν while the atomic transition frequency is ω_a

Assume that initially the electronic system and vibrational motional are in the ground state, $|\psi(0)\rangle = |0\rangle \otimes |g\rangle$. In the first step, we apply a carrier pulse (so $\chi = 0$) for a time, T, such that $\omega T = \pi/2$. This gives the state transformation

$$|0\rangle \otimes |g\rangle \overset{\pi/2}{\Longrightarrow} |0\rangle \otimes \frac{1}{\sqrt{2}}(|g\rangle + |e\rangle) \qquad (17.15)$$

In the second step we turn off the carrier pulse and turn on the displacement Raman pulse for a time τ. Only the $|e\rangle$ component sees the displacement, according to (17.14) so the state is transformed as

$$\frac{1}{\sqrt{2}}(|0\rangle \otimes |g\rangle + |0\rangle \otimes |e\rangle) \overset{\text{displace}}{\Longrightarrow} \frac{1}{\sqrt{2}}(|0\rangle \otimes |g\rangle + |\alpha\rangle \otimes |e\rangle) \qquad (17.16)$$

In the third step we apply a $\pi = \Omega T$ carrier pulse that flips the electronic states and inserts a π phase shift

$$\frac{1}{\sqrt{2}}(|0\rangle \otimes |g\rangle + |\alpha\rangle \otimes |e\rangle) \overset{\pi}{\Longrightarrow} \frac{1}{\sqrt{2}}(|\alpha\rangle \otimes |g\rangle - |0\rangle \otimes |e\rangle) \tag{17.17}$$

In the fourth step we apply another state selective displacement with a relative phase ϕ,

$$\frac{1}{\sqrt{2}}(|\alpha\rangle \otimes |g\rangle - |0\rangle \otimes |e\rangle) \overset{\text{displace}}{\Longrightarrow} \frac{1}{\sqrt{2}}(|\alpha\rangle \otimes |g\rangle - |\alpha e^{i\phi}\rangle \otimes |e\rangle) \tag{17.18}$$

In the fifth and final step, we apply another $\pi/2$ pulse to give

$$\frac{1}{\sqrt{2}}(|\alpha\rangle \otimes |g\rangle - |\alpha e^{i\phi}\rangle \otimes |e\rangle) \overset{\pi/2}{\Longrightarrow} \left(\frac{|\alpha\rangle - |\alpha e^{i\phi}\rangle}{2}\right) \otimes |g\rangle + \left(\frac{|\alpha\rangle + |\alpha e^{i\phi}\rangle}{2}\right) \otimes |e\rangle$$

$$\equiv |\alpha_-\rangle \otimes |g\rangle + |\alpha_+\rangle \otimes |e\rangle \tag{17.19}$$

If we now readout the state of the ion, the conditional states are highly non classical superpositions of two different coherent states of vibrational motion, $|\alpha_\pm\rangle$ known in quantum optics as *cat states*.

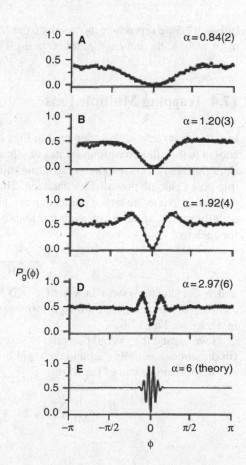

Fig. 17.5 The probability to find the ion in the ground state as a function of the phase of displacements for different choices of the magnitude of displacement. From Fig. 4 of Monroe et al., Science, **272**, 1135 (1996)

We thus have correlated different motional states with each of the electronic states. This kind of entangled state is reminiscent of Schrödinger's famous thought experiment in which two different metabolic (and thus macroscopic) states of a cat are correlated with a two level system in just this way. Indeed if we stopped after the second step the cat state analogy could be sustained with the identification $|\alpha\rangle \rightarrow |\text{alive}\rangle$ and $|0\rangle \rightarrow |\text{dead}\rangle$. However pursuing the analogy to the final state at the end of step 5 produces the rather disturbing prospect (for the cat at least) of correlating different superposition of metabolic states with the internal electronic states. It is for this reason that superpositions of coherent states are called cat states in quantum optics.

In the experiment of Monroe at al. [9], the presence of an entangled state of different coherent states was demonstrated by measuring the electronic state at the end of step 5. Repeated measurements enabled a sampling of the distribution $P_g(\phi)$, for different values of ϕ. This is given by

$$
\begin{aligned}
P_g(\phi) &= \langle \alpha_- | \alpha_- \rangle \\
&= \frac{1}{2}\left[1 - e^{-|\alpha|^2(1-\cos\phi)} \cos\left(|\alpha|^2 \sin\phi\right) \right]
\end{aligned}
\tag{17.20}
$$

In Fig. 17.5 we reproduce the results from Monroe et al. [9] comparing the experiment with the theoretical prediction in Eq. 17.20. The agreement is remarkable.

17.4 Trapping Multiple Ions

In a linear ion trap such as depicted in Fig. 17.1, multiple ions may be trapped and cooled to the collective ground states of vibrational motion. Each ion has an equilibrium position, $\bar{x}_i$, corresponding to a minimum in the total potential comprising the trap plus Coulomb potential for each ion. These equilibrium points are analogous to the atomic ions at the lattice points of a crystal, however unlike a crystal they are not equally spaced. In terms of a natural length scale given by the Coulomb potential for each ion,

$$
l = \left(\frac{Z^2 e^2}{4\pi\varepsilon_0 M v^2} \right)^{1/3},
\tag{17.21}
$$

and a coordinate system in which $z = 0$ is in the middle of the trapped ions, *James* [10] has computed the equilibrium positions for different numbers of ions in a trap, see Fig. 17.6

If we expand the overall potential to second order in the small oscillations, $q_n(t)$ (in dimensionless units), around the equilibrium points we obtain a simple coupled oscillator Hamiltonian of the form,

$$
H = \frac{1}{2M} \sum_{n=1}^{N} p_m^2 + \frac{Mv^2}{2} \sum_{n,m=1}^{N} A_{nm} q_n q_m
\tag{17.22}
$$

N	Scaled equilibrium positions									
2					-0.62996	0.62996				
3				-1.0772	0	1.0772				
4			-1.4368	-0.45438	0.45438	1.4368				
5			-1.7429	-0.8221	0	0.8221	1.7429			
6		-2.0123	-1.1361	-0.36992	0.36992	1.1361	2.0123			
7		-2.2545	-1.4129	-0.68694	0	0.68694	1.4129	2.2545		
8	-2.4758	-1.6621	-0.96701	-0.31802	0.31802	0.96701	1.6621	2.4758		
9	-2.6803	-1.8897	-1.2195	-0.59958	0	0.59958	1.2195	1.8897	2.6803	
10	-2.8708	-2.10003	-1.4504	-0.85378	-0.2821	0.2821	0.85378	1.4504	2.10003	2.8708

Fig. 17.6 The equilibrium positions for varying numbers of ions in the trap in units of the length scale given in 17.21. From Fig. 1 James et al. Appl. Phys. B, 66, 181, (1998)

where p_n is the canonical momentum to q_n. Explicit expressions for the coefficients A_{nm} are given in [10]. This Hamiltonian represents a linear array of N simple harmonic oscillators with quadratic coupling. We can now make a change of variable to normal-mode coordinates (sometimes called collective or global coordinates). The transformation is chosen to diagonalise the real symmetric $N \times N$ matrix with entries A_{nm}. The eigenvalue equation is

$$\sum_{n=1}^{N} A_{nm}\beta_n^{(p)} = \mu_p \beta_m^{(p)} \quad (p = 1, \ldots, N) \tag{17.23}$$

where the eigenvalues are $\mu_p > 0$ and the eigenvectors $\vec{\beta}^{(p)}$ are assumed to be numbered in order of increasing eigenvalue and are normalised so that

$$\sum_{p=1}^{N} \beta_n^{(p)} \beta_m^{(p)} = \delta_{nm}$$

$$\sum_{n=1}^{N} \beta_n^{(p)} \beta_n^{(q)} = \delta_{pq}$$

For example, when $N = 3$ the eigenvalues are $\mu_1 = 1, \mu_2 = 3, \mu_3 = 29/5$. The normal modes are then given in terms of the small oscillations as

$$Q_p(t) = \sum_{m=1}^{N} \beta_m^{(p)} q_m(t) \tag{17.24}$$

Note the number of normal modes is equal to the number of ions. Of course we can equally well write the *local coordinates* q_n as

$$q_m(t) = \sum_{p=1}^{N} \beta_m^{(p)} Q_p(t) \tag{17.25}$$

The first normal mode, Q_1 represents the centre of mass mode in which all the ions oscillate as if they were train wagons linked together. The second mode Q_2 represents a breathing mode in which each ion oscillates with an amplitude proportional to is displacement form the trap centre. In terms of the normal mode coordinates, Q_p and conjugate momenta P_p, the Hamiltonian is

$$H = \frac{1}{2M} \sum_{p=1}^{N} P_p^2 + \frac{M}{2} \sum_{p=1}^{N} v_p^2 Q_p^2 \qquad (17.26)$$

where the frequencies of each of the normal modes is given by

$$v_p = v\sqrt{\mu_p} \qquad (17.27)$$

This is the Hamiltonian of N independent simple harmonic oscillators. Thus we introduce raising and lowering operators for each normal mode as

$$Q_p = \sqrt{\frac{\hbar}{2Mv_p}}(b_p + b_p^\dagger) \qquad (17.28)$$

$$P_p = -i\sqrt{\frac{\hbar M v_p}{2}}(b_p - b_p^\dagger) \qquad (17.29)$$

with $[b_p, b_q^\dagger] = \delta_{pq}$.

Let us now assume that each ion in the trap can be addressed with a separate laser beam. For example in a linear ion trap for $^{40}Ca^+$ built in Innsbruck, the average spacing for 4 ions was greater than $5\,\mu m$, which is above the diffraction limit for the laser beams. This spacing is also sufficient for the fluorescence (at readout) of each ion to be separately imaged.

The interaction picture Hamiltonian describing how the ith ion is coupled to small oscillations around equilibrium is given by an obvious generalisation of Eq. 17.4

$$H_I^{(i)} = \hbar \frac{\Omega_i}{2} \left(\sigma_-^{(i)} e^{-ik_L(q_i(t)} e^{-i(\omega_A - \omega_L)t} + \text{h.c} \right) \qquad (17.30)$$

where we have taken the Rabi frequency for the ith ion to be Ω_i and $\sigma_\pm^{(i)}$ are the Pauli raising and lowering operators for the ith ion, while $q_i(t)$ are local coordinates of the ith ion. If we now again assume that the Lamb–Dicke parameter for each ion is small, the interaction between the electronic and vibrational degree of freedom is

$$H_I^{(i)} = -i\hbar \frac{k_L \Omega_i}{2} \left(\sigma_-^{(i)} q_i(t) e^{-i(\omega_A - \omega_L)t} - \text{h.c} \right) \qquad (17.31)$$

This may be written in terms of the global modes as

$$H_I^{(I)} = -i\hbar \frac{\eta \Omega_i}{2\sqrt{N}} \sum_{p=1}^{N} s_i^{(p)} \left(b_p e^{-iv_p t} + b_p^\dagger e^{iv_p t} \right) e^{-i(\omega_A - \omega_L)t} \sigma_-^{(i)} - \text{h.c} \qquad (17.32)$$

where $s_p^{(i)} = \sqrt{N}\mu_p^{-1/4}\beta_i^{(p)}$.

We now assume that we can tune the external laser to address only a single global vibrational mode (a particular normal mode), say the centre of mass mode at frequency, $\mu_1 = v$ and $s_i^{(1)} = 1$ with $\omega_A - \omega_L = v$. Then we can ignore all the other modes and approximate the Hamiltonian as

$$H_I^{(i)} = -i\hbar\frac{\eta\Omega_i}{2\sqrt{N}}\left(\sigma_+^{(i)}b_1 + \sigma_-^{(i)}b_1^\dagger\right) \tag{17.33}$$

This is the Cirac–Zoller Hamiltonian [1] and was used by these authors in a scheme for quantum computing using trapped ions (see below). If there are many ions in the trap this may not be a good approximation. In that case there are many normal modes and it is difficult to resolve individual normal mode frequencies as they become very closely spaced. To some extent this may be mitigated by cooling all the normal modes to their ground states. Further discussion of the validity of this approximation may be found in [10] and also [11].

There is an interesting interpretation of the Hamiltonian for local modes, (17.31). Define

$$\sigma_1^{(i)}(t) = \sigma_- e^{-i\Delta t} + \sigma_+ e^{i\Delta t} \tag{17.34}$$

with $\Delta = \omega_A - \omega_L$. Now define a quantum field $\hat{\psi}(x,t) = (2Mv/\hbar)^{1/2}q_i(t)$ by replacing the discrete index, i, with a position variable $x = id_i$ where d_i is the position of the ith ion form the centre of the trap. This field is a scalar field operator describing the small oscillations of the ion at x from equilibrium. The interaction Hamiltonian then takes the form

$$H_I = \hbar\chi\,\hat{\psi}(x,t)\sigma_1(x,t) \tag{17.35}$$

which is in the form of local field dipole detector interaction Hamiltonian, with $\chi = \eta\Omega/2$.

17.5 Ion Trap Quantum Information Processing

In 1995 Cirac and Zoller [1] proposed the first scheme for implementing quantum logic gates for trapped ions. In a quantum computer information is stored in the states of a collection of two level systems, generically referred to as qubits. In the Cirac–Zoller (CZ) scheme, the qubits are the two-level electronic states of the trapped ions. Arbitrary transformations of the state of a single qubit are easily implemented by external laser fields. For universal computation we also need to have access to two qubit interactions. However the electronic states of different ions do not interact. CZ proposed to overcome this by using the collective vibrational mode of the ions to implement a virtual interaction between the qubits.

We will discuss a way to implement a particular two qubit interaction, known as a controlled NOT (CNOT) gate which is a universal two-qubit gate. If we denote the two states of a qubit as $|x\rangle, x = 0, 1$, the CNOT gate is defined by the unitary transformation

$$U_{CN}|x\rangle \otimes |y\rangle = |x\rangle \otimes |x \oplus y\rangle \tag{17.36}$$

Fig. 17.7 The electronic level
scheme for each ion in the
CNOT gate scheme of Cirac
and Zoller

where $x \oplus y$ is addition modulo two. In words, the state of the second qubit, called
the target, is flipped if and only if the state of the first qubit, the control, is $|1\rangle$. In all
cases the state of the control qubit is unchanged.

In the CZ scheme each ion has a set of internal electronic levels, $|0\rangle, |1\rangle$ and $|e\rangle$,
depicted in Fig. 17.7. The mapping between physical electronic states and logical
states is $|0\rangle \leftrightarrow |0\rangle$, $|1\rangle \leftrightarrow |1\rangle$. Note that in addition to the qubit states, there is an
additional auxiliary state, $|e\rangle$ which helps implement the degree of control required.
We suppose that is possible to direct a laser onto a particular ion inducing electronic transitions in that ion alone. This can couple the qubit state of the ion to its
vibrational degree of freedom. In the Lamb–Dicke limit and for carefully chosen
detunings, it is possible to couple a single qubit state to a chosen collective state of
vibrational motion of all the trapped ions. In the CZ scheme the collective vibrational modes are all prepared in the ground state by a prior sideband cooling step.

Let us suppose the laser is directed towards the nth ion and tuned to the first red
sideband of the collective centre of mass mode described by raising and lowering
operators, $a^\dagger, a$. The Hamiltonian for this is

$$H = \hbar \frac{\eta \Omega}{2\sqrt{N}} (a|1\rangle_n \langle 0| e^{-i\phi} + a^\dagger |0\rangle_n \langle 1| e^{i\phi}) \tag{17.37}$$

If this laser is on for a time T such that $\eta \Omega T \sqrt{N} = k\pi$ (a $k\pi$-pulse), the unitary
operator,

$$U_{01}^{k,n}(\phi) = \exp\left[-i\frac{\pi}{2}(a|1\rangle_n \langle 0| e^{-i\phi} + a^\dagger |0\rangle_n \langle 1| e^{-i\phi})\right] \tag{17.38}$$

is implemented. This unitary interaction couples the electronic states to the vibra-
tional phonon number states $|0\rangle, |1\rangle$;

$$U_{01}^{k,n}(\phi)|0\rangle_n|1\rangle = \cos(k\pi/2)|0\rangle_n|1\rangle - ie^{i\phi}\sin(k\pi/2)|1\rangle_n|0\rangle \tag{17.39}$$

$$U_{01}^{k,n}(\phi)|1\rangle_n|0\rangle = \cos(k\pi/2)|1\rangle_n|0\rangle - ie^{-i\phi}\sin(k\pi/2)|0\rangle_n|1\rangle \tag{17.40}$$

We will also need to implement $k\pi$ pulse between the ground state $|0\rangle$ and the aux-
iliary excited state $|e\rangle$. This implements the unitary transformation

$$U_{oe}^{k,n}(\phi) = \exp\left[-i\frac{\pi}{2}(a|e\rangle_n \langle 0| e^{-i\phi} + a^\dagger |0\rangle_n \langle e| e^{-i\phi})\right] \tag{17.41}$$

and can be done by changing the polarisation of the exciting laser.

We now consider a three pulse sequence: on the mth ion implement $U_{01}^{1,m}(0)$,
then on the nth ion a 2π pulse between the ground state and the auxillary excited
state, $U_{oe}^{2,n}(0)$, finally, again on the m'th ion, $U_{01}^{1,m}(0)$. The three pulse sequence thus
implements the product unitary, $U^{m,n} = U_{01}^{1,m}(0)U_{oe}^{2,n}(0)U_{01}^{1,m}(0)$. Acting on each of

the four possible two qubit states of the ions we find that all states remain unchanged except when both ions are initially excited;

$$|1\rangle_m |1\rangle_n \rightarrow -|1\rangle_m |1\rangle_n \qquad (17.42)$$

This is a two qubit gate known as the CSIGN gate. To use this to implement a CNOT gate we now choose one of the ions to be the control qubit, say mth ion, and first use a laser pulse to put it into a superposition of the logical states, $|0\rangle_m + |1\rangle_m$. This can be done by tuning the laser to the carrier frequency so that it is resonant with the $|0\rangle_m \leftrightarrow |1\rangle_m$ transition, adjusting the phase and pulse area to implement the unitary

$$V_m = \exp\left[-\frac{\pi}{4}(|1\rangle_m\langle 0| - \text{h.c.})\right] \qquad (17.43)$$

After we implement the V single qubit unitary on the mth ion we implement a CSIGN between the mth ion and the nth ion, target ion. Finally we again act with a V pulse on the mth ion. The net effect is to implement a CNOT gate between the mth ion as the control on the nth ion as the target. Clearly the ions do not need to be adjacent. Furthermore we can implement a number of CNOT gates between different pairs in parallel so long as we can individually resolve the ions with the control lasers. The Cirac–Zoller scheme was first implemented by the Innsbruck group led by Blatt in 2003 [12]. They used two $^{40}\text{Ca}^+$ ions held in a linear trap and individually addressed by focussed laser beams.

Other schemes have been proposed for implementing quantum gates in ion traps. *Sørenson* and *Mølmer* [13] developed a scheme which mitigates to some extent the deleterious effects of noise entering via the vibrational degree of freedom (e.g. patch potential heating) and implemented by the Wineland group in NIST [14]. A related scheme [15] uses far off-resonance optical dipole forces to implement a geometric phase gate, also first implemented by the NIST group [16]. The basic idea of a geometric phase gate is to use a sequence of laser pulse sequences, applied to two ions, that move the vibrational degree of freedom of the ion through a loop in phase space that depends on the internal states of the two ions. A simple, though impractical, way to achieve this is to use phase space displacements that move around a rectangle in phase space, starting at the origin, in a direction that depends on the internal state of the ion. For example, the unitary operators

$$U_j(\kappa_x) = e^{-i\kappa_x \hat{p} \sigma_{z,n}/\hbar} \qquad (17.44)$$

$$U_j(\kappa_p) = e^{-i\kappa_p \hat{x} \sigma_{z,n}/\hbar} \qquad (17.45)$$

give conditional displacements of the vibrational degree of freedom along the x-axis for the jth ion in the case of $U_j(\kappa_x)$ and along the p-axis in the case of $U_j(\kappa_p)$. If we use the commutation relation $[\hat{x}, \hat{p}] = i\hbar$ we can show that the following

$$U_k(\kappa_p)U_j(-\kappa_x)U_k(-\kappa_p)U_j(\kappa_x) = e^{i\kappa_x \kappa_p \sigma_{z,j} \sigma_{z,k}} \qquad (17.46)$$

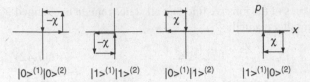

Fig. 17.8 The conditional phase space displacements of the vibrational degrees of freedom of two ions. Four cases are shown corresponding to the four distinct states of two qubits. The path followed is different for each case but the enclosed area is the same

This is an Ising-like two qubit unitary interaction between the j, kth qubits. Note that there is no dependance on the vibrational degree of freedom at all. Inspection of the various phase space orbits, see Fig. 17.8 indicates why this is called a geometric phase gate. The effective conditional phase between the two qubits is proportional to the area of the rectangle, $\chi = \kappa_x \kappa_p$ and the sign is given by the sense of rotation. It is clear that the actual shape of the closed orbit in phase space does not matter: only the area and sense of rotation matter. In an experiment the phase space rotations are done by a time varying driving fields, with both amplitude and phase modulation (see Exercise 17.3). The idea of conditional phase space displacements opens up a path to fast quantum gates for two ions [17].

Exercises

17.1 A laser is tuned to the first read sideband transition for a single two level transition, $|g\rangle \leftrightarrow |e\rangle$, with a spontaneous emission rate of γ. Ignoring all but the spontaneous emission decay channel, the master equation (in the interaction picture) describing this system is

$$\frac{d\rho}{dt} = \frac{\eta\Omega}{2}[a\sigma_+ - a^\dagger\sigma_-, \rho] + \gamma\mathscr{D}[\sigma_-]\rho \qquad (17.47)$$

where η is the Lamb–Dicke parameter, Ω is the Rabi frequency for the transition and $a, a^\dagger$ are the lowering and raising operators for the vibrational motion of the ion in the trap. Obtain equations of motion for $\bar{n} = \langle a^\dagger a\rangle$, $\langle a\rangle$, $\langle\sigma_\pm\rangle$, $\langle\sigma_z\rangle$ by factorising all higher order moments in the equations of motion. Assuming that the spontaneous emission rate is large enough so that the average polarisation $\langle\sigma_\pm\rangle$ is stationary and the vibrational motion is slaved to the atomic motion, show that the rate of change of $\bar{n}$ is given by (17.13).

17.2 A simple mode for the heating of a trapped ion due to fluctuation potentials may be given in terms of the Hamiltonian

$$H(t) = \hbar v a^\dagger a + \hbar\varepsilon(t)(a + a^\dagger) \qquad (17.48)$$

where $\varepsilon(t)$ is fluctuating force term with the following classical moments

$$\bar{\varepsilon} = \mathscr{E}(\varepsilon(t)) = 0$$

$$G(\tau) = \mathscr{E}(\varepsilon(t)\varepsilon(t+\tau)) = \frac{D}{2\gamma}e^{-\gamma|\tau|}$$

Show that the heating rate is given by

$$\frac{d\langle a^\dagger a\rangle}{dt} = \frac{\pi}{2}S(v) \qquad (17.49)$$

where the noise power spectrum for the fluctuating force is defined by

$$S(\omega) = \frac{1}{2\pi}\int_{-\infty}^{\infty} e^{-i\omega t}G(\tau) \qquad (17.50)$$

17.3 Show that if a harmonic oscillator in its ground state is subjected to a sequence of displacements in phase space that form a closed loop, the state is returned to the ground state up to an overall phase proportional to the area of the loop.

References

1. J.I. Cirac, P. Zoller: Phys. Rev. Letts. **74**, 4091 (1995)
2. H.G. Dehmelt: Bull. Am. Phys. Soc. **20**, 60 (1975)
3. J.C. Bergquist, R.G. Hulet, W.M. Itano, D.J. Wineland: Phys. Rev. Lett. **57**, 1699 (1986): W. Nagourney, J. Sandberg, H.G. Dehmelt: Phys. Rev. Lett. **56**, 2797 (1986); T. Sauter, W. Neuhauser, R. Blatt, P. E. Toschek: Phys. Rev. Lett. **57**, 1696 (1986)
4. Ch. Roos, Th. Zeiger, H. Rohde, H.C. Nägerl, J. Eschner, D. Leibfried. F. Schmidt-kaler, R. Blatt, Phys. Rev. Letts., **83**, 4713 (1999)
5. D. Leibfried, R. Blatt, C. Monroe, D. Wineland: Rev. Mod. Phys. **75**, 281 (2003)
6. J.I. Cirac, R. Blatt, P. Zoller, W.D. Philips: Phys. Rev. A **46**, 2668 (1992)
7. C. Monroe, D.M. Meekhof, B.E. King, S.R. Jefferts, W.M. Itano, D.J. Wineland, P.L. Gould: Phys. Rev. Lett. **75**, 4011 (1995)
8. M.J. Gagen, G.J. Milburn: Phys. Rev. A **45**, 5228 (1992)
9. C. Monroe, D.M. Meekhof, B.E. King, D.J. Wineland: Science, **272**, 1131 (1996)
10. D.F.V. James: Appl. Phys. B **66**, 181, (1998)
11. D.J. Wineland, C. Monroe, W.M. Itano, D. Liebfried, B.E. King, D.M. Meekhof: J. Res. Natl. Stand. Technol. **103**, 259 (1998)
12. F. Schmidt-Kaler, H. Häffner, M. Riebe, S. Gulde, G.P.T. Lancaster, T. Deuschle, C. Becher, C.F. Roos, J. Eschner, R. Blatt: Nature, **422**, 408, (2003)
13. A. Sørensen, K. Mølmer: Phys. Rev. A **62**, 022311 (2000)
14. C.A. Sackett, D. Kielpinski, B.E. King, C. Langer, V. Meyer, C.J. Myatt, M. Rowe, Q.A. Turchette, W.M. Itano, D.J. Wineland, C. Monroe: Nature, **404**, 256 (2000)
15. G.J. Milburn, S. Schneider, D.F.V. James: Fortschritte der Physik, **48**, 801 (2000)
16. D. Leibfried, B. De MArco, V. Meyer, D. Lucas, M. Barrett, J. Britton, W.M. Itano, B. Jelenkovic, C. Langer, T. Rosenband, D.J. Wineland: Nature, **422**, 412 (2003)
17. J.J. Garcia-Ripoll, P. Zoller, J.I. Cirac: Phys. Rev. Lett. **91**, 157901 (2004); L.-M. Duan: Phys. Rev. Lett. **93**, 100502 (2004)

Chapter 18
Light Forces

Abstract In recent years it has become possible to manipulate atoms with light beams, to trap them and cool them to temperatures of milliKelvin and below. The first proposal of laser cooling by Hänsch and Schawlow [1] was based upon Doppler cooling in a two-level atom. Consider an atom irradiated by counterpropagating laser beams tuned to the low frequency side of atomic resonance. The beam counter-propagating with the atom will be Doppler shifted towards resonance, thus increasing the probability of photon absorption. The beam co-propagating with the atom will be frequency-shifted away from resonance, so there will be a net absorption of photons opposing the motion of the atom. The absorbed photon gives the atom a momentum impulse $\Delta p = \hbar k = h/\lambda$ in the direction of the beam. The atom re-emits a photon by spontaneous emission in a random scattering direction. Thus, the net force of the time averaged emitted photons is zero. The resultant force due to the absorption of the photons opposes the atom's motion. By surrounding the atom with three pairs of counter-propagating beams along the x, y and z axes, a drag force opposing the velocity of the atom can be generated. The term "optical molasses" was coined to describe this situation. For two level atoms the minimum temperature achievable, the so-called Doppler limit, was predicted to be $k_B T = \hbar\gamma/2$, where γ is the atomic decay rate. Optical molasses in sodium was first observed in 1985 by *Chu* et al. [2] with a temperature $\sim 240\,\mu$K, close to the Doppler limit.

In order to trap the cold atoms, techniques including purely optical forces, the dipole trap, or a combination of optical and magnetic forces, the magneto-optical trap (MOT), were developed. Experiments performed by W. Phillips [3] and colleagues cooling sodium atoms in a MOT recorded temperatures considerably lower than expected, i.e. $\sim 40\,\mu$K as opposed to $T_D \sim 240\,\mu$K. The explanation for the lower temperatures observed experimentally was given by *Dalibard* and *Cohen-Tannoudji* [4] and *Chu* [5] and colleagues. They showed that optical pumping between atomic magnetic sublevels could result in lower temperatures (sub-Doppler cooling) with a limit close to the recoil energy,

$$k_B T = \hbar\omega_{\text{Rec}},$$

with $\hbar\omega_{\text{Rec}} = \hbar^2 k^2 / 2M$.

More recently, sub recoil cooling schemes have been proposed and implemented, using, for example, the accumulation of ultracold atoms in dark states, which do not couple to the light fields provided the atomic kinetic energy is less than the recoil energy.

In 1997 the Nobel Prize in Physics was awarded to Steven Chu, Claude Cohen-Tannoudji and William Phillips for their work on atom trapping and cooling. In this chapter we shall present a theoretical description of light forces on a two level atoms. We shall follow closely the treatment developed by *Gordon* and *Ashkin* [6] and *Cohen-Tannoudji* [7].

18.1 Radiative Forces in the Semiclassical Limit

We begin with the Hamiltonian for a two–level atom coupled to the electromagnetic field and driven by a near–resonant laser field:

$$\mathcal{H} = \mathcal{H}_A + \mathcal{H}_V + \mathcal{H}_{AL} + \mathcal{H}_{AV} \,. \tag{18.1}$$

Here, $\mathcal{H}_A$ is the atomic Hamiltonian,

$$\mathcal{H}_A = \frac{\vec{P}^2}{2m} + \frac{\hbar \omega_a}{2} \sigma_z \tag{18.2}$$

with $\sigma_z = (|e\rangle\langle e| - |g\rangle\langle g|)$, $\mathcal{H}_V$ is the Hamiltonian of the (vacuum) radiation field, $\mathcal{H}_{AV}$ is the coupling of the atom to this field, and $\mathcal{H}_{AL}$ describes the atom–laser coupling,

$$\mathcal{H}_{AL}(\vec{R}) = -\vec{d} \cdot \vec{E}_L(\vec{R}, t) = \hbar \Omega_L(\vec{R}) \left[\sigma_+ e^{-i(\omega_L t - \Phi(\vec{R}))} + \text{h.c.} \right] , \tag{18.3}$$

where $\sigma_+ = |e\rangle\langle g|$ and the rotating wave approximation has been made. The interaction with the vacuum field is

$$\mathcal{H}_{AV}(\vec{R}) = \hbar \left[\sigma_+ \Gamma(\vec{R}) + \sigma_- \Gamma^\dagger(\vec{R}) \right] , \tag{18.4}$$

where $\Gamma(\vec{R})$ is the bath operator for the vacuum defined as

$$\Gamma(\vec{R}) = \sum_k g_k b_k e^{-i(\omega_k t - \vec{k} \cdot \vec{R})} \tag{18.5}$$

(see Sect. 10.1).

The force acting, $\vec{F}(\vec{R})$, on the atom is given by

$$\vec{F}(\vec{R}) = -\nabla \mathcal{H}_{AL}(\vec{R}) - \nabla \mathcal{H}_{AV}(\vec{R})$$

The coupling of the atom to the vacuum radiation field is responsible for spontaneous emission. This process introduces friction and damping to the system— necessary for Doppler cooling – but it also introduces some of the fluctuations which lead to momentum diffusion, limiting the final temperatures.

In a classical description the atom has a well–defined position and a well–defined momentum. For a semiclassical description to be valid, one therefore requires that the atomic wave packet be sufficiently well localised in position space and in momentum space.

Denote the spatial and momentum spreads of the atomic wave packet by ΔR and ΔP, which must satisfy $(\Delta R)(\Delta P) \geq \hbar$. The force exerted by the laser on the atom varies on a distance determined by the laser wavelength λ_L. If the atom is very well localised on this scale we can neglect the fluctuations of the force due to the spread in atom positions and simply evaluate the force at the average position of the wave packet. Furthermore, as atoms moving with different velocities see a varying Doppler shift, the force due to the laser will also fluctuate due to momentum fluctuations. However if the initial momentum fluctuations are small enough we can neglect fluctuations in Doppler shifts and assume all atoms see a single Doppler shift determined by the mean velocity of the atomic wave packet. We are thus led to two conditions for a semiclassical description to be valid (i) that the position spread be small compared to λ_L,

$$\Delta R \ll \lambda_L \quad \Leftrightarrow \quad k_L \Delta R \ll 1 \tag{18.6}$$

and
(ii) that the velocity spread be small enough that the corresponding spread of Doppler shifts be negligible compared to the natural linewidth γ,

$$\frac{k_L \Delta P}{m} \ll \gamma. \tag{18.7}$$

Combining (18.6) and (18.7) with the uncertainty relation gives

$$\frac{\hbar k_L^2}{m} \ll \gamma \quad \text{or} \quad \hbar \omega_{\text{rec}} \ll \hbar \gamma. \tag{18.8}$$

It turns out that this is equivalent to the condition that the timescale for internal atomic evolution ($\sim \gamma^{-1}$) is much shorter than the timescale for external evolution $[\sim (\hbar k_L^2/m)^{-1}]$, i.e., for damping of the motion.

We now denote the *mean* atomic position and momentum by $\vec{r} = \langle \vec{R} \rangle$ and $\vec{p} = \langle \vec{P} \rangle$. The semiclassical force is given by

$$\vec{f} = -\langle \nabla \mathcal{H}_{\text{AL}}(\vec{r},t) \rangle|_{\vec{r}=\vec{r}_0+\vec{v}_0 t} \,. \tag{18.9}$$

Note that contribution from the atom–vacuum–field coupling term, $\mathcal{H}_{\text{AV}}$, vanishes, due to the symmetry of spontaneous emission. Now

$$-\nabla \mathcal{H}_{\text{AL}}(\vec{r},t) = -\hbar \sigma_+ \exp^{-i\omega_L t} \nabla \left[\Omega(\vec{r}) \exp^{-i\Phi(\vec{r})} \right] + \text{h.c.}, \tag{18.10}$$

and we can write

$$\nabla\left[\Omega_L(\vec{r})\exp^{-i\Phi(\vec{r})}\right] = \Omega_L(\vec{r})\exp^{-i\Phi(\vec{r})}\left[\vec{\alpha}(\vec{r}) - i\vec{\beta}(\vec{r})\right] \qquad (18.11)$$

where

$$\vec{\alpha}(\vec{r}) \equiv \frac{\nabla\Omega_L(\vec{r})}{\Omega_L(\vec{r})}, \qquad \vec{\beta}(\vec{r}) = \nabla\Phi(\vec{r}). \qquad (18.12)$$

$\vec{\alpha}(\vec{r})$ – characterises spatial variation of the Rabi frequency.

$\vec{\beta}(\vec{r})$ – characterises spatial variation of the phase.

This gives the following expression for the mean force,

$$\vec{f}(\vec{r},t) = -2\mathrm{Re}\left\{\langle\sigma_+(t)\rangle\,\hbar\Omega_L(\vec{r})\exp^{-i[\omega_L t + \Phi(\vec{r})]}\left[\vec{\alpha}(\vec{r}) - i\vec{\beta}(\vec{r})\right]\right\}$$

$$= -2\hbar\Omega_L(\vec{r})\left[u(t)\vec{\alpha}(\vec{r}) + v(t)\vec{\beta}(\vec{r})\right], \qquad (18.13)$$

where $\langle\sigma_+(t)\rangle = \mathrm{Tr}[\sigma_+\rho_{\mathrm{int}}(t)]$, with $\rho_{\mathrm{int}}(t)$ the internal atomic density operator, and

$$u(t) = \mathrm{Re}\left\{\langle\sigma^+(t)\rangle\exp^{-i[\omega_L t + \Phi(\vec{r})]}\right\} \qquad (18.14)$$

$$v(t) = \mathrm{Im}\left\{\langle\sigma^+(t)\rangle\exp^{-i[\omega_L t + \Phi(\vec{r})]}\right\}. \qquad (18.15)$$

The average values of the internal atomic operators are computed from the optical Bloch equations, which, in terms of the variables defined above, take the form

$$\dot{u} = -(\gamma/2)u + (\delta + \dot{\Phi})v$$

$$\dot{v} = -(\gamma/2)v - (\delta + \dot{\Phi})u - 2\Omega_L w$$

$$\dot{w} = -(\gamma/2) - \gamma w + 2\Omega_L v \qquad (18.16)$$

where $w = [\langle\sigma^+\sigma^-\rangle - \langle\sigma^-\sigma^+\rangle]/2$ is the inversion, $\delta = \omega_L - \omega_A$ is the laser–atom detuning, and

$$\dot{\Phi} = \vec{v}\cdot\nabla\Phi = \vec{v}\cdot\beta. \qquad (18.17)$$

18.2 Mean Force for a Two–Level Atom Initially at Rest

Some physical insight into the nature of the forces acting on the atom can be gained by considering the zero velocity case. Consider a two–level atom initially at rest at the origin:

$$\vec{r} = 0, \quad \vec{p} = 0. \qquad (18.18)$$

The steady state solutions of the optical Bloch equations are:

$$u_{ss} = \frac{\delta}{2\Omega_L} \frac{s}{1+s} \tag{18.19}$$

$$v_{ss} = \frac{\gamma}{4\Omega_L} \frac{s}{1+s} \tag{18.20}$$

$$w_{ss} = -\frac{1}{2(1+s)} \tag{18.21}$$

where s is the saturation parameter,

$$s = \frac{2\Omega_L^2}{\delta^2 + (\gamma^2/4)}. \tag{18.22}$$

Note that the population of the upper state is given by

$$P_{ss}^e = w_{ss} + 1/2 = \frac{1}{2} \frac{s}{1+s}. \tag{18.23}$$

Substituting the steady-state values u_{ss} and v_{ss} in the expression (18.13) for the force gives the average force in the case where the internal degrees of freedom have reached their steady state (adiabatic approximation). The mean force $\vec{f} f$ is the sum of two contributions proportional to v and u, respectively. These two contributions are known as the *spontaneous force* and the *dipole force* (or *dissipative force* and *reactive force*).

The spontaneous force is due to the radiation pressure. It results from the absorption of photons from a traveling wave laser from which the atom receives a transfer of momentum. The subsequent spontaneous emission of photons does not contribute to the average force since spontaneous emission occurs with equal probabilities in all directions. The dissipative force is zero for a stationary atom in a standing wave since absorption of photons from both directions cancel. The spontaneous force is given by the component proportional to v_{ss},

$$\vec{f}_{spon} = -2\hbar\Omega_L v_{ss} \vec{\beta}. \tag{18.24}$$

Note that $-2\Omega_L v_{ss} = \gamma\rho_{ss}^e$ which is simply the mean number of photons spontaneously emitted per unit time.

For a plane wave,

$$\vec{E}_L(\vec{r},t) = \vec{E}_0 \cos\left(\omega t - \vec{k}_L \cdot \vec{R}\right) \tag{18.25}$$

The phase of the field is $\Phi(\vec{r}) = -\vec{k}_L \cdot \vec{R}$ which gives

$$\beta = \nabla\Phi|_{\vec{r}=0} = -\vec{k}_L, \tag{18.26}$$

and the force,

$$f_{\mathrm{spon}} = \hbar k_{\mathrm{L}} \left(\frac{\gamma}{2}\right) \frac{2\Omega_{\mathrm{L}}^2}{\delta^2 + \frac{\gamma^2}{4} + 2\Omega_{\mathrm{L}}^2}. \tag{18.27}$$

For low intensity $(s \ll 1)$ $f_{\mathrm{spon}} \propto \Omega_{\mathrm{L}}^2 \sim I_{\mathrm{L}}$ (laser intensity). For high intensity laser fields $f_{\mathrm{spon}} \to \hbar k_{\mathrm{L}} \frac{\gamma}{2}$, that is the maximum force is limited by the spontaneous emission rate. The maximum acceleration imparted to the atom by the dissipative force is,

$$\vec{a}_{\mathrm{max}} = \frac{\hbar \vec{k}_{\mathrm{L}}}{m} \frac{\gamma}{2}. \tag{18.28}$$

While the recoil velocity $v_{\mathrm{recoil}} = \frac{\hbar k_{\mathrm{L}}}{m}$ due to absorption of a single photon is small $(\sim 3 \, \mathrm{cm/sec}$ for sodium, $\sim 3 \, \mathrm{mm/sec}$ for cesium), the number of fluorescent cycles/sec is high for intense fields. For sodium $\gamma^{-1} = 1.6 \times 10^{-8}$ s, which gives $a_{\mathrm{max}} \sim 10^6 \, \mathrm{m/s^2}$ which is 10^5 times the acceleration due to gravity.

The dipole force is given by the component proportional to u_{ss}:

$$\vec{f}_{\mathrm{dip}} = -2\hbar \Omega_{\mathrm{L}} u_{\mathrm{ss}} \vec{\alpha} \tag{18.29}$$

In a plane travelling wave, $\vec{\alpha} = 0$ as the amplitude is independent of $\vec{r}$ and hence $\vec{f}_{\mathrm{dip}} = 0$. The dipole force is only nonzero if the laser field is a superposition of plane waves, e.g., a *laser standing wave*. The dipole force results from the absorption of a photon with momentum $\hbar k$ from a standing wave and reemission of a photon in the opposite direction, that is with momentum $-k$. This results in a total momentum gain for the atom $2\hbar k$. For a travelling wave laser the dipole force is zero since photons can only be reemitted into the same direction from which they where absorbed.

Using the solution for u_{ss},

$$\vec{f}_{\mathrm{dip}} = -\hbar \delta \frac{\nabla(\Omega_{\mathrm{L}}^2)}{\delta^2 + (\gamma^2/4) + 2\Omega_{\mathrm{L}}^2} \tag{18.30}$$

For $\delta < 0$ $(\omega_{\mathrm{L}} < \omega_{\mathrm{A}})$, the dipole force pushes the atom towards regions of higher intensity since $\mathrm{sgn}\{\vec{f}_{\mathrm{dip}}\} = \mathrm{sgn}\{\nabla I_{\mathrm{L}}\}$ (and vice–versa for $\delta > 0$). For each value of Ω_{L}^2 (with $\Omega_{\mathrm{L}} \gg \gamma$), $\vec{f}_{\mathrm{dip}}$ is optimised for $\delta \sim \Omega_{\mathrm{L}}$, and

$$\left(\vec{f}_{\mathrm{dip}}\right)_{\mathrm{max}} \simeq \frac{\hbar \nabla(\Omega_{\mathrm{L}}^2)}{\Omega_{\mathrm{L}}} \simeq \hbar \nabla \Omega_{\mathrm{L}} \tag{18.31}$$

Hence, $\vec{f}_{\mathrm{dip}}$ increases with laser intensity and is not bounded like $\vec{f}_{\mathrm{spon}}$ so that much greater acceleration is possible with the dipole force.

The dipole force can be derived from an effective potential U as

$$\vec{F}_{\mathrm{dip}} = -\nabla U, \tag{18.32}$$

with

$$U(\vec{r}) = \frac{\hbar\delta}{2} \ln\left[1 + \frac{2\Omega_L^2(\vec{r})}{\delta^2 + (\gamma^2/4)}\right].$$

(18.33)

For $\delta < 0$, a region of maximum intensity appears as an attractive *potential well* for the atom. For a given Ω_L, the maximum depth of the potential well occurs for a saturation parameter $s \simeq 4$ giving

$$|U_{max}| \simeq 0.6\left|\hbar\Omega_L^{max}\right|.$$

(18.34)

18.3 Friction Force for a Moving Atom

Now suppose that the atom is moving with a velocity $\vec{v}$, such that

$$\vec{r} = \vec{v}t \quad (\vec{r} = 0 \text{ at } t = 0).$$

(18.35)

As previously we assume also that we are in the semiclassical limit so that fluctuations in velocity due to momentum dispersion in the wave packet can be ignored. The velocity, $\vec{v}$, is determined by the average momentum of the wave packet. We assume a laser plane wave with wave vector $\vec{k}_L$, so that the Rabi frequency

$$\Omega_L(\vec{r} = \vec{v}t) = \Omega_0 = \text{constant}$$

(18.36)

while

$$\Phi(\vec{r}) = -\vec{k}_L \cdot \vec{r} \quad \Rightarrow \quad \dot{\Phi} = \frac{d\vec{r}}{dt} \cdot \nabla\Phi = \vec{v} \cdot \nabla\Phi = -\vec{k}_L \cdot \vec{v}.$$

(18.37)

Since Ω_L and $\dot{\Phi}$ are time independent (and hence the coefficients of the optical Bloch equations are time independent), the steady state solutions are as before (for a stationary atom), but with $\omega_L \rightarrow \omega_L - \vec{k}_L \cdot \vec{v}$, i.e., an atom moving with velocity $\vec{v}$ "sees" the laser frequency shifted by the *Doppler shift* $-\vec{k}_L \cdot \vec{v}$. Hence, the force is

$$\vec{f} = \hbar\vec{k}_L \frac{\gamma}{2} \frac{2\Omega_0^2}{(\delta - \vec{k}_L \cdot \vec{v})^2 + (\gamma^2/4) + 2\Omega_0^2}.$$

(18.38)

We assume $\vec{k}_L = -k_L\vec{e}_x$ and consider motion along the x axis. For $\delta < 0$ (and $v_x > 0$), the force is negative and has a maximum where $\delta = -k_L v_x$, i.e., where the apparent laser frequency $\omega_L + k_L v_x = \omega_A$ (see Fig. 18.1). Near $v_x = 0$, one can make an expansion in the velocity, writing where the term linear in v_x is a friction force, since it is proportional to v_x. The coefficient of proportionality, α, is called the *friction coefficient* and is derived as

$$\alpha = -\hbar k_L^2 \frac{s}{(1+s)^2} \left(\frac{\delta\gamma}{\delta^2 + (\gamma^2/4)}\right)$$

(18.39)

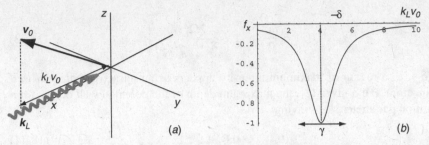

Fig. 18.1 (a) Atom moving with velocity v_0 in a laser travelling wave with wave vector $\vec{k}_L$. (b) Mean force experienced by the atom versus $k_L v_0$ in units of $\hbar k_L \gamma$

where $s = (2\Omega_0^2)/[\delta^2 + (\gamma^2/4)]$. The friction coefficient is optimised for $\delta = -\gamma/2$ and $s = 1$ ($\Omega_0 = \gamma/2$), giving

$$\alpha_{\max} = \frac{\hbar k_L^2}{4}, \tag{18.40}$$

so that the atomic velocity is damped at a rate

$$\frac{\alpha}{m} = \frac{\hbar k_L^2}{4m} = \frac{\varepsilon_{\mathrm{rec}}}{2\hbar} \tag{18.41}$$

where $\varepsilon_{\mathrm{rec}}$ is the recoil energy, i.e., the external atomic variables, such as the velocity, have a characteristic damping time on the order of $\hbar/\varepsilon_{\mathrm{rec}}$ (typically between $1 - 100\,\mu s \gg \gamma^{-1}$).

18.3.1 Laser Standing Wave—Doppler Cooling

The interaction with a standing wave can be considered as the interaction with a superposition of two counter-propagating plane waves with the same amplitude E_0. However the force exerted by the two standing waves is not just the sum of the radiation pressures of the two counter-propagating plane waves. Interference terms play an important role.

We now consider a laser standing wave along the x axis, linearly polarised along the z axis, so that

$$\begin{aligned}\vec{E}_L(\vec{r},t) &= \vec{\varepsilon}_z \mathscr{E}_0 \left[\cos(\omega_L t - k_L x) + \cos(\omega_L t + k_L x)\right] \\ &= 2\vec{\varepsilon}_z \mathscr{E}_0 \cos(k_L x) \cos(\omega_L t)\end{aligned} \tag{18.42}$$

The phase of the field is the same everywhere ($\vec{\beta} = 0$), while the Rabi frequency is position dependent and can be written

$$\Omega_L(x) = 2\Omega_0 \cos(k_L x) \tag{18.43}$$

where $\Omega_0 = -(d\mathscr{E}_0)/(2\hbar)$, and

$$\vec{\alpha} = -k_{\mathrm{L}}\tan(k_{\mathrm{L}}x)\vec{\varepsilon}_x \ . \tag{18.44}$$

For a moving atom, one can replace x with $v_x t$ in the optical Bloch equations, then $\Omega_{\mathrm{L}}(x)$ becomes a sinusoidal function of t. In general, it is impossible to solve these equations analytically.

For small velocities we can use an approximation scheme, first introduced by *Gordon* and *Ashkin* [6], in which one makes an expansion in powers of $k_{\mathrm{L}}v_x/\gamma$. The zeroth order term represents the "adiabatic" solution, corresponding to the situation where the motion is so slow that the internal state of an atom passing the position x is the same as if it were at rest at x. The first order term gives the first correction to the adiabatic approximation. Note that $k_{\mathrm{L}}v_x/\gamma$ is equal to the ratio between the distance $v_x\gamma^{-1}$ over which the atom travels during the internal response time γ^{-1} and the laser wavelength k^{-1}.

The expansion proceeds as follows. We write

$$\dot{\mathbf{u}} = \left(\frac{\partial}{\partial t} + v\frac{\partial}{\partial x}\right)\mathbf{u} = \mathscr{B}\mathbf{u} - \mathbf{s} \tag{18.45}$$

where

$$\mathbf{u} = \begin{pmatrix} u \\ v \\ w \end{pmatrix}, \quad \mathbf{s} = \begin{pmatrix} 0 \\ 0 \\ \gamma/2 \end{pmatrix}, \quad \mathscr{B} = \begin{bmatrix} -\gamma/2 & \delta & 0 \\ -\delta & -\gamma/2 & -2\Omega_{\mathrm{L}}(x) \\ 0 & 2\Omega_{\mathrm{L}}(x) & -\gamma \end{bmatrix} \tag{18.46}$$

and the "hydrodynamic derivative" $\mathrm{d}/\mathrm{d}t = (\partial/\partial t) + v_x(\partial/\partial x)$ has been used. In the steady state, we write

$$v\frac{\partial}{\partial x}\mathbf{u} = \mathscr{B}\mathbf{u} - \mathbf{s} \tag{18.47}$$

and insert the expansion

$$\mathbf{u} = \mathbf{u}^{(0)} + \mathbf{u}^{(1)} + \ldots \tag{18.48}$$

of $\mathbf{u}$ in powers of $k_{\mathrm{L}}v_x/\gamma$. To order 0,

$$\mathbf{u}^{(0)} = \mathscr{B}^{-1}\mathbf{s} \ , \tag{18.49}$$

which is the steady state Bloch vector for an atom at rest in x. To order 1, we get

$$v\frac{\partial}{\partial x}\mathbf{u}^{(0)} = \mathscr{B}\mathbf{u}^{(1)} \quad \Rightarrow \quad \mathbf{u}^{(1)} = \mathscr{B}^{-1}v\frac{\partial}{\partial x}\mathbf{u}^{(0)} \ . \tag{18.50}$$

Inserting the expansion for u into the expression for the force, one obtains

$$f_x(v_x, x) = \hbar\frac{s}{1+s}\,\delta k_{\mathrm{L}}\tan(k_{\mathrm{L}}x)\left\{1 + \frac{\gamma^2(1-s) - 2s^2[\delta^2 + (\gamma^2/4)]}{\gamma[\delta^2 + (\gamma^2/4)](1+s)^2}\,k_{\mathrm{L}}v_x\tan(k_{\mathrm{L}}x)\right\} \ ,$$
$$\tag{18.51}$$

where

$$s = \frac{8\Omega_0^2 \cos^2(k_L x)}{\delta^2 + (\gamma^2/4)} . \tag{18.52}$$

For weak intensities, the interference effects average to zero over a wavelength, hence the friction force averaged over a wavelength coincides with the sum of the two friction forces exerted by the two counterpropagating plane waves. In particular,

$$\bar{f} = -\alpha v \quad \text{with} \quad \alpha = -4\hbar k_L^2 \Omega_0^2 \frac{\delta \gamma}{[\delta^2 + (\gamma^2/4)]^2} . \tag{18.53}$$

Hence, because of the Doppler effect, for $\delta < 0$ the atom gets closer to resonance with the wave opposing its motion and further away from resonance with the other wave, so that the two forces exerted by the two waves become unbalanced and the net force opposes the motion of the atom. This is the mechanism for Doppler cooling.

18.4 Dressed State Description of the Dipole Force

If the Rabi frequency is large compared to the spontaneous emission rate the dressed atom picture provides an intuitive interpretation. We may write the interaction Hamiltonian for a two level atom, with frequency ω_a, interacting with a single mode standing wave laser field with frequency $\Omega_L = \omega_a + \Delta$

$$H_I(x) = \frac{\hbar \Delta}{2} \sigma_z + \hbar g(x) \left(a\sigma_+ + a^\dagger \sigma_- \right) , \tag{18.54}$$

where $g(x) = gf(x)$ is the spatially dependent vacuum Rabi frequency, $\sigma_+ = \sigma_-^\dagger = |e\rangle\langle g|$ and $a, a^\dagger$ are the annihilation and creation operators for the field. The dressed states (see section 10.2) for this interaction are:

$$|n, +\rangle = \sin\theta_n(x)|n, e\rangle + \cos\theta_n(x)|n+1, g\rangle \tag{18.55}$$
$$|n, -\rangle = \cos\theta_n(x)|n, e\rangle - \sin\theta_n(x)|n+1, g\rangle \tag{18.56}$$

where $|n, e\rangle = |n\rangle \otimes |e\rangle$, $|n, g\rangle = |n\rangle \otimes |g\rangle$ with $|n\rangle$ a photon number eigenstate. The coefficients are given by

$$\cos 2\theta_n(x) = -\frac{\Delta/2}{\Omega_n(x)} , \tag{18.57}$$

$$\sin 2\theta_n(x) = \frac{g(x)}{\Omega_n(x)} , \tag{18.58}$$

and $\Omega_n(x) = \sqrt{g^2(x)(n+1) + \Delta^2/4}$ is the effective Rabi frequency. The corresponding energy levels are $U_{n,\pm}(x) = \pm\hbar\Omega_n(x)$. This is a generalisation of the dressed state picture discussed in Sect. 10.2. since the expansion coefficients are now position dependent. The dressed states form doublets separated the photon energy $\hbar\omega_L$.

Typically the optical field is well approximated by a coherent state of large amplitude. The Poisson nature of the photon number distribution in such a state enables us to replace $\Omega_n(x)$ by its average value $\Omega(x) = \sqrt{g^2 I(x) + \Delta^2/4}$ where $I(x) = f^2(x)\bar{n}$ and $\bar{n}$ is the average photon number is the field. Thus $I(x)$ is the spatially varying intensity of the field.

The variation in x of the energies of the dressed states are shown in Fig. 18.2 for two manifolds $\varepsilon(N)$ and $\varepsilon(N-1)$. The manifolds are connected by spontaneous emission. Outside the laser beam the energy levels of the dressed states tend to the uncoupled states and are separated by the detuning Δ. Within the laser beam $I(x)$ is nonzero and the splitting

$$\hbar\Omega(x) = \hbar\sqrt{4g^2 I(x) + \Delta^2} \tag{18.59}$$

between the two dressed levels of the same manifold increases with increasing values of $I(x)$.

We shall now use this dressed state picture to provide a physical description of the dipole force. Initially we shall neglect spontaneous emission. We assume the system is either in state $|n,+\rangle$ or $|n,-\rangle$ and satisfies the semiclassical limit (7). Assuming the atom velocity is sufficiently slow so that nonadiabatic transitions from one level to another can be neglected, the system will follow adiabatically the level $|n,+\rangle$ or $|n,-\rangle$ in which it is found initially. The energy curves in Fig. 18.2 are then potential energy level curves $V_{e,n}(x)$ and $V_{g,n}(x)$. The dressed atom therefore experiences a force

$$F_g = -\nabla V_{g,n}(x) = -\frac{\hbar}{2}\nabla\Omega(x) , \tag{18.60}$$

if it is in level $|g\rangle$ and a force

$$F_e = -\nabla V_{e,n}(x) = \frac{\hbar}{2}\nabla\Omega(x) = -F_g, \tag{18.61}$$

if it is in level $|e\rangle$.

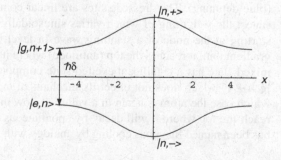

Fig. 18.2 The dependance of the dressed state energy levels on position due to a spatially varying intensity, a Gaussian beam profile

Fig. 18.3 A slow moving
atom in a standing wave
is cooled by spontaneous
emission: Sisyphus cooling

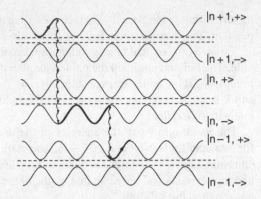

This dependence of the force on the internal state is very similar to the Stern–Gerlach effect that occurs for a spin $\frac{1}{2}$ particle in an inhomogeneous magnetic field. This effect is known as the optical Stern–Gerlach effect will be described in more detail in Sect. 18.6. It also provides a mechanism for atom mirrors where atoms in one of the dressed states are repelled by a light field.

Spontaneous emission causes the dressed atom to make a transition to a lower manifold. In doing so it may change the type of state it is in e.g. a transition $|g, n+1> \rightarrow |g, n\rangle$ by emitting a photon. This causes the sign of the force to change abruptly. The time intervals τ_1 and τ_2 spent in the different dressed levels between two successive quantum jumps are Poisson random variables. This gives rise to the fluctuations in the dipole force.

The mean dipole force in the steady state is given by the mean of the forces F_1 and F_2 weighted by the proportion of time spent in the type 1 and 2 levels, which are simply the steady state populations of the dressed levels π_1 and π_2. Thus

$$\langle F_{\text{dip}} \rangle = F_1 \pi_1 + F_2 \pi_2 = -\frac{\hbar^2}{2} \nabla \Omega(x)(\pi_1 - \pi_2) \ . \tag{18.62}$$

This corresponds to the expression derived earlier.

The dressed state picture gives a clear illustration of the mechanism for cooling a moving atom in a strong standing wave, known as Sisyphus cooling. In this case $I(x) = I_0 \sin^2(k_L x)$. Figure 18.3 represents the dressed states for a positive detuning (blue detuning). The dressed states are linear combinations of ground and excited states, the weighting of which varies sinusoidally in space. If we follow an atom starting at the node of a standing wave in level $|g, n+1\rangle$, it climbs up the light gradient until it reaches the top (antinode) where its decay rate is a maximum as the mixed state has a significant excited state component. It may then jump into level $|g, n+1\rangle$ which does not effect its mechanical motion or jump into level $|e, n\rangle$ in which case the atom is again in a valley. It now must climb against the gradient to reach the top where it will decay by spontaneous emission again. The mechanism has been named Sisyphus cooling by analogy with the story from Greek mythology.

Between two spontaneous emissions, the total energy (kinetic plus potential) of the atom is conserved. When the atom climbs up hill, its kinetic energy is transformed into potential energy by stimulated emission processes which redistribute photons between the two counterpropagating waves at rate Ω. Atomic momentum is therefore transferred to laser photons. The total atomic energy is then dissipated by spontaneous emission which carries away part of the atomic potential energy. The force reaches its maximum value for velocities such that the atom travels over a distance of the order of a wavelength between two spontaneous emissions ($k_L v_0 \sim \gamma$). The magnitude of this friction force is directly related to the modulated depth (Ω) of the dressed energy levels and hence increases indefinitely with laser intensity.

18.5 Atomic Diffraction by a Standing Wave

We shall consider the deflection of a beam of two-level atoms by a classical standing wave light field in a cavity. The atomic beam is normally incident on the standing wave and experiences an exchange of momentum with the photons in the light wave. We shall assume that the frequency of the light field is well detuned from the atomic resonance so that we may neglect spontaneous emission.

The Hamiltonian describing the interaction is

$$\mathcal{H} = \hbar \frac{\omega_0}{2} \sigma_z + \frac{p^2}{2m} + \hbar\Omega(\sigma_- e^{-i\omega t} + \sigma_+ e^{i\omega t})\cos kx\,, \qquad (18.63)$$

where p is the centre of mass momentum of the atom along the transverse (x direction), m is the atomic mass, σ_z and $\sigma_\pm$ are the pseudo spin operators for the atom, ω_0 and ω are the atomic and field frequencies, $k = \omega/c$ is the wave number of the standing wave, and $\Omega = \mu\varepsilon_0/\hbar$ the Rabi frequency with μ being the dipole moment and ε_0 the maximum field amplitude. We shall assume that the interaction time is sufficiently small that the transverse kinetic energy absorbed by the atom during the interaction can be neglected. This is known as the Raman–Nath regime and requires $t < 2\pi/\omega_r$, where the recoil energy is $\hbar\omega_r = (2r\hbar k)^2/2m$ where r is an integer. This is equivalent to neglecting the term $p^2/2m$ in the Hamiltonian.

Transforming to the interaction picture with $\mathcal{H}_0 = \hbar\omega\sigma_z$ the Hamiltonian may be written in the form

$$\mathcal{H} = \hbar\frac{\delta\omega}{2}\sigma_z + 2\hbar\Omega\sigma_x \cos kx\,, \qquad (18.64)$$

where $\delta\omega = \omega_a - \omega$ and $2\sigma_x = \sigma_+ + \sigma_-$. This Hamiltonian may be diagonalised and written in the form

$$\mathcal{H} = V(x)[\cos\theta(x)\sigma_z + \sin\theta(x)\sigma_x] \qquad (18.65)$$

where

$$V(x) = \hbar\sqrt{(\delta\omega)^2 + (2\Omega\cos kx)^2},$$

$$\cos\theta(x) = \frac{\delta\omega}{\sqrt{(\delta\omega)^2 + (2\Omega\cos kx)^2}},$$

$$\sin\theta(x) = \frac{2\Omega\cos kx}{\sqrt{(\delta\omega)^2 + (2\Omega\cos kx)^2}}.$$

In the limit of large detuning ($\delta\omega \gg 2\Omega\cos kx$),

$$\cos\theta \approx 1, \quad \text{and} \quad \sin\theta \approx 0,$$

so that

$$V(x) \approx \hbar\delta\omega\left(1 + \frac{2\Omega^2\cos^2 kx}{\delta\omega^2}\right). \tag{18.66}$$

This leads to the effective Hamiltonian

$$\mathcal{H}_{\text{eff}} = \hbar\frac{\delta\omega}{2}\sigma_z + \left(\frac{2\hbar\Omega^2\cos^2 kx}{\delta\omega}\right)\sigma_z. \tag{18.67}$$

The atomic state vector in the coordinate representation may be written as

$$\langle x|\psi(t)\rangle = a(x,t)|a\rangle + b(x,t)|b\rangle, \tag{18.68}$$

where $|a\rangle$ and $|b\rangle$ are the upper and lower atomic states, and $a(x,t)(b(x,t))$ are the probability amplitudes for the atom to be in the upper (lower) state at the transverse coordinate x at time t.

We assume that the atoms are initially in their ground state with a Gaussian wavefunction

$$a(\xi,0) = 0,$$

$$b(\xi,0) = (\pi\sigma^2)^{-1/4}\exp\left(-\frac{\xi^2}{2\sigma^2}\right), \tag{18.69}$$

where $\xi = kx$ and σ is proportional to the r.m.s. transverse position spread of the input beam. This may be written as

$$b(\xi,0) = \left(\frac{2\sigma_\eta^2}{\pi}\right)^{1/4}\exp[-(\sigma_\eta\xi)^2], \tag{18.70}$$

where σ_η is the r.m.s. transverse momentum spread of the input beam scaled to the photon momentum $\hbar k(\eta = p/\hbar k)$. The Schrödinger equation in the large detuning limit is

$$\frac{\partial}{\partial\tau}\begin{pmatrix} a \\ b \end{pmatrix} = \begin{pmatrix} -i\Delta - \frac{i\cos^2\xi}{2\Delta} & 0 \\ 0 & i\Delta + \frac{i\cos^2\xi}{2\Delta} \end{pmatrix}\begin{pmatrix} a \\ b \end{pmatrix} \tag{18.71}$$

where $\tau = \Omega t$ and $\Delta = \delta\omega/2\Omega$. We shall assume that the atom interacts with a field of constant amplitude for a time τ. The Rabi frequency is Ω.

The solution for $b(\varepsilon, \tau)$ may be written in the form [8]

$$b(\xi, \tau) = \exp\left[i\left(\Delta + \frac{1}{4\Delta}\right)\tau\right] \sum_{n=-\infty}^{\infty} i^n J_n\left(\frac{\tau}{4\Delta}\right) \exp(2in\xi)b(\xi, 0). \qquad (18.72)$$

Taking the Fourier transform of this relationship shows the effect in momentum space is a convolution

$$\tilde{b}(\eta, \tau) = \exp\left[i\left(\Delta + \frac{1}{4\Delta}\right)\tau\right] \sum_{n=-\infty}^{\infty} i^n J_n\left(\frac{\tau}{4\Delta}\right) \delta(\eta - 2n) * \tilde{b}(\eta, 0)$$

$$= \exp\left[i\left(\Delta + \frac{1}{4\Delta}\right)\tau\right] \sum_{n=-\infty}^{\infty} i^n J_n\left(\frac{\tau}{4\Delta}\right) \tilde{b}(\eta - 2n, 0) \qquad (18.73)$$

where $\tilde{b}$ denotes the Fourier transform of b.

The scattered ground state wavefunction is a superposition of Gaussian modulated plane waves with momenta $p = 2n\hbar k$. The momentum transferred from the field to the atom is in even multiplies of $\hbar k$ corresponding to the absorption of a photon from the $(+k)$ component, followed by induced emission into the $(-k)$ component of the standing wave. The final output momentum probability distribution is composed of a comb of images of the initial momentum distribution. In order to resolve these peaks it is necessary to have a narrow initial momentum spread such that $\Delta p \ll 2\hbar k$ or $\sigma_\eta \ll 1$. The output momentum distribution is shown in Fig. 18.4 for $\sigma_\eta = 0.1$, where the propagation time after the interaction is assumed short so that further spreading has been neglected. The above result holds for large atomic detuning. For smaller atomic detunings spontaneous emission becomes important. Since the recoil imparted to an atom by a spontaneously emitted photon occurs in a random direction, exchanges of momentum in non-integral multiplies of $\hbar k$ may occur and the diffractive peaks will be smeared out. That is, as the laser is tuned closer

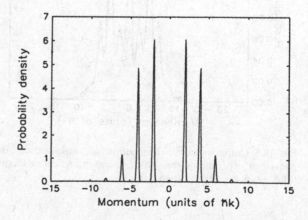

Fig. 18.4 Output momentum distribution for atoms scattered from a standing optical wave in the Kapitza-Dirac regime in the large detuning limit. Initial RMS momentum uncertainty is $\sigma_\eta = 0.1$ (units of $\hbar k$) and the normalised interaction time $\tau/\Delta = 10$

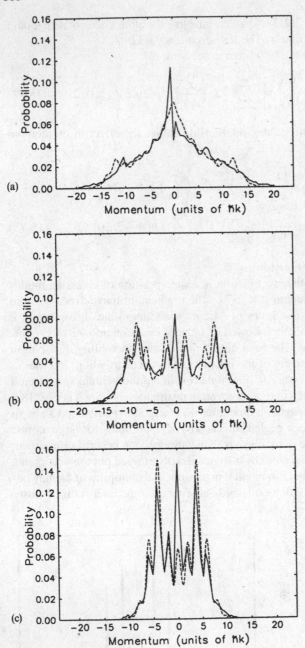

Fig. 18.5 Output momentum distribution for atoms scattered from a standing optical wave. Comparison of experimental data [20] (—) and theoretical predictions [21] (- - -) for (a) $\Delta = 0$, (b) $\Delta = 0.6$, (c) $\Delta = 1.2$ (from [10])

to the atomic resonance one moves from the diffractive to the diffusive regime. The transition from the diffractive to the diffusive regime has been demonstrated in an experiment by *Gould* et al. [11]. A fit to their data using a calculation which includes spontaneous emission has been given by *Tan* and *Walls* [10] and is shown in Fig. 18.5.

18.6 Optical Stern–Gerlach Effect

In the previous section, we discussed the diffraction of atoms in their ground state from a standing wave light field. In order to resolve the diffraction peaks in the momentum distribution it was necessary that the initial atomic spatial wave packet be larger than a wavelength. In this regime, the momentum transfer is symmetrical about $\Delta p = 0$ and is the same for input atoms in either the ground or excited states. As the spatial extent of the input wave packet is reduced to a fraction of the optical wavelength, the momentum transfer becomes asymmetrical and dependent upon the initial atomic state. Figure 18.6 shows the outgoing probability density of the atomic momentum for an initial atomic beam width $\sigma = 0.3$. The solid curve is for atoms in the ground state while the dotted curve is for the excited state.

In the limit when the spatial extent of the input wave packet is very small compared to the optical wavelength, an input beam is split into two beams depending on its atomic state. This is called the optical Stern–Gerlach limit in which the atom interacts with only a small part of the light wave ($\Delta x^{IN} \ll 1/k$) and the individual photon exchanges are not resolvable ($\Delta p \gg \hbar k$). In the large-detuning limit the momentum transfer, which can be many times the photon momentum, depends on the intensity gradient of the optical field. An experimental demonstration of the optical Stern–Gerlach effect has been given by *Sleator* et al. [12] using a beam of metastable He atoms (Fig. 18.7).

Fig. 18.6 Output momentum distribution for atoms scattered from a standing optical wave in the Stern-Gerlach regime. Initial rms momentum uncertainty is $\sigma_\eta = 2.4$ (units of $\hbar k$). The atomic beam is incident midway between a node and an antinode and the normalised interaction time $\tau/\Delta = 20$. (**a**) *solid curve*, ground state atoms, (**b**) *dotted curve*, excited state atoms

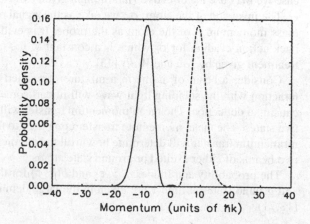

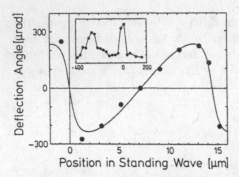

Fig. 18.7 Deflection of an atomic beam by a standing wave as a function of the atomic beam position in the standing wave. The detuning from resonance was $\Delta/2\pi = 160\,\text{MHz}$. Position 0 on the horizontal scale was arbitrarily chosen to be at a node. Inset: Atomic intensity profile at the detector for the atomic beam at a position of $-11\,\mu\text{m}$ in the standing wave. The peak at zero angle is due to undeflected atoms [12]

We shall demonstrate how the standing wave optical field encodes information about the atomic state in the centre of mass momenta of the outgoing atoms. A non-destructive determination of the atomic state may be useful in conjunction with micromaser experiments in which the photon statistics within the micromaser cavities are indirectly measured via their effects on the states of the Rydberg atoms which pass through them. The ability to spatially separate atoms according to their state without destroying them also makes it possible to consider their subsequent coherent recombination in an atomic interferometer.

In order to measure their atomic state we shall require that the momentum transfer be well correlated with the initial atomic state and that the process of splitting the input beam should not cause this atomic state to change. We shall demonstrate how one may make a QND measurement of the atomic inversion. We consider the case where the standing wave field is far detuned from the atomic resonance. In this case we may use the effective Hamiltonian given by (18.67).

The inversion of the atom, σ_z, we take as the signal observable and the centre of mass momentum p of the atom as the probe. It is evident that the Hamiltonian is back action evading for σ_z which is a constant of the motion. We shall follow the treatment given by *Tan* and *Walls* [13].

Consider a beam of atoms in a mixture of excited and ground states. The interaction with the standing light wave will impart some momentum to these atoms causing a deflection. The mean momentum transfer will have opposite signs for the two states. The mean momentum transfer compared to the standard deviation of the momentum transfer will determine how well the atomic beam can be separated into two beams of either excited or ground state atoms.

The probability amplitudes $a(\xi, \tau)$ and $b(\xi, \tau)$ for the atom to be in the excited and ground states at time τ are given in the large detuning limit by the solution of (18.71).

After passage through the field, the mean momentum transfer to this beam is given by

$$\langle \eta^{\mathrm{OUT}} \rangle = \int_{-\infty}^{\infty} a(\xi,\tau)^* \left(\frac{-i\partial}{\partial \xi} \right) a(\xi,\tau) \mathrm{d}\xi + \int_{-\infty}^{\infty} b(\xi,\tau)^* \left(\frac{-i\partial}{\partial \xi} \right) b(\xi,\tau) \mathrm{d}\xi$$

(18.74)

where $\eta = p/\hbar k$ is a normalised momentum. Similarly the mean squared momentum transfer is

$$\sigma_\eta^2 = \langle \eta^{\mathrm{OUT}^2} \rangle = \int_{-\infty}^{\infty} a^*(\xi,\tau) \left(\frac{-\partial^2}{\partial \xi^2} \right) a(\xi,\tau) \mathrm{d}\xi$$
$$+ \int b^*(\xi,\tau) \left(\frac{-\partial^2}{\partial \xi^2} \right) b(\xi,\tau) \mathrm{d}\xi .$$

(18.75)

In the large detuning limit we find for a Gaussian beam with width σ centred at ξ_0

$$\langle \eta^{\mathrm{OUT}}_\pm \rangle = \pm \frac{\tau}{2\Delta} e^{-\sigma^2} \sin(2\xi_0)$$

(18.76)

$$\langle \eta^{\mathrm{OUT}^2}_\pm \rangle = \sigma_\eta^2 = \frac{1}{2\sigma^2} + \frac{\tau^2}{8\Delta^2} [1 - 2e^{-2\sigma^2} \sin^2(2\xi_0) - e^{-4\sigma^2} \cos(4\xi_0)]$$

(18.77)

where the $+$ sign is for initial state $|e\rangle$ and the $-$ sign for initial state $|g\rangle$. In Fig. 18.8 we plot these quantities as a function of the width of the atomic beam σ for $\tau/\Delta = 20$. The mean momentum transfer depends on the gradient of the intensity of the light field where the atom crosses and so the atomic beam must be narrow

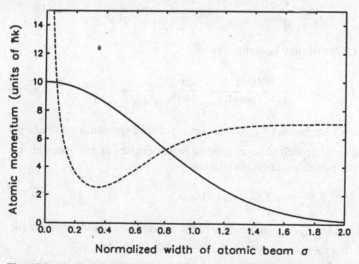

Fig. 18.8 Mean (*solid curve*) and standard deviation (*dashed curve*) of the atomic momentum plotted against the normalised width of the atomic beam. The atomic beam is incident midway between a node and an antinode and the normalised interaction time $\tau/\Delta = 20$ (from [13])

in order for the momentum transfer to be well defined. As the width of the atomic beam increases, different portions of the beam are deflected differently, reducing the mean momentum transfer. The variance of the output momentum arises from two effects: The term involving $(2\sigma^2)^{-1}$ represents the original momentum uncertainty of the input beam, which increases rapidly as σ is reduced. If, however, σ is increased so that it covers a significant portion of an optical wavelength, the variation in momentum imparted to different parts of the beam again increases the variance. Consequently there is a minimum in the standard deviation, as seen in Fig. 18.8. The condition for a good measurement is that the mean momentum transfer to the beam prepared in an eigenstate is larger than the spread of the outgoing momentum. From Fig. 18.8 it is clear that there is an optimal width of the input beam which gives the best quality of measurement. This optimum width depends on the interaction time since the mean momentum transfer rises with τ/Δ, and for larger interaction times this will exceed the intrinsic momentum uncertainty for a narrower initial beam. We may use the QND correlation coefficient introduced in Chap. 14 to evaluate the quality of the measurement. In this case the signal is the atomic inversion σ_z and the probe is the centre-of-mass momentum η of the atom.

These correlation coefficients depend on expectation values which have to be taken over some initial state. We choose the state which is a statistical mixture of ground and excited states with equal probability. In the large detuning limit the measurement correlation may be written as

$$
C^2_{A^{\mathrm{IN}}_{\mathrm{s}} A^{\mathrm{OUT}}_{\mathrm{P}}} = \frac{|\langle \eta^{\mathrm{OUT}} \sigma^{\mathrm{IN}}_z \rangle|^2}{\langle \eta^{\mathrm{OUT}^2} \rangle \langle \sigma^{\mathrm{IN}^2}_z \rangle}
$$

$$
= \frac{|\frac{1}{2}[\langle \eta^{\mathrm{OUT}}_+ \rangle (\frac{1}{2}) - \langle \eta^{\mathrm{OUT}}_- \rangle (\frac{-1}{2})]|^2}{\frac{1}{2}(\langle \eta^{\mathrm{OUT}^2}_+ \rangle + \langle \eta^{\mathrm{OUT}^2}_- \rangle) \frac{1}{2}[(\frac{1}{2})^2 + (\frac{-1}{2})^2]} , \tag{18.78}
$$

from (18.76 and 18.77) this may be written as

$$
C^2_{A^{\mathrm{IN}}_{\mathrm{s}} A^{\mathrm{OUT}}_{\mathrm{P}}} = \frac{2\sigma^2 \tau^2 e^{-2\sigma^2} \sin^2(2\xi_0)}{4\Delta^2 + \sigma^2 \tau^2 [1 - e^{-4\sigma^2} \cos(4\xi_0)]} . \tag{18.79}
$$

The state preparation correlation $C^2_{A^{\mathrm{IN}}_{\mathrm{s}} A^{\mathrm{OUT}}_{\mathrm{P}}}$ has an identical expression. In the large detuning limit an atom prepared in an eigenstate of σ_z remains in an eigenstate of σ_z. Consequently, the non-demolition correlation

$$
C^2_{A^{\mathrm{IN}}_{\mathrm{s}} A^{\mathrm{OUT}}_{\mathrm{s}}} = 1 . \tag{18.80}
$$

In Fig. 18.9 we plot $C^2_{A^{\mathrm{IN}}_{\mathrm{s}} A^{\mathrm{OUT}}_{\mathrm{P}}}$ as a function of the width of the atomic beam σ for a range of different interaction times $\tau/\Delta = 2, 5, 10, 20, 50$.

The position of the beam centre is $\xi_0 = \pi/4$ corresponding to the point midway between a node and antinode where the intensity gradient is greatest. We see that for sufficiently large interaction times and an atomic beam of optimal width the

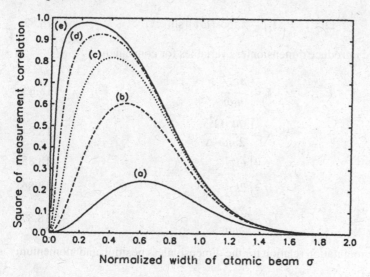

Fig. 18.9 Measurement correlation coefficient squared *plotted* against the normalised width of the atomic beam for $\tau/\Delta =$ (a) 2, (b) 5, (c) 10, (d) 20, (e) 50 (from [13])

measurement correlation may approach unity. Thus the deflection of an atomic beam by a standing wave field may be used to give a good QND measurement of the atomic inversion σ_z in the detuning limit. For smaller detuning a QND measurement of an operator $dx\sigma_x + dy\sigma_y + dz\sigma_z$ can be made. For example, for zero detuning the appropriate QND observable is the atomic polarisation σ_x.

18.7 Quantum Chaos

A cold atom moving in a well detuned laser field sees a dipole potential that has the same spatial dependence as the field intensity. These potentials are usually non linear and anharmonic. They thus provide an ideal means to investigate anharmonic quantum non linear dynamics. Given the ease with which the optical field can be modulated in time, we also have the possibility of investigating non adiabatic dynamics of time dependent potentials and quantum chaos.

We can describe both spatial and temporal modulation through the Rabi frequency $\Omega(x,t)$. As we have seen a two-level atom moving in one dimension (the x-direction) in an off-resonant field can be well described by the effective Hamiltonian

$$H_{\text{eff}} = \frac{p_x^2}{2m} + \frac{|\Omega(x,t)|^2}{2\Delta}\sigma_z \qquad (18.81)$$

where $\Delta = \omega_L - \omega_a$. We will typically be concerned with a standing wave laser field with wave vector k, with modulated intensity. In that case

$$\Omega(x,t) = \Omega[1 - 2\varepsilon\cos(\omega\tau)]\sin(kx) \tag{18.82}$$

Following [14] we introduce dimensionless variables for convenience:

$$\bar{k} = \frac{4\hbar k^2}{m\omega} \tag{18.83}$$

$$\kappa = \frac{\hbar k^2\Omega^2}{2m\omega^2\Delta} \tag{18.84}$$

$$t = \omega\tau \tag{18.85}$$

$$q = 2kx \tag{18.86}$$

$$p = \frac{2kp_x}{m\omega} \tag{18.87}$$

Noting that the commutation relation for the dimensionless position and momentum is

$$[q,p] = i\bar{k} \tag{18.88}$$

we see that $\bar{k}$ has the interpretation of a dimensionless Planck's constant.

We first consider the case of a time independent standing wave. In this case the Hamiltonian in dimensionless variables is

$$H = \frac{p^2}{2} - \kappa\cos q \tag{18.89}$$

This is a standard problem in classical nonlinear dynamics and is fully integrable by a canonical transformation to action-angle variables. If at time $t = 0$ the atom has initial conditions $(q(t = 0), p(t = 0)) = (q_0, p_0)$ it will move in the phase space so as to conserve energy so it must remain on the curve $E = p(t)^2/2 - \kappa\cos(q(t)) = p_0^2/2 - \kappa\cos(q_0)$. The motion is periodic if the initial energy is such that $-\kappa < E < 0$, in which case the atom remains localised in one well. For $E > \kappa$ the motion is unbounded. The curve $E = 0$ is called the spearatrix. The frequency of the oscillatory motion depends on energy, and tends to zero as the initial energy approaches zero on the separatrix. This is shown in Fig. 18.10.

The nonlinear dependance of oscillator frequency on energy is important when we add a periodic temporal modulation to the potential. For a simple harmonic oscillator, resonance is only possible at one frequency and this does not depend on initial conditions. For a non linear oscillator this is no longer the case and a complex hierarchy of resonances is possible leading to chaos. In fact, periodically driven systems with a nonlinear potential is generically chaotic even in one dimension. Consider the case of an atom in a standing wave with modulated amplitude. The Hamiltonian is

$$H(t) = \frac{p^2}{2} - \kappa(1 - 2\varepsilon\cos t)\cos q \tag{18.90}$$

As the Hamiltonian is periodic in time, the most appropriate way to describe the dynamics is in terms of a Poincaré section with respect to time. This means that we

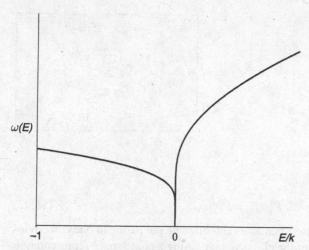

Fig. 18.10 The energy dependance of the oscillation frequency for an atom moving in a sinusoidal potential, $\omega(E)$ versus E/κ. Note that for energies at the bottom of the wells, the energy is almost independent of frequency, and the motion is simple harmonic

only need to view the dynamics at discrete times which are multiples of the driving period (see [15]). This defines a non linear map on phase space which is sometimes called a *stroboscopic map*. In figure 18.11 we illustrate this for a variety of initial conditions. A number of islands of regular motion are apparent surrounded by a sea of chaotic orbits.

18.7.1 Dynamical Tunnelling

The regular orbits are near elliptic fixed points of period-one of the map. The two inner regular orbits apparent in Fig. 18.11 are two distinct period-one orbits. If the system is prepared with initial conditions close to one of these period-one orbits it must remain within the island of stability surrounding the fixed point. A distribution of phase space points initially localised in this region will remain localised. As we now discuss, this is no longer the case quantum mechanically.

The quantum equivalent of a stroboscopic map is the unitary operator corresponding to the dynamics integrated over one period of modulation. This is called the Floquet operator, $\hat{F}$. Iterations of the quantum map are thus defined by

$$|\psi_{n+1}\rangle = \hat{F}|\psi_n\rangle \qquad (18.91)$$

As the dynamics is not integrable even classically the operator $\hat{F}$ is difficult find and we must resort to numerical methods. See [16] for further discussion. Once it is found, we proceed by first finding its eigenstates and eigenvalues (which must all lie on the unit circle)

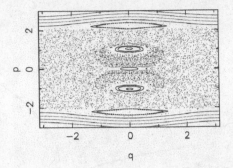

$$\hat{F}|\phi_\alpha\rangle = e^{i\phi_\alpha}|\phi_\alpha\rangle \tag{18.92}$$

It is a remarkable fact that the eigenstates can often be put in one to one correspondence with particular orbits of the classical map, using a phase space representation such as the Q-function. For example there are pair of nearly degenerate eigenstates associated with the period one fixed points near the origin in Fig. 18.11. Given the solution to the eigenvalue problem the iteration of a given initial state is found by expanding the state over the Floquet eigenstates, $|\psi_0\rangle = \sum_\alpha c_\alpha |\phi_\alpha\rangle$

$$|\psi_n\rangle = \sum_\alpha c_\alpha e^{in\phi_\alpha}|\phi_\alpha\rangle \tag{18.93}$$

We now take the initial state to be well localised inside the island of stability surrounding one of the period one orbits. We iterate the state and at each iteration step calculate the average and variance of the position and momentum. The results are plotted in Fig. 18.12. We see a very different situation to what would be expected classically (also shown in the figure). The quantum state does not remain localised near the period one fixed point at which it started. Rather it appears to tunnel across classically forbidden regions of phase space to the symmetric partner of the fixed

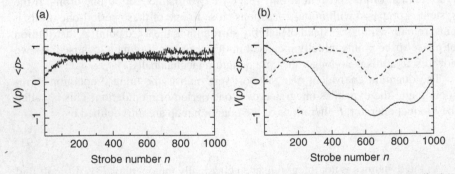

Fig. 18.12 The average and variance of position and momentum for an initial classical distribution (**a**) localised near a period one fixed point and (**b**) an initial quantum state localised near a period one fixed point. The quantum system demonstrates tunneling form one period one fixed point to another

point. This is called dynamical tunneling [17]. Dynamical tunneling was observed by *Hensinger* et al. [18] using a Bose Einstein condensate to prepare the initial localised state in phase space.

18.7.2 Dynamical Localisation

More complicated anharmonic modulation of the optical dipole potential can be easily implemented. An extreme case corresponds to a kicked system in which the standing wave is pulsed on and off very quickly compared to the period of free dynamics between pulses. In the limit of infinitely short pulses we obtain the delta kicked rotor described by the Hamiltonian

$$H(t) = \frac{p^2}{2} - \kappa \cos q \sum_n \delta(t - n) \qquad (18.94)$$

This is a well studied system that classically is described by a stroboscopic map that is an example of the *standard map*. If the kicking strength, κ, is sufficiently large the phase space is dominated by large sea of chaotic orbits. In this region an initial well localised distribution of points spreads very rapidly in position but only slowly diffuses in momentum. In fact to a very good approximation the spread in momentum is indeed described by a diffusion process with diffusion constant proportional to κ. See [19] for further detail.

The quantum description of this system gives a very different result. An initial state well localised in momentum will follow the classical diffusion until a time know as the *break time*. At that time the diffusion in momentum will cease and the momentum variance saturates. This is illustrated in Fig. 18.13. Dynamical localisation using laser cooled atoms was first observed by Raizen's group in 1995 [20].

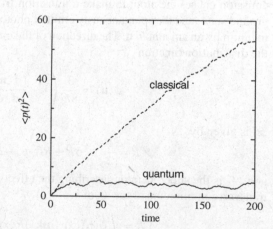

Fig. 18.13 Classical *(dashed)* and quantum *(solid)* of the average momentum squared versus kick number. In this example $k = 0.24, \kappa = 1.2$

18.8 The Effect of Spontaneous Emission

In this section we give a brief presentation of the effect of spontaneous emission on the nonlinear quantum dynamics of atoms. Even in a far detuned optical dipole potential there is a non zero probability that the atom will be excited. In that case it suffers a recoil induced momentum kick. Once in the excited state it can relax to the ground state by either stimulated emission or spontaneous emission. The best way to deal with the stochastic momentum kicks resulting from absorption, or stimulated and spontaneous emission is via the stochastic Schrödinger equation discussed in Chap. 6.

We will follow the approach of Dyrting [21]. The coherent coupling between the internal and center-of-mass variables is through the position dependent Rabi frequency. We can write the Rabi frequency in the general form

$$\Omega(x) = \Omega f(kx,t) . \tag{18.95}$$

$k = c/\omega_{\mathrm{L}}$ is called the wavenumber and ω_{L} is the frequency of the laser. Ω is a measure of the intensity of the field and without loss of generality it can be chosen to be real and may be explicitly time dependent. The quantum mechanical atom can be described by its internal state $|\sigma\rangle$, where σ represents either g or e, and a centre-of-mass state $|\psi\rangle$.

The master equation for a two-level atom interacting with the field including spontaneous emission is

$$\frac{\mathrm{d}\hat{R}}{\mathrm{d}\tau} = -\frac{\mathrm{i}}{\hbar}\left[\hat{H},\hat{R}\right] - \gamma\mathscr{L}\hat{R} . \tag{18.96}$$

The Hamiltonian $\hat{H}$ generates the coherent dynamics for the center-of-mass and internal states of the atom. The superoperator $\mathscr{L}$ describes incoherent evolution due to coupling with the vacuum field modes at a rate γ. The effect of a spontaneous emission causes the atom to make a transition from its internal excited state to its ground state, and the spontaneously emitted photon changes its center-of-mass momentum by an amount $\hbar k\mathbf{n}$. The direction of the emitted photon $\mathbf{n}$ is random and has the distribution function

$$\phi(\mathbf{n}) = \frac{3}{8\pi}\left(1 - \frac{(\mathbf{d}\cdot\mathbf{n})^2}{\mathbf{d}\cdot\mathbf{d}}\right) . \tag{18.97}$$

$\mathscr{L}$ is given by

$$\mathscr{L}\hat{R} = \frac{1}{2}\left(\hat{\sigma}^\dagger\hat{\sigma}\hat{R} + \hat{R}\hat{\sigma}^\dagger\hat{\sigma} - 2\hat{\sigma}\mathscr{N}\hat{R}\hat{\sigma}^\dagger\right) , \tag{18.98}$$

and $\mathscr{N}$ is the superoperator describing the effect of a spontaneous emission on the transverse momentum of the atom

$$\mathscr{N}\hat{R} = \int \phi(\mathbf{n})\exp\left(\mathrm{i}nk\hat{x}\right)\hat{R}\exp\left(-\mathrm{i}nk\hat{x}\right)d\mathbf{n} . \tag{18.99}$$

Here the integral is done over the surface of the unit sphere $\mathbf{n} \cdot \mathbf{n} = 1$.

In the limit that the detuning Δ is much larger than the Rabi frequency Ω and the spontaneous emission rate γ the system can be described by the effective master equation

$$
\frac{\mathrm{d}\hat{R}}{\mathrm{d}\tau} = -\frac{\mathrm{i}}{\hbar}\left[\hat{H}_0,\hat{R}\right] - \gamma\mathscr{L}\hat{R} + \frac{\mathrm{i}}{\Delta}\left[\Omega(\hat{x})\hat{\sigma}^\dagger, \left(1-\mathrm{i}\frac{\gamma}{\Delta}\mathscr{L}\right)^{-1}\left[\Omega(\hat{x})^\dagger\hat{\sigma},\hat{R}\right]\right]
$$

$$
-\frac{\mathrm{i}}{\Delta}\left[\Omega(\hat{x})^\dagger\hat{\sigma}, \left(1+\mathrm{i}\frac{\gamma}{\Delta}\mathscr{L}\right)^{-1}\left[\Omega(\hat{x})\hat{\sigma}^\dagger,\hat{R}\right]\right] , \tag{18.100}
$$

where $\hat{H}_0$. The evolution of the reduced center-of-mass density operators $\hat{\rho}_g = \langle g|\hat{R}|g\rangle$ and $\hat{\rho}_e = \langle e|\hat{R}|e\rangle$ is given by the coupled equations

$$
\frac{\mathrm{d}\hat{\rho}_g}{\mathrm{d}t} = -\frac{\mathrm{i}}{k}\left[\hat{H}_g,\hat{\rho}_g\right] + \Gamma\mathscr{N}\hat{\rho}_e
$$

$$
-\frac{\eta}{2}\left[\{f(\hat{q}/2,t)^\dagger f(\hat{q}/2,t),\hat{\rho}_g\} - 2f(\hat{q}/2,t)^\dagger\hat{\rho}_e \bar{f}(\hat{q}/2,t)\right] , \tag{18.101}
$$

$$
\frac{\mathrm{d}\hat{\rho}_e}{\mathrm{d}t} = -\frac{\mathrm{i}}{k}\left[\hat{H}_e,\hat{\rho}_e\right] - \Gamma\hat{\rho}_e
$$

$$
-\frac{\eta}{2}\left[\{f(\hat{q}/2,t)^\dagger f(\hat{q}/2,t),\hat{\rho}_e\} - 2f(\hat{q}/2,t)\hat{\rho}_g f(\hat{q}/2,t)^\dagger\right] , \tag{18.102}
$$

where

$$
\hat{H}_g = \frac{\hat{p}^2}{2} + \frac{2\kappa}{|v|^2}f(\hat{q}2,t)^\dagger f(\hat{q}/2,t) , \tag{18.103}
$$

$$
\hat{H}_e = \frac{\hat{p}^2}{2} - \frac{2\kappa}{|v|^2}f(\hat{q}/2,t)^\dagger f(\hat{q}/2,t) , \tag{18.104}
$$

and $v = 1 - \mathrm{i}\gamma/2\Delta$ and $\{\ ,\ \}$ denotes the anti-commutator. We have again used the dimensionless variables for convenience: $k = \frac{4\hbar k^2}{m\omega}$, $\kappa = \frac{\hbar k^2\Omega^2}{2m\omega^2\Delta}$, $\Gamma = \frac{\gamma}{\omega}$, $t = \omega\tau$, $\eta = \frac{\gamma\Omega^2}{4\omega\Delta^2|v|^2}$, $q = 2kx$, and $p = \frac{2kp_x}{m\omega}$.

The master equation may be unravelled as a stochastic Schrödinger equation as follows. The internal state changes according to the two jump processes N_1, and N_2 which have the following actions:

$$
g \xrightarrow{N_1} e, \quad \text{absorption} \tag{18.105}
$$

$$
e \xrightarrow{N_1} g, \quad \text{stimulated emission} \tag{18.106}
$$

$$
g \xrightarrow{N_2} e \quad \text{spontaneous emission} . \tag{18.107}
$$

These two jump processes proceed at the rates

$$E\left[dN_1\right] = \eta\langle\psi|f(\hat{q}/2,t)^2|\psi\rangle dt\,, \qquad (18.108)$$

$$E\left[dN_2\right] = \Gamma dt\,. \qquad (18.109)$$

Here $E[\,]$ denotes an ensemble average. thus represents spontaneous emission while $N_2(t)$ represents stimulated emission. The centre-of-mass state evolves according to the un-normalised stochastic Schrödinger equation

$$d|\psi\rangle = -\frac{i}{\hbar}dt\hat{K}_\sigma|\psi\rangle + dN_1\left(\frac{f(\hat{q}/2,t)}{\sqrt{\langle f(\hat{q}/2,t)^2\rangle}} - 1\right)|\psi\rangle$$

$$+ dN_2\left(\frac{\exp(i\bar{p}\hat{q}/\hbar)}{\sqrt{\langle\psi|\psi\rangle}} - 1\right)|\psi\rangle\,, \qquad (18.110)$$

where $\langle f(\hat{q}/2,t)^2\rangle = \langle\psi|f(\hat{q}/2,t)^2|\psi\rangle$. This equation does not preserve the normalisation of the state $|\psi\rangle$. This will be important when we come to generate the times for the jump N_1. The jump terms determine the state after a jump $|\psi_{\text{after}}\rangle$ in terms of the state before $|\psi_{\text{before}}\rangle$ by

$$N_1 : \quad |\psi_{\text{after}}\rangle = \frac{f(\hat{q}/2,t)|\psi_{\text{before}}\rangle}{\sqrt{\langle\psi_{\text{before}}|f(\hat{q}/2,t)^2|\psi_{\text{before}}\rangle}}\,, \qquad (18.111)$$

$$N_2 : \quad |\psi_{\text{after}}\rangle = \frac{\exp(i\bar{p}\hat{q}/\hbar)|\psi_{\text{before}}\rangle}{\sqrt{\langle\psi_{\text{before}}|\psi_{\text{before}}\rangle}}\,. \qquad (18.112)$$

where $\bar{p}$ is the random kick in momentum due to spontaneous recoil which satisfies

$$\text{Prob}(\bar{p},\bar{p}+d\bar{p}) = \phi(\bar{p})d\bar{p} \qquad (18.113)$$

The operator $\hat{K}_\sigma$ is non-Hermitian and depends on the internal state σ as follows:

$$\hat{K}_\sigma = \begin{cases} \hat{p}^2/2 + V(\hat{q},t)/v^* & \sigma = g \\ \hat{p}^2/2 - V(\hat{q},t)/v & \sigma = e. \end{cases} \qquad (18.114)$$

where $v = 1 - i\gamma/2\Delta$ and $V(\hat{q},t) = f^\dagger(\hat{q},t)f(\hat{q},t)$. Between jumps the state evolves in a complex potential and the imaginary part of the complex potential causes the normalisation of $|\psi\rangle$ to decay. The effect of an N_1 jump is to change the internal state and to change the centre-of-mass state. The cumulative distribution function for the stimulated jump N_1 is given by

$$P_{\text{stim}}(t) = 1 - |\langle\psi(t)|\psi(t)\rangle|\,. \qquad (18.115)$$

We generate a random number z_{stim} which has a uniform distribution on the interval and provided no spontaneous emission has occurred in the meantime we integrate the wave equation with generator $\hat{K}_\sigma$ to the time t such that $z_{\text{stim}} = 1 - |\langle\psi(t)|\psi(t)\rangle|$. In this way we compute when an atom makes a stimulated transition.

When the atom makes a transition to state b it is easy to generate the random number t_{spont} equal to the time the atom spontaneously emits using the cumulative distribution function $P_{spont}(t) = 1 - \exp(-\Gamma t)$. The effect of spontaneous emission N_2 is to change the momentum of the state by the amount $\bar{p}$ given by

$$\bar{p} = k(\cos\zeta\cos\theta + \sin\phi\sin\theta\sin\zeta)/2 \,, \tag{18.116}$$

where ζ is the angle between the dipole moment and the x-axis. The angle $\phi \in [0, 2\pi]$ is random with a uniform distribution and θ is given by

$$\theta = \arccos\left[2\cos\left(\frac{\arccos(2y-1)+4\pi}{3}\right)\right] \,, \tag{18.117}$$

where $y \in [0,1]$ is a random number with a uniform distribution. In our numerical calculations we have chosen $\zeta = \pi/2$.

To recover the centre-of-mass density operator one takes the ensemble average of the conditioned operators

$$\hat{\rho} = \hat{\rho}_g + \hat{\rho}_e = E\left[\frac{|\psi\rangle\langle\psi|}{\langle\psi|\psi\rangle}\right] \tag{18.118}$$

In the simulations of [16] the state $|\psi\rangle$ was evolved forward for a time δt using the first order split operator method. Then the norm of $|\psi(t + \delta t)\rangle$ is calculated to determine if a stimulated jump has occurred and whether the spontaneous emission time t_{spont} is reached. If a jump occurs the appropriate transformation (18.8) is

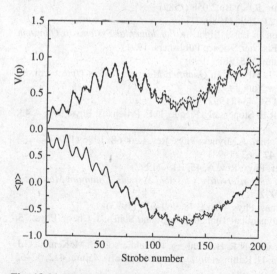

Fig. 18.14 Tunneling between second order resonances as reflected in the momentum mean $\langle p \rangle$ and variance $V(p)$. Solid line, coherent motion; dashed line with spontaneous emission included and $\Gamma = 1.525, \eta = 6.1 \times 10^{-4}$

applied in the momentum representation and then the new state is evolved forward δt and so on until it has been evolved forward the desired time. The whole process is repeated over $1,000$ trajectories.

To test whether spontaneous emission obscures coherent tunneling simulations were made using the atomic system of Ytterbium with $\bar{k} = 0.25$ and with an initial wavepacket localised in phase space at $(q_0, p_0) = (0.0, 1.0)$ and a momentum variance of $V(p) = 0.04$. In Fig. 18.14 we show quantum Monte Carlo simulations with and without spontaneous emission. We still see definite sign of coherent tunneling even when spontaneous emission is included.

For the initial state a semi-classical an estimate, based on the simulation, for the coherence damping rate gives $\Gamma_{\rm coh} \approx 0.06\eta$. The time taken to tunnel to the opposite second order resonance is $T \approx 125 \times 2\pi$. This implies that the value $\eta \approx 0.02$ is required for the critical damping of the tunneling oscillations. Dynamic localisation is a coherent effect and noise due to spontaneous emission must be kept low. For the spontaneous decay rate of Ytterbium of $\gamma/2\pi = 183$ kHz, the spontaneous and stimulated rates are $\Gamma = 1.5$ and $\eta = 4.5 \times 10^{-4}$ respectively.

References

1. T.W. Hänsch, A.L. Schawlow: Opt. Comm. **13**, 68 (1975)
2. S. Chu, L. Hollberg, J.E. Bjorkholm, A. Cable, A. Ashkin: Phys. Rev. Lett. **55**, 48 (1985)
3. P.D. Lett, R.N. Watts, C.I. Westbrook, W.D. Phillips, P.L. Gould, H.J. Metcalf: Phys. Rev. Lett. **62**, 1118 (1988)
4. J. Dalibard, C. Cohen-Tannoudji: JOSA B**6**, 2023 (1989)
5. P.J. Ungar, D.S. Weiss, E. Ris, S. Chu: JOSA B**6**, 2058 (1989)
6. J. Gordon, A. Ashkin: Phys. Rev. A **21**, 1606 (1980)
7. C. Cohen-Tannoudji: Atomic Motion in Laser Light, in *Fundamental Systems in Quantum Optics*, edited by J. Dalibard et al. (Elsevier Science Publishers, 1991)
8. E.M. Wright, P. Meystre: Opt. Commun. **75**, 388 (1990)
9. T. Sleator, O. Carnal, A. Faulstich, J. Mlynek: In *Quantum Measurement in Optics*, edi. by P. Tombesi, D.F. Walls (Plenum, New York 1992)
10. S.M. Tan, D.F. Walls: Appl. Phys. B **54**, 434 (1992)
11. P.L. Gould, P.J. Martin, G.A. Ruff, R.E. Stoner, J.L. Pieque, D.E. Pritchard: Phys. Rev. A **43**, 585 (1991)
12. T. Sleator, T. Pfau, V. Balykin, O. Carnal, J. Mlynek: Phys. Rev. Lett. **68**, 1996 (1992)
13. S.M. Tan, D.F. Walls: Phys. Rev. A **47**, 663 (1993)
14. R.Graham, M. Schlautman, P. Zoller: Phys. Rev. A, **45**, R19, (1992)
15. L. Reichl: *The Transition to Chaos, in Conservative Classical Systems: Quantum Manifestations* (Springer Verlag, New York 1992)
16. S. Dyrting, G.J. Milburn, C.A. Holmes: Phys. Rev. E 48, 969–978 (1993)
17. M.J. Davis, E.J. Heller: Quantum Dynamical Tunneling in Bound States: J. Chem. Phys. **75**, 246 (1981)
18. W.K. Hensinger, H. Hffner, A. Browaeys, N.R. Heckenberg, K. Helmerson, C. McKenzie, G.J. Milburn, W.D. Phillips, S.L. Rolston, H. Rubinsztein-Dunlop, B. Upcroft: Nature, **412**, 52–55 (2001)
19. H.-J. Stöckman: *Quantum Chaos: An Introduction* (Cambridge University Press, 2001)
20. F.L. Moore, J.C. Robinson, C.F. Barucha, B. Sunduram, M.G. Raizen: Phys. Rev. Lett. **75**, 4598 (1995)

21. S. Dyrting, G.J. Milburn: Phys. Rev. A **51**, 3136 (1995)
22. T. Dittrich, R. Graham: Ann. Phys. **200**, 363 (1990)

Further Reading

S. Chu: Laser Manipulation of Atoms and Particles, Science **253**, 861 (1991)

E. Arimondo, W.D. Phillips, F. Strumia: *Laser Manipulation of Atoms and Ions* (North Holland, 1991)

H.J. Metcalf, P. van der Stratten: *Laser Cooling and Trapping* (Springer, New York 1999)

P. Meystre: *Atom Optics* (Springer, New York, 2001)

A.P. Kazantsev, G.I. Surdutovich, V.P. Yakovlev: *Mechanical Action of Light on Atoms* (World Scientific, 1990)

Chapter 19
Bose-Einstein Condensation

Abstract Bose-Einstein condensation (BEC) refers to a prediction of quantum statistical mechanics (Bose [1], Einstein [2]) where an ideal gas of identical bosons undergoes a phase transition when the thermal de Broglie wavelength exceeds the mean spacing between the particles. Under these conditions, bosons are stimulated by the presence of other bosons in the lowest energy state to occupy that state as well, resulting in a macroscopic occupation of a single quantum state. The condensate that forms constitutes a macroscopic quantum-mechanical object. BEC was first observed in 1995, seventy years after the initial predictions, and resulted in the award of 2001 Nobel Prize in Physics to Cornell, Ketterle and Weiman. The experimental observation of BEC was achieved in a dilute gas of alkali atoms in a magnetic trap. The first experiments used ^{87}Rb atoms [3], ^{23}Na [4], ^{7}Li [5], and H [6] more recently metastable He has been condensed [7]. The list of BEC atoms now includes molecular systems such as Rb_2 [8], Li_2 [9] and Cs_2 [10]. In order to cool the atoms to the required temperature ($\sim$200 nK) and densities (10^{13}–10^{14} cm^{-3}) for the observation of BEC a combination of optical cooling and evaporative cooling were employed. Early experiments used magnetic traps but now optical dipole traps are also common. Condensates containing up to 5×10^9 atoms have been achieved for atoms with a positive scattering length (repulsive interaction), but small condensates have also been achieved with only a few hundred atoms. In recent years Fermi degenerate gases have been produced [11], but we will not discuss these in this chapter.

BECs are now routinely produced in dozens of laboratories around the world. They have provided a wonderful test bed for condensed matter physics with stunning experimental demonstrations of, among other things, interference between condensates, superfluidity and vortices. More recently they have been used to create optically nonlinear media to demonstrate electromagnetically induced transparency and neutral atom arrays in an optical lattice via a Mott insulator transition.

Many experiments on BECs are well described by a semiclassical theory discussed below. Typically these involve condensates with a large number of atoms, and in some ways are analogous to describing a laser in terms of a semiclassical mean field. More recent experiments however have begun to probe quantum properties of the condensate, and are related to the fundamental discreteness of the field and nonlinear quantum dynamics. In this chapter, we discuss some of these quantum properties of the condensate. We shall make use of "few mode" approximations which treat only essential condensate modes and ignore all noncondensate modes. This enables us to use techniques developed for treating quantum optical systems described in earlier chapters of this book.

19.1 Hamiltonian: Binary Collision Model

The effects of interparticle interactions are of fundamental importance in the study of dilute–gas Bose–Einstein condensates. Although the actual interaction potential between atoms is typically very complex, the regime of operation of current experiments is such that interactions can in fact be treated very accurately with a much–simplified model. In particular, at very low temperature the de Broglie wavelengths of the atoms are very large compared to the range of the interatomic potential. This, together with the fact that the density and energy of the atoms are so low that they rarely approach each other very closely, means that atom–atom interactions are effectively *weak* and dominated by (elastic) *s*–wave scattering. It follows also that to a good approximation one need only consider *binary* collisions (i.e., three–body processes can be neglected) in the theoretical model.

The *s*–wave scattering is characterised by the *s*–wave scattering length, *a*, the sign of which depends sensitively on the precise details of the interatomic potential [$a > 0$ ($a < 0$) for repulsive (attractive) interactions]. Given the conditions described above, the interaction potential can be approximated by

$$U(\mathbf{r} - \mathbf{r}') = U_0 \delta(\mathbf{r} - \mathbf{r}') \, , \tag{19.1}$$

(i.e., a hard sphere potential) with U_0 the interaction "strength," given by

$$U_0 = \frac{4\pi\hbar^2 a}{m} \, , \tag{19.2}$$

and the Hamiltonian for the system of weakly interacting bosons in an external potential, $V_{\mathrm{trap}}(\mathbf{r})$, can be written in the second quantised form as

$$\hat{H} = \int \mathrm{d}^3 r \, \hat{\Psi}^\dagger(\mathbf{r}) \left[-\frac{\hbar^2}{2m} \nabla^2 + V_{\mathrm{trap}}(\mathbf{r}) \right] \hat{\Psi}(\mathbf{r})$$

$$+ \frac{1}{2} \int \mathrm{d}^3 r \int \mathrm{d}^3 r' \hat{\Psi}^\dagger(\mathbf{r}) \hat{\Psi}^\dagger(\mathbf{r}') U(\mathbf{r} - \mathbf{r}') \hat{\Psi}(\mathbf{r}') \hat{\Psi}(\mathbf{r}) \tag{19.3}$$

where $\hat{\Psi}(\mathbf{r})$ and $\hat{\Psi}^\dagger(\mathbf{r})$ are the boson field operators that annihilate or create a particle at the position $\mathbf{r}$, respectively.

To put a quantitative estimate on the applicability of the model, if ρ is the density of bosons, then a necessary condition is that $a^3\rho \ll 1$ (for $a > 0$). This condition is indeed satisfied in the alkali gas BEC experiments [3, 4], where achieved densities of the order of $10^{12} - 10^{13}$ cm^{-3} correspond to $a^3\rho \simeq 10^{-5} - 10^{-6}$.

19.2 Mean–Field Theory — Gross-Pitaevskii Equation

The Heisenberg equation of motion for $\hat{\Psi}(\mathbf{r})$ is derived as

$$i\hbar\frac{\partial \hat{\Psi}(\mathbf{r},t)}{\partial t} = \left[-\frac{\hbar^2}{2m}\nabla^2 + V_{\text{trap}}(\mathbf{r})\right]\hat{\Psi}(\mathbf{r},t) + U_0\hat{\Psi}^\dagger(\mathbf{r},t)\hat{\Psi}(\mathbf{r},t)\hat{\Psi}(\mathbf{r},t)\,, \quad (19.4)$$

which cannot in general be solved. In the mean–field approach, however, the expectation value of (19.4) is taken and the field operator decomposed as

$$\hat{\Psi}(\mathbf{r},t) = \Psi(\mathbf{r},t) + \tilde{\Psi}(\mathbf{r},t)\,, \qquad (19.5)$$

where $\Psi(\mathbf{r},t) = \langle\hat{\Psi}(\mathbf{r},t)\rangle$ is the "condensate wave function" and $\tilde{\Psi}(\mathbf{r})$ describes quantum and thermal fluctuations around this mean value. The quantity $\Psi(\mathbf{r},t)$ is in fact a classical field possessing a well–defined phase, reflecting a broken gauge symmetry associated with the condensation process. The expectation value of $\tilde{\Psi}(\mathbf{r},t)$ is zero and, in the mean–field theory, its effects are assumed to be small, amounting to the assumption of the thermodynamic limit, where the number of particles tends to infinity while the density is held fixed. For the effects of $\tilde{\Psi}(\mathbf{r})$ to be *negligibly small* in the equation for $\Psi(\mathbf{r})$ also amounts to an assumption of zero temperature (i.e., pure condensate). Given that this is so, and using the normalisation

$$\int d^3r\,|\Psi(\mathbf{r},t)|^2 = 1\,, \qquad (19.6)$$

one is lead to the nonlinear Schrödinger equation, or "Gross–Pitaevskii equation" (GP equation), for the condensate wave function $\Psi(\mathbf{r},t)$ [13],

$$i\hbar\frac{\partial \Psi(\mathbf{r},t)}{\partial t} = \left[-\frac{\hbar^2}{2m}\nabla^2 + V_{\text{trap}}(\mathbf{r}) + NU_0|\Psi(\mathbf{r},t)|^2\right]\Psi(\mathbf{r},t)\,, \qquad (19.7)$$

where N is the mean number of particles in the condensate. The nonlinear interaction term (or mean–field pseudo–potential) is proportional to the number of atoms in the condensate and to the s–wave scattering length through the parameter U_0.

A stationary solution for the condensate wavefunction may be found by substituting $\psi(\mathbf{r},t) = \exp\left(\frac{-i\mu t}{\hbar}\right)\psi(\mathbf{r})$ into (19.7) (where μ is the chemical potential of the condensate). This yields the time independent equation,

$$\left[\frac{-\hbar^2}{2m}\nabla^2 + V_{\text{trap}}(\mathbf{r}) + NU_0|\psi(\mathbf{r})|^2\right]\psi(\mathbf{r}) = \mu\psi(\mathbf{r}) \ . \tag{19.8}$$

The GP equation has proved most successful in describing many of the mean field properties of the condensate. The reader is referred to the review articles listed in further reading for a comprehensive list of references. In this chapter we shall focus on the quantum properties of the condensate and to facilitate our investigations we shall go to a single mode model.

19.3 Single Mode Approximation

The study of the quantum statistical properties of the condensate (at $T = 0$) can be reduced to a relatively simple model by using a mode expansion and subsequent truncation to just a single mode (the "condensate mode"). In particular, one writes the Heisenberg atomic field annihilation operator as a mode expansion over single–particle states,

$$\hat{\Psi}(\mathbf{r},t) = \sum_\alpha a_\alpha(t)\psi_\alpha(\mathbf{r})\exp^{-i\mu_\alpha t/\hbar}$$

$$= a_0(t)\psi_0(\mathbf{r})\exp^{-i\mu_0 t/\hbar} + \tilde{\Psi}(\mathbf{r},t) \ , \tag{19.9}$$

where $[a_\alpha(t), a_\beta^\dagger(t)] = \delta_{\alpha\beta}$ and $\{\psi_\alpha(\mathbf{r})\}$ are a complete orthonormal basis set and $\{\mu_\alpha\}$ the corresponding eigenvalues. The first term in the second line of (19.9) acts only on the condensate state vector, with $\psi_0(\mathbf{r})$ chosen as a solution of the stationary GP equation (19.8) (with chemical potential μ_0 and mean number of condensate atoms N). The second term, $\tilde{\Psi}(\mathbf{r},t)$, accounts for non–condensate atoms. Substituting this mode expansion into the Hamiltonian

$$\hat{H} = \int d^3r\, \hat{\Psi}^\dagger(\mathbf{r})\left[-\frac{\hbar^2}{2m}\nabla^2 + V_{\text{trap}}(\mathbf{r})\right]\hat{\Psi}(\mathbf{r})$$

$$+ (U_0/2)\int d^3r\, \hat{\Psi}^\dagger(\mathbf{r})\hat{\Psi}^\dagger(\mathbf{r})\hat{\Psi}(\mathbf{r})\hat{\Psi}(\mathbf{r}) \ , \tag{19.10}$$

and retaining only condensate terms, one arrives at the single–mode effective Hamiltonian

$$\hat{H} = \hbar\tilde{\omega}_0 a_0^\dagger a_0 + \hbar\kappa a_0^\dagger a_0^\dagger a_0 a_0 \ , \tag{19.11}$$

where

$$\hbar\tilde{\omega}_0 = \int d^3r\, \psi_0^*(\mathbf{r})\left[-\frac{\hbar^2}{2m}\nabla^2 + V_{\text{trap}}(\mathbf{r})\right]\psi_0(\mathbf{r}) \ , \tag{19.12}$$

and

$$\hbar\kappa = \frac{U_0}{2}\int d^3r\, |\psi_0(\mathbf{r})|^4 \ . \tag{19.13}$$

We have assumed that the state is prepared slowly, with damping and pumping rates vanishingly small compared to the trap frequencies and collision rates. This means that the condensate remains in thermodynamic equilibrium throughout its preparation. Finally, the atom number distribution is assumed to be sufficiently narrow that the parameters $\tilde{\omega}_0$ and κ, which of course depend on the atom number, can be regarded as constants (evaluated at the mean atom number). In practice, this proves to be a very good approximation.

19.4 Quantum State of the Condensate

A Bose-Einstein condensate (BEC) is often viewed as a coherent state of the atomic field with a definite phase. The Hamiltonian for the atomic field is independent of the condensate phase (see Exercise 19.1) so it is often convenient to invoke a symmetry breaking Bogoliubov field to select a particular phase. In addition, a coherent state implies a superposition of number states, whereas in a single trap experiment there is a fixed number of atoms in the trap (even if we are ignorant of that number) and the state of a simple trapped condensate must be a number state (or, more precisely, a mixture of number states as we do not know the number in the trap from one preparation to the next). These problems may be bypassed by considering a system of two condensates for which the total number of atoms N is fixed. Then, a general state of the system is a superposition of number difference states of the form,

$$|\psi\rangle = \sum_{k=0}^{N} c_k |k, N - k\rangle \tag{19.14}$$

As we have a well defined superposition state, we can legitimately consider the relative phase of the two condensates which is a Hermitian observable. We describe in Sect. 19.6 how a particular relative phase is established due to the measurement process.

The identification of the condensate state as a coherent state must be modified in the presence of collisions except in the case of very strong damping.

19.5 Quantum Phase Diffusion: Collapses and Revivals of the Condensate Phase

The macroscopic wavefunction for the condensate for a relatively strong number of atoms will exhibit collapses and revivals arising from the quantum evolution of an initial state with a spread in atom number [21]. The initial collapse has been described as quantum phase diffusion [20]. The origins of the collapses and revivals may be seen straightforwardly from the single–mode model. From the Hamiltonian

$$\hat{H} = \hbar\tilde{\omega}_0 a_0^\dagger a_0 + \hbar\kappa a_0^\dagger a_0^\dagger a_0 a_0 , \tag{19.15}$$

the Heisenberg equation of motion for the condensate mode operator follows as

$$\dot{a}_0(t) = -\frac{i}{\hbar}[a_0, H]$$

$$= -i\left(\tilde{\omega}_0 a_0 + 2\kappa a_0^\dagger a_0 a_0\right) , \tag{19.16}$$

for which a solution can be written in the form

$$a_0(t) = \exp\left[-i\left(\tilde{\omega}_0 + 2\kappa a_0^\dagger a_0\right)t\right] a_0(0) . \tag{19.17}$$

Writing the initial state of the condensate, $|i\rangle$, as a superposition of number states,

$$|i\rangle = \sum_n c_n |n\rangle , \tag{19.18}$$

the expectation value $\langle i|a_0(t)|i\rangle$ is given by

$$\langle i|a_0(t)|i\rangle = \sum_n c_{n-1}^* c_n \sqrt{n} \, \exp\{-i[\tilde{\omega}_0 + 2\kappa(n-1)]t\}$$

$$= \sum_n c_{n-1}^* c_n \sqrt{n} \, \exp\left(-\frac{i\mu t}{\hbar}\right) \exp\{-2i\kappa(n-N)t\} , \tag{19.19}$$

where the relationship

$$\mu = \hbar\tilde{\omega}_0 + 2\hbar\kappa(N-1) , \tag{19.20}$$

has been used [this expression for μ uses the approximation $\langle n^2 \rangle = N^2 + (\Delta n)^2 \approx N^2$]. The factor $\exp(-i\mu t/\hbar)$ describes the deterministic motion of the condensate mode in phase space and can be removed by transforming to a rotating frame of reference, allowing one to write

$$\langle i|a_0(t)|i\rangle = \sum_n c_{n-1}^* c_n \sqrt{n} \, \{\cos[2\kappa(n-N)t] - i\sin[2\kappa(n-N)t]\} . \tag{19.21}$$

This expression consists of a weighted sum of trigonometric functions with different frequencies. With time, these functions alternately "dephase" and "rephase," giving rise to collapses and revivals, respectively, in analogy with the behaviour of the Jaynes–Cummings Model of the interaction of a two–level atom with a single electromagnetic field mode described in Sect. 10.2. The period of the revivals follows directly from (19.21) as $T = \pi/\kappa$. The collapse time can be derived by considering the spread of frequencies for particle numbers between $n = N + (\Delta n)$ and $n = N - (\Delta n)$, which yields $(\Delta\Omega) = 2\kappa(\Delta n)$; from this one estimates $t_{\text{coll}} \simeq 2\pi/(\Delta\Omega) = T/(\Delta n)$, as before.

From the expression $t_{\text{coll}} \simeq T/(\Delta n)$, it follows that the time taken for collapse depends on the statistics of the condensate; in particular, on the "width" of the initial distribution. This dependence is illustrated in Fig. 19.1, where the real part of $\langle a_0(t) \rangle$

Fig. 19.1 The real part of the condensate amplitude versus time, $\mathrm{Re}\{\langle a_0(t)\rangle\}$ for an amplitude–squeezed state, (**a**) and a coherent state (**b**) with the same mean number of atoms, $N = 25$

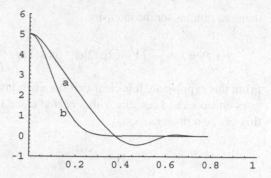

is plotted as a function of time for two different initial states: (a) an amplitude–squeezed state, (b) a coherent state. The mean number of atoms is chosen in each case to be $N = 25$.

The timescales of the collapses show clear differences; the more strongly number–squeezed the state is, the longer its collapse time. The revival times, however, are independent of the degree of number squeezing and depend only on the interaction parameter, κ. For example, a condensate of Rb 2,000 atoms with the $\omega/2\pi = 60\,\mathrm{Hz}$, has revival time of approximately 8 s, which lies within the typical lifetime of the experimental condensate (10–20 s).

One can examine this phenomenon in the context of the interference between a pair of condensates and indeed one finds that the visibility of the interference pattern also exhibits collapses and revivals, offering an alternative means of detecting this effect. To see this, consider, as above, that atoms are released from two condensates with momenta k_1 and k_2 respectively. Collisions within each condensate are described by the Hamiltonian (neglecting cross–collisions)

$$\hat{H} = \hbar\kappa\left[\left(a_1^\dagger a_1\right)^2 + \left(a_2^\dagger a_2\right)^2\right], \qquad (19.22)$$

from which the intensity at the detector follows as

$$I(x,t) = I_0\langle[a_1^\dagger(t)\exp^{ik_1 x} + a_2^\dagger(t)\exp^{ik_2 x}][a_1(t)\exp^{-ik_1 x} + a_2(t)\exp^{-ik_2 x}]\rangle$$

$$= I_0\left\{\langle a_1^\dagger a_1\rangle + \langle a_2^\dagger a_2\rangle\right.$$

$$\left. + \langle a_1^\dagger \exp\left[2i\left(a_1^\dagger a_1 - a_2^\dagger a_2\right)\kappa t\right]a_2\rangle\exp^{-i\phi(x)} + \mathrm{h.c.}\right\}, \qquad (19.23)$$

where $\phi(x) = (k_2 - k_1)x$.

If one assumes that each condensate is initially in a coherent state of amplitude $|\alpha|$, with a relative phase ϕ between the two condensates, i.e., assuming that

$$|\varphi(t=0)\rangle = |\alpha\rangle|\alpha e^{-i\phi}\rangle, \qquad (19.24)$$

then one obtains for the intensity

$$I(x,t) = I_0 \frac{|\alpha|^2}{2} \left\{ 1 + \exp\left[2|\alpha|^2 \left(\cos(2\kappa t) - 1\right)\right] \cos\left[\phi(x) - \phi\right] \right\} . \qquad (19.25)$$

From this expression, it is clear that the visibility of the interference pattern undergoes collapses and revivals with a period equal to π/κ. For short times $t \ll 1/2\kappa$, this can be written as

$$I(x,t) = I_0 \frac{|\alpha|^2}{2} \left[1 + \exp\left(-|\alpha|^2 \kappa^2 t^2\right) \right] , \qquad (19.26)$$

from which the collapse time can be identified as $t_{\mathrm{coll}} = 1/\kappa|\alpha|$.

An experimental demonstration of the collapse and revival of a condensate was done by the group of Bloch in 2002 [12]. In the experiment coherent states of ^{87}Rb atoms were prepared in a three dimensional optical lattice where the tunneling is larger than the on-site repulsion. The condensates in each well were phase coherent with constant relative phases between the sites, and the number distribution in each well is close to Poisonnian. As the optical dipole potential is increased the depth of the potential wells increases and the inter-well tunneling decreases producing a sub-Poisson number distribution in each well due to the repulsive interaction between the atoms. After preparing the states in each well, the well depth is rapidly increased to create isolated potential wells. The nonlinear interaction of (19.15) then determines the dynamics in each well. After some time interval, the *hold time*, the condensate is released from the trap and the resulting interference pattern is imaged. As the mean field amplitude in each well undergoes a collapse the resulting interference pattern visibility decreases. However as the mean field revives, the visibility of the interference pattern also revives. The experimental results are shown in Fig. 19.2.

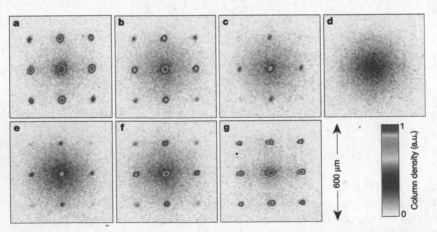

Fig. 19.2 The interference pattern imaged from the released condensate after different hold times. In (**d**) the interference fringes have entirely vanished indicating a complete collapse of the amplitude of the condensate. In (**g**), the wait time is now close to the complete revival time for the coherent amplitude and the fringe pattern is restored. From Fig. 2 of [12]

19.6 Interference of Two Bose–Einstein Condensates and Measurement–Induced Phase

The standard approach to a Bose–Einstein condensate assumes that it exhibits a well–defined *amplitude*, which unavoidably introduces the condensate *phase*. Is this phase just a formal construct, not relevant to any real measurement, or can one actually observe something in an experiment? Since one needs a phase reference to observe a phase, two options are available for investigation of the above question. One could compare the condensate phase to itself at a different time, thereby examining the condensate phase dynamics, or one could compare the phases of two distinct condensates. This second option has been studied by a number of groups, pioneered by the work of Javanainen and Yoo [23] who consider a pair of statistically independent, physically–separated condensates allowed to drop and, by virtue of their horizontal motion, overlap as they reach the surface of an atomic detector. The essential result of the analysis is that, even though no phase information is initially present (the initial condensates may, for example, be in number states), an interference pattern may be formed and a relative phase established as a result of the measurement. This result may be regarded as a constructive example of spontaneous symmetry breaking. Every particular measurement produces a certain relative phase between the condensates; however, this phase is random, so that the symmetry of the system, being broken in a *single measurement*, is restored if an *ensemble of measurements* is considered.

The physical configuration we have just described and the predicted interference between two overlapping condensates was realised in a beautiful experiment performed by Andrews et al. [18] at MIT. The observed fringe pattern is shown in Fig. 19.8.

19.6.1 Interference of Two Condensates Initially in Number States

To outline this effect, we follow the working of Javanainen and Yoo [23] and consider two condensates made to overlap at the surface of an atom detector. The condensates each contain $N/2$ (noninteracting) atoms of momenta k_1 and k_2, respectively, and in the detection region the appropriate field operator is

$$\hat{\psi}(x) = \frac{1}{\sqrt{2}}\left[a_1 + a_2 \exp^{i\phi(x)}\right], \tag{19.27}$$

where $\phi(x) = (k_2 - k_1)x$ and a_1 and a_2 are the atom annihilation operators for the first and second condensate, respectively. For simplicity, the momenta are set to $\pm\pi$, so that $\phi(x) = 2\pi x$. The initial state vector is represented simply by

$$|\varphi(0)\rangle = |N/2, N/2\rangle. \tag{19.28}$$

Assuming *destructive* measurement of atomic position, whereby none of the atoms interacts with the detector twice, a direct analogy can be drawn with the theory of absorptive photodetection and the joint counting rate R^m for m atomic detections at positions $\{x_1, \cdots, x_m\}$ and times $\{t_1, \cdots, t_m\}$ can be defined as the normally–ordered average

$$R^m \ (x_1, t_1, \ldots, x_m, t_m)$$
$$= K^m \langle \hat{\psi}^\dagger(x_1, t_1) \cdots \hat{\psi}^\dagger(x_m, t_m) \hat{\psi}(x_m, t_m) \cdots \hat{\psi}(x_1, t_1) \rangle \ . \quad (19.29)$$

Here, K^m is a constant that incorporates the sensitivity of the detectors, and $R^m = 0$ if $m > N$, i.e., no more than N detections can occur.

Further assuming that all atoms are in fact detected, the *joint probability density* for detecting m atoms at positions $\{x_1, \cdots, x_m\}$ follows as

$$p^m (x_1, \cdots, x_m) = \frac{(N-m)!}{N!} \langle \hat{\psi}^\dagger(x_1) \cdots \hat{\psi}^\dagger(x_m) \hat{\psi}(x_m) \cdots \hat{\psi}(x_1) \rangle \quad (19.30)$$

The *conditional probability density*, which gives the probability of detecting an atom at the position x_m given $m-1$ previous detections at positions $\{x_1, \cdots, x_{m-1}\}$, is defined as

$$p(x_m | x_1, \cdots, x_{m-1}) = \frac{p^m(x_1, \cdots, x_m)}{p^{m-1}(x_1, \cdots, x_{m-1})} \ , \quad (19.31)$$

and offers a straightforward means of directly simulating a sequence of atom detections [23, 24]. This follows from the fact that, by virtue of the form for $p^m(x_1, \cdots, x_m)$, the conditional probabilities can all be expressed in the simple form

$$p(x_m | x_1, \cdots, x_{m-1}) = 1 + \beta \cos(2\pi x_m + \varphi) \ , \quad (19.32)$$

where β and φ are parameters that depend on $\{x_1, \cdots, x_{m-1}\}$. The origin of this form can be seen from the action of each measurement on the previous result,

$$\langle \varphi_m | \hat{\psi}^\dagger(x) \hat{\psi}(x) | \varphi_m \rangle = (N - m) + 2A \cos[\theta - \phi(x)] \ , \quad (19.33)$$

with $A \exp^{-i\theta} = \langle \varphi_m | a_1^\dagger a_2 | \varphi_m \rangle$.

So, to simulate an experiment, one begins with the distribution $p^1(x) = 1$, i.e., one chooses the first random number (the position of the first atom detection), x_1, from a uniform distribution in the interval $[0, 1]$ (obviously, before any measurements are made, there is no information about the phase or visibility of the interference). After this "measurement," the state of the system is

$$|\varphi_1\rangle = \hat{\psi}(x_1)|\varphi_0\rangle$$
$$= \sqrt{N/2} \left\{ |(N/2) - 1, N/2\rangle + |N/2, (N/2) - 1\rangle \exp^{i\phi(x_1)} \right\} \ . \quad (19.34)$$

That is, one now has an entangled state containing phase information due to the fact that one does not know from which condensate the detected atom came. The corresponding conditional probability density for the second detection can be derived as

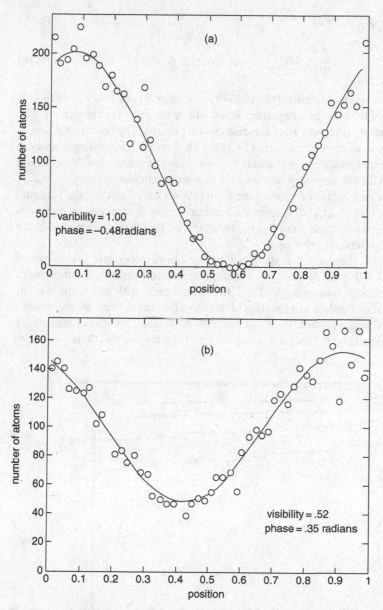

Fig. 19.3 (a) Numerical simulation of 5,000 atomic detections for $N = 10,000$ (circles). The solid curve is a least-squares fit using the function $1 + \beta \cos(2\pi x + \varphi)$. The free parameters are the visibility β and the phase φ. The detection positions are sorted into 50 equally spaced bins. (b) Collisions included ($\kappa = 2\gamma$ giving a visibility of about one-half of the no collision case. From Wong et al. [24]

$$p(x|x_1) = \frac{p^2(x_1,x)}{p^1(x_1)} = \frac{1}{N-1}\frac{\langle \hat{\psi}^{\dagger}(x_1)\hat{\psi}^{\dagger}(x)\hat{\psi}(x)\hat{\psi}(x_1)\rangle}{\langle \hat{\psi}^{\dagger}(x_1)\hat{\psi}(x_1)\rangle} \qquad (19.35)$$

$$= \frac{1}{2}\left\{1+\frac{N}{2(N-1)}\cos\left[\phi(x)-\phi(x_1)\right]\right\} . \qquad (19.36)$$

Hence, after just one measurement the visibility (for large N) is already close to $1/2$, with the phase of the interference pattern dependent on the first measurement x_1. The second position, x_2, is chosen from the distribution (19.36). The conditional probability $p(x|x_1)$ has, of course, the form (19.32), with β and φ taking simple analytic forms. However, expressions for β and φ become more complicated with increasing m, and in practice the approach one takes is to simply calculate $p(x|x_1,\cdots,x_{m-1})$ numerically for two values of x [using the form (19.30) for $p^m(x_1,\ldots,x_{m-1},x)$, and noting that $p^{m-1}(x_1,\ldots,x_{m-1})$ is simply a number already determined by the simulation] and then, using these values, solve for β and φ. This then defines exactly the distribution from which to choose x_m.

The results of simulations making use of the above procedure are shown in Figs 19.3 – 19.4. Figure 19.3 shows a histogram of 5,000 atom detections from condensates initially containing $N/2 = 5,000$ atoms each with and without collisions. From a fit of the data to a function of the form $1+\beta\cos(2\pi x+\varphi)$, the visibility of the interference pattern, β, is calculated to be 1. The conditional probability distributions calculated before each detection contain what one can define as a *con-*

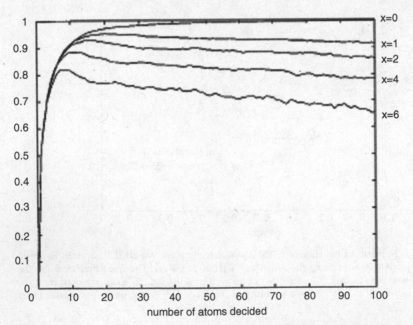

Fig. 19.4 Averaged conditional visibility as a function of the number of detected atoms. From Wong et al. [13]

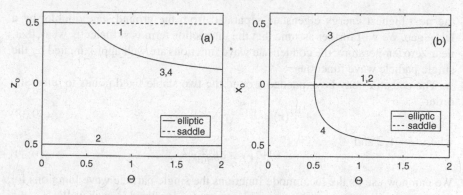

Fig. 19.5 Fixed point bifurcation diagram of the two mode semiclassical BEC dynamics. (**a**) z^*, (**b**) x^*. *Solid line* is stable while *dashed line* is unstable.

ditional visibility. Following the value of this conditional visibility gives a quantitative measure of the buildup of the interference pattern as a function of the number of detections. The conditional visibility, averaged over many simulations, is shown as a function of the number of detections in Fig. 19.4 for $N = 200$. One clearly sees the sudden increase to a value of approximately 0.5 after the first detection, followed by a steady rise towards the value 1.0 (in the absence of collisions) as each further detection provides more information about the phase of the interference pattern.

One can also follow the evolution of the *conditional phase* contained within the conditional probability distribution. The final phase produced by each individual simulation is, of course, random but the trajectories are seen to stabilise about a particular value after approximately 50 detections (for $N = 200$).

19.7 Quantum Tunneling of a Two Component Condensate

A two component condensate in a double well potential is a non trivial nonlinear dynamical model. Suppose the trapping potential in (19.3) is given by

$$V(\mathbf{r}) = b(x^2 - q_0^2)^2 + \frac{1}{2}m\omega_t^2(y^2 + z^2) \tag{19.37}$$

where ω_t is the trap frequency in the y–z plane. The potential has elliptic fixed points at $\mathbf{r}_1 = +q_0\mathbf{x}$, $\mathbf{r}_2 = -q_0\mathbf{x}$ near which the linearised motion is harmonic with frequency $\omega_0 = q_o(8b/m)^{1/2}$. For simplicity we set $\omega_t = \omega_0$ and scale the length in units of $r_0 = \sqrt{\hbar/2m\omega_0}$, which is the position uncertainty in the harmonic oscillator ground state. The barrier height is $B = (\hbar\omega/8)(q_0/r_0)^2$. We can justify a two mode expansion of the condensate field by assuming the potential parameters are chosen so that the two lowest single particle energy eigenstates are below the barrier, with the next highest energy eigenstate separated from the ground state doublet by a large gap. We will further assume that the interaction term is sufficiently weak that, near zero temperature, the condensate wave functions are well approximated by the single particle wave functions.

The potential may be expanded around the two stable fixed points to quadratic order

$$V(\mathbf{r}) = \tilde{V}^{(2)}(\mathbf{r} - \mathbf{r}_j) + \dots \tag{19.38}$$

where $j = 1, 2$ and

$$\tilde{V}^{(2)}(\mathbf{r}) = 4bq_0^2|\mathbf{r}|^2 \tag{19.39}$$

We can now use as the local mode functions the single particle wave functions for harmonic oscillators ground states, with energy E_0, localised in each well,

$$u_j(\mathbf{r}) = \frac{-(-1)^j}{(2\pi r_0^2)^{3/4}} \exp\left[-\frac{1}{4}((x - q_0)^2 + y^2 + z^2)/r_0^2\right] \tag{19.40}$$

These states are almost orthogonal, with the deviation from orthogonality given by the overlap under the barrier,

$$\int d^3 r\, u_j^*(\mathbf{r}) u_k(\mathbf{r}) = \delta_{j,k} + (1 - \delta_{j,k})\varepsilon \tag{19.41}$$

with $\varepsilon = e^{-\frac{1}{2} q_0^2/r_0^2}$.

The localised states in (19.40) may be used to approximate the single particle energy (and parity) eigenstates as

$$u_\pm \approx \frac{1}{\sqrt{2}}[u_1(\mathbf{r}) \pm u_2(\mathbf{r})] \tag{19.42}$$

corresponding to the energy eigenvalues $E_\pm = E_0 \pm \mathscr{R}$ with

$$\mathscr{R} = \int d^3 r\, u_1^*(\mathbf{r})[V(\mathbf{r}) - \tilde{V}(\mathbf{r} - \mathbf{r}_1)]u_2(\mathbf{r}) \tag{19.43}$$

A localised state is thus an even or odd superposition of the two lowest energy eigenstates. Under time evolution the relative phase of the superposition can change sign after a time $T = 2\pi/\Omega$, the tunneling time, where the tunneling frequency is given by

$$\Omega = \frac{2\mathscr{R}}{\hbar} = \frac{3}{8}\omega_0 \frac{q_0^2}{r_0^2} e^{-q_0^2/2r_0^2} \tag{19.44}$$

We now make the two-mode approximation by expanding the field operator as

$$\hat{\psi}(\mathbf{r}, t) = c_1(t)\, u_1(\mathbf{r}) + c_2(t)\, u_2(\mathbf{r}) \tag{19.45}$$

where

$$c_j(t) = \int d^3\mathbf{r}\, u_1^*(\mathbf{r})\hat{\psi}(\mathbf{r},t) \tag{19.46}$$

and $[c_i, c_k^\dagger] = \delta_{j,k}$. The two mode approximation is good provided ε is small, equivalently if $\Omega << \omega_0$. Numerical calculations indicate that the two mode approximation can be acceptable even for such small values as $q_0/r_0 = 3$. With eh two mode expansion the full many body Hamiltonian may be approximated by [25]

$$\hat{H}_2 = E_0(c_1^\dagger c_1 + c_2^\dagger c_2) + \frac{\hbar\Omega}{2}(c_1 c_2^\dagger + c_2 c_1^\dagger) + \hbar\kappa\left((c_1^\dagger c_1)^2 + (c_2^\dagger c_2)^2\right) \tag{19.47}$$

where $\kappa = U_0/2\hbar V_{\text{eff}}$ and $V_{\text{eff}}^{-1} = \int d^3\mathbf{r}|u_0(\mathbf{r})|^4$ is the inverse effective mode volume of each well. Neglected terms are of order ε^2.

This approximate Hamiltonian is expected to be valid so long as the atomic interactions are not so large as to cause large deviations between the single particle localised states and the true stationary state of the condensate in each well. In practice this means a restriction on atomic number such that $N << \frac{r_0}{|a_0|}$. If we use $r_0 = 5\,\mu m$ and $a_0 = 5\,nm$, an atom number as $N = 100$ satisfies the condition. Recently a number of experiments have begun to explore this low atomic number region where quantum fluctuations in the field are dominant, as we discuss on more detail below.

19.7.1 Semiclassical Dynamics

Before proceeding to the full quantum analysis of the two-mode Hamiltonian we first consider the mean-field approximation. For this we employ the Hartree approximation for a fixed number of atoms N, and write the atomic state vector as

$$|\Psi_N(t)\rangle = \frac{1}{\sqrt{N!}}\left[\int d^3\mathbf{r}\phi_N(\mathbf{r},t)\hat{\psi}^\dagger(\mathbf{r},0)\right]^N |0\rangle, \tag{19.48}$$

where $|0\rangle$ is the vacuum. The self-consistent nonlinear Schrödinger equation or Gross-Pitaevskii equation for the condensate wave function $\phi_N(\mathbf{r},t)$ follows from the Schrödinger equation $i\hbar|\dot{\Psi}_N(t)\rangle = \hat{H}(0)|\Psi_N(t)\rangle$, and is given by

$$i\hbar\frac{\partial\phi_N}{\partial t} = \left[-\frac{\hbar^2}{2m}\nabla^2 + V(\mathbf{r}) + NU_0|\phi_N|^2\right]\phi_N. \tag{19.49}$$

For a particular choice of the global potential $V(\mathbf{r})$, (19.49) can be solved numerically for a given initial condition. In particular, this equation allows simulations of condensate tunnelling to be performed without the limitations imposed by the two-mode approximation.

In the two-mode approximation we use the local modes described above and write

$$\phi_N(\mathbf{r},t) = e^{-iE_0 t/\hbar}[b_1(t)u_1(\mathbf{r}) + b_2(t)u_2(\mathbf{r})]. \tag{19.50}$$

Then, to first-order in ε we obtain the coupled-mode equations

$$\frac{db_j}{dt} = -\frac{i\Omega}{2}b_{3-j} - 2i\kappa N|b_j|^2 b_j \ , \tag{19.51}$$

The number of atoms in the jth well is given by

$$N_j(t) = \langle \Psi_N(t)|\hat{c}_j^\dagger \hat{c}_j|\Psi_N(t)\rangle = N|b_j(t)|^2 \ , \tag{19.52}$$

and this provides the link between the coupled-mode amplitudes and the expectation values of the quantum problem.

The coupled-mode (19.51) have an exact solution [15]. For the case that all N atoms are initially localised in well 1, $N_1(0) = N|b_1(0)|^2 = N$, the number of atoms in well 1 varies in time as

$$N_1(t) = \frac{N}{2} \left[1 + cn(\Omega t|N^2/N_c^2)\right] \ , \tag{19.53}$$

with $N_1(t) + N_2(t) = N$. Here $cn(\phi|m)$ is a Jacobi elliptic function, and N_c is the critical number of atoms given by

$$N_c = \frac{\Omega}{\kappa} \ . \tag{19.54}$$

For $N < N_c$ this solution exhibits complete and periodic oscillations between the two condensates with a period $K(N^2/N_c^2)$ which depends on the number of atoms, where $K(m)$ is a complete elliptic integral of the first kind. For $N << N_c$, cn becomes cos, and the oscillations are precisely like those in the Josephson effect. As the number of atoms is increased the oscillation period increases, until at $N = N_c$ the period is infinite. This marks a bifurcation in the nonlinear system and at this point the system asymptotically evolves to equal number of atoms $N/2$ in each well. For $N > N_c$ the period of oscillation reduces again but the exchange between the wells is no longer complete. That is, the coherent tunnelling oscillations are inhibited at high numbers of atoms, and this is the analogue of the self-trapping transition [15] for the double-well BEC. Note that this result arises even for a fixed number of atoms N, and does not therefore rely on coherence between different number states. It does, however, require there to be a well defined relative phase between the amplitudes $b_{1,2}$ of the two potential wells.

We can equally well write the solution in terms of three real variables x, y, z defined by

$$x = \frac{1}{2}(|b_2|^2 - |b_1|^2) \tag{19.55}$$

$$y = -\frac{i}{2}(b_1^* b_2 - b_2^* b_1) \tag{19.56}$$

$$z = \frac{1}{2}(b_1^* b_2 + b_1 b_2^*) \tag{19.57}$$

with $x^2 + y^2 + z^2 = 1/4$, so that the dynamics is constrained to the surface of a sphere. We can gain some further insight into the nature of the nonlinear dynamics by considering the fixed points of this dynamical system, that is, those points for which the velocity vanishes ($\dot{x} = \dot{y} = \dot{z} = 0$). There are four fixed points, (x^*, y^*, z^*) given by

$$(x^*, y^*, z^*) = \begin{cases} (0, 0, \frac{1}{2}) & (1) \\ (0, 0, -\frac{1}{2}) & (2) \\ \frac{1}{4}(\sqrt{4\Theta^2 - 1}, 0, 1)\Theta & (3) \\ \frac{1}{4}(-\sqrt{4\Theta^2 - 1}, 0, 1)\Theta & (4) \end{cases} \qquad (19.58)$$

where $\Theta = \kappa N / \Omega$. If we linearise the motion around each fixed point we see that the resultant eigenvalues for each of the four fixed points above are

$$\lambda_1 = \pm\sqrt{2\Theta - 1}$$
$$\lambda_2 = \pm i\sqrt{2\Theta + 1}$$
$$\lambda_{3,4} = \pm i\sqrt{4\Theta^2 - 1}$$

The bifurcation at $\Theta = 1/2$ is thus a pitchfork bifurcation. For $\Theta < 1/2$ there are two elliptic fixed points at the top and bottom of the sphere. Above the bifurcation, $\Theta > 1/2$, the elliptic fixed point at the top of there sphere becomes a saddle point, giving rise to two elliptic fixed points that move from the north pole towards the equator at $x = \pm 1/2$. The bifurcation diagram is shown in Fig.19.5.

An experimental demonstration of the transition between tunneling and self trapping was published by the Oberthaler group in 2006 [16]. The double well system was created using an optical dipole standing wave potential superimposed on a magnetic harmonic trap. The potential formed in one dimension has the form

$$V_{dw} = \frac{1}{2}m\omega_x^2(x - \Delta x)^2 + V_0 \cos^2\left(\frac{\pi x}{d_l}\right) \qquad (19.59)$$

where Δx is a relative offset that controls the asymmetry of the potential: when $\Delta x = 0$, a symmetric double well is obtained. The parameters of the system were carefully determined by independent measurement to be $\omega_x = 2\pi \times 78\,\text{Hz}$, $V_0 = h \times 412\,\text{Hz}$ and $d_l = 5.18\,\mu\text{m}$ There is harmonic confinement in the other two dimensions. The atoms used were ^{87}Rb. The atomic number varies from one preparation to another but is of the order of $N = 1000$. At the start of a run the parameter Δx is set to initialise a particular population difference $z = (N_l - N_r)/(N_l + N_r)$ between a left and right well. In the experiment the self trapping region of phase space requires a critical population difference $z_c = 0.39$, that is to say Josephson-like tunneling between the wells is apparent when the initial condition is set at $z < z_c$, while self trapping results for $z > z_c$. After the preparation of the BEC in the asymmetric double well, the offset is changed rapidly (faster than the typical tuneling time of 50 ms) to zero to produce a symmetric double well.

The experimental results were described by an extended two-mode model developed by Ananikian and Bergeman [17]. A comparison of the extended two-mode

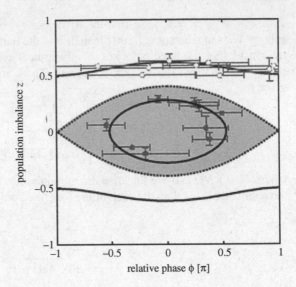

Fig. 19.6 A comparison of theoretically determined phase-portrait of the extended two mode model and the experimental conditions for self trapping and tunneling oscillations for a BEC in a double well potential. The Josephson oscillation region is shaded. From Gati et al. Applied Physics B 82, 207 (2006), Fig. 5

model and the experimental results is shown in Fig. 19.6 in terms of the theoretically determined phase space portrait and the experimentally determined parameters for motion inside the separatrix(tunneling) and motion outside the separatrix (self-trapping).

19.7.2 Quantum Dynamics

The two mode Hamiltonian given in (19.47) can be written in terms of the generators of $su(2)$ as

$$H_2 = \hbar\Omega\hat{J}_z + 2\hbar\kappa\hat{J}_x^2 \tag{19.60}$$

with

$$\hat{J}_x = \frac{1}{2}(c_2^\dagger c_2 - c_1^\dagger c_1) \tag{19.61}$$

$$\hat{J}_y = \frac{i}{2}(c_2^\dagger c_1 - c_1^\dagger c_2) \tag{19.62}$$

$$\hat{J}_z = \frac{1}{2}(c_2^\dagger c_1 + c_1^\dagger c_2) \tag{19.63}$$

The corresponding Casimir invariant is related to number conservation:

$$\hat{J}^2 = \frac{\hat{N}}{2}\left(\frac{\hat{N}}{2}+1\right) \tag{19.64}$$

We will generally work in a Hilbert space subspace in which all states are number eigenstates, and thus we can use the $N/2(N/2 + 1)$ dimensional representation of su(2). For this reason we have dropped terms that commute with $\hat{N}$ from the Hamiltonian. In this form the Hamiltonian describes a nonlinear top, with a linear precession around the z-axis and a nonlinear precession around the x axis. These operators are of course the operator equivalents to the semiclassical variables defined in the previous section.

From the Heisenberg equations of motion we see that the semiclassical equations of motion are found by taking scaled moments of the operator equations and factorising all quadratic product averages, for example $\langle \hat{J}_z \hat{J}_x \rangle / N^2 = \langle \hat{J}_z \rangle \langle \hat{J}_x \rangle / N^2 = xz$. In Exercise 19.2 we show that this approximation becomes good for condensates with $N \gg 1$.

The su(2) operators have an obvious interpretations. The operator $\hat{J}_x$ corresponds to particle number difference between localised states. In Exercise 19.2 you are asked to show that in fact it is simply the occupation number representation of the condensate position operator in the two-mode approximation. Likewise we can show that $\hat{J}_y$ represents the condensate momentum while $\hat{J}_z$ represents the particle number difference between the two lowest energy eigenstates of the potential.

We can contrast the quantum and classical dynamics of the two mode condensate by solving the Schrödinger equation in the representation that diagonalises $\hat{J}_z$. This is shown in Fig. 19.7. The semiclassical oscillations are modified by a periodic collapse and revival envelope. For small condensates considered here, the collapse occurs after only a few tunneling oscillations. However in the semiclassical limit of large atomic number, and the collapse and revival times are much larger.

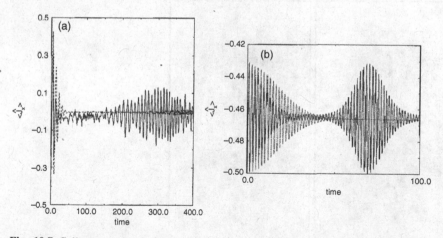

Fig. 19.7 Collapses and revivals in the tunneling oscillations of condensates containing 100 (*solid line*) and 400 atoms (*dashed line*). In (**a**) the number of atoms is such that we are below the critical number $N = 0.9N_c$, while in (**b**) we are above, $N = 2.0N_c$. The time axis has been scaled by $t_0 = 1/\Omega$. From Milburn et al. Phys. Rev. A 55, 4318–4324 (1997) [25]

19.8 Coherence Properties of Bose–Einstein Condensates

The coherence properties of a Bose condensate may be determined in a similar fashion to those for laser light. In a laser first order optical coherence is established via interference experiments and second and higher order optical coherence via intensity correlation measurements.

19.8.1 1st Order Coherence

First order coherence in a Bose-Einstein condensate was established in an experiment demonstrating interference between two condensates [18]. The interference was obtained between two condensates created in a double well trap which were released from the trap and allowed to expand and overlap. The interference fringes observed by absorption imaging are shown in Fig. 19.8. The fringe spacing may be established by considering two point-like condensates with separation d. The relative speed between the two condensates at any point in space is d/t where t is the delay between switching of the trap and observation. The fringe spacing is the de Broglie wavelength λ associated with the relative motion of atoms with mass m,

$$\lambda = \frac{\hbar t}{md}.$$

(19.65)

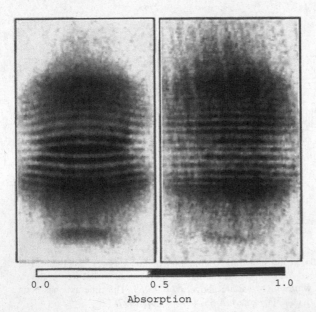

0.0   0.5   1.0

Absorption

Fig. 19.8 The interference pattern of two expanding condensates released from a trap (Fig. 2 from [18]

Their observation confirmed that the fringe spacing became smaller for larger values of d.

The observed contrast of the atomic interference was between 50 and 100% . Since the condensates are much larger than the observed fringe spacing they must have a high degree of spatial coherence. These measurements established the long range order of the condensate. The theoretical calculations of Sect. 19.6 predicted that two independent condensates will exhibit interference fringes with a phase that varies from run to run. This was not possible to verify in the experiment since mechanical instabilities were sufficient to generate a random phase.

Spatial interference fringes have also been observed between condensate atoms outcoupled from a trap demonstrating that the coherence is preserved by the output coupler. This may be considered as the first prototype of an "atom laser".

19.8.2 Higher Order Coherence

Evidence for *higher–order* coherence, strengthening the analogy between condensates and optical laser photons, has also been provided through careful interpretation of some fundamental condensate properties, in particular, of the loss rate of atoms from the condensate via three–body recombination and of the mean field energy of the condensate.

The atom loss rate due to three–body recombination is directly related to the probability of finding three atoms close to each other [26], and can therefore act as a probe of the third–order correlation function

$$g^{(3)}(\mathbf{r},\mathbf{r},\mathbf{r}) = \frac{\langle \hat{\Psi}^\dagger(\mathbf{r})\hat{\Psi}^\dagger(\mathbf{r})\hat{\Psi}^\dagger(\mathbf{r})\hat{\Psi}(\mathbf{r})\hat{\Psi}(\mathbf{r})\hat{\Psi}(\mathbf{r})\rangle}{n(\mathbf{r})^3} , \qquad (19.66)$$

where $n(\mathbf{r}) = \langle \hat{\Psi}^\dagger(\mathbf{r})\hat{\Psi}(\mathbf{r})\rangle$ is the atomic density. Importantly, the value of this function differs between condensates and thermal clouds by a factor of $3! = 6$; in particular, the value of $g^{(3)}(\mathbf{r},\mathbf{r},\mathbf{r})$ for a thermal cloud is a factor of six larger than that for a condensate, implying an atom loss rate due to three–body recombination six times larger. The ratio of the noncondensate to the condensate rate constants for this loss process was found by Burt et al. [27] to be 7.4 ± 2.0, confirming the presence of at least third–order coherence in their condensates.

Similarly, Ketterle and Miesner [28] have pointed out that the mean–field energy of a condensate, $\langle U \rangle$, provides a direct measure of the second–order correlation function,

$$g^{(2)}(\mathbf{r},\mathbf{r}) = \frac{\langle \hat{\Psi}^\dagger(\mathbf{r})\hat{\Psi}^\dagger(\mathbf{r})\hat{\Psi}(\mathbf{r})\hat{\Psi}(\mathbf{r})\rangle}{n(\mathbf{r})^2} , \qquad (19.67)$$

through the relationship (see 19.5)

$$\langle U \rangle = \left(\frac{2\pi\hbar^2 a}{m} \right) g^{(2)}(0) \int \mathrm{d}^3 r \, [n(\mathbf{r})]^2 , \qquad (19.68)$$

where $g^{(2)}(0) \equiv g^{(2)}(\mathbf{r},\mathbf{r})$, assuming that $g^{(2)}(\mathbf{r},\mathbf{r}')$ depends only on $\mathbf{r} - \mathbf{r}'$. Re–analysing condensate data from earlier experiments, they obtain values of $g^{(2)}(0)$ close to 1, as expected for a condensate and differing from that of a thermal cloud, for which $g^{(2)}(0) = 2$. The reduced value of $g^{(2)}(0)$ for a condensate reflects reduced density fluctuations, in direct analogy with the reduced intensity fluctuations of a (photon) laser in comparison with a thermal light source.

We note that $g^{(2)}(0) = 1, g^{(3)}(0) = 1$, while consistent with a coherent state does not distinguish from a number state for $g^{(2)}(0) = 1 - 1/n, g^{(3)}(0) = 1 - 3/n$ for $n \sim 10^6$.

A direct experimental determination of the atom counting statistics may be made by out coupling atoms from the condensate and letting them fall under the action of gravity: that is an *atom laser*. The atoms fall through an optical cavity and modulate the transmission of a coherent laser beam through the cavity by changing the

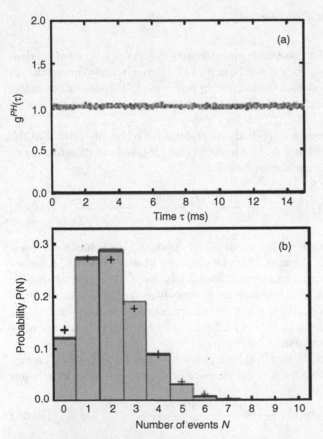

Fig. 19.9 The atom number counting statistics for an out-coupled BEC. In (a) the second order correlation functions is shown. (b) the full counting statistics for the atomic number counted in a time $T - 1.5\,\mathrm{ms}$. The symbol + indicates the probability for a Poisson distribution with the same mean number $\bar{n} = 1.99$ is shown (Fig. 3 from [22])

absorption and refractive index of the cavity. In the experiment of Öttl et al. [22], using an out-coupled ^{87}Rb BEC, single atoms transiting the driven cavity resulted in a drop in transmission. In this way a Hanbury-Brown Twiss experiment can be done to determine not only $g^{(2)}$ for the atoms but the full counting statistics. The results are shown in Fig. 19.9. We can clearly see the expected value of 1 indicating a coherent beam.

Exercises

19.1. Show that the condensate Hamiltonian is invariant under the gauge transformation $\hat{\psi}(r) \rightarrow \hat{\psi}(r)e^{i\phi(r)}$.

19.2. The spin coherent states are defined by (see Exercise 15.3)

$$|\alpha\rangle = (1+|\alpha|^2)^{-j} \sum_{m=-j}^{j} \binom{N}{m}^{1/2} \alpha^{(m+j)}|j,m\rangle \tag{19.69}$$

where $|j,m\rangle$ is the simultaneous eigenstate of $\hat{J}^2$ and $\hat{J}_z$ and α is a complex number. These states have the same form as the general total number eigenstate for a two-mode condensate given in (19.14) and with the operator correspondence given in Sect. 19.7.2. Compute the moments $\langle \hat{J}_k \rangle$ $k = x, y, z$ in terms of α, and show that α lies in the complex plane of the stereographic projection of the Bloch sphere. Also compute the second order moment $\langle \hat{J}_z \hat{J}_x \rangle / N^2$ and show that for $N \gg 1$ it may be factorised.

19.3. Show that the occupation number representation of the position operator $\hat{x}$ in the two-mode approximation is given by

$$\hat{x} = \frac{2q_0}{N} \hat{J}_x \tag{19.70}$$

References

1. S.N. Bose: Z. Phys. **26**, 178 (1924)
2. A. Einstein: Sitzber. Kgl. Preuss. Akad. Wiss. 3 (1925)
3. M.H. Anderson, J.R. Ensher, M.R. Matthews, C.E. Wieman, E.A. Cornell: 1995, Science **269**, 198 (1995)
4. K.B. Davis, M.-O. Mewes, M.R. Andrews, N.J. van Druten, D.S. Durfee, D.M. Kurn, W. Ketterle: Phys. Rev. Lett. **75**, 3969 (1995)
5. C.C. Bradley, C.A. Sackett, J.J. Tollett, R.G. Hulet: Phys. Rev. Lett. **75**, 1687 (1995)
6. Fried, D.G., T.C. Killian, L. Willmann, D. Landhuis, S.C. Moss, D. Kleppner, T.J. Greytak: Phys. Rev. Lett. **81**, 3811 (1998)
7. A. Robert, O. Sirjean, A. Browaeys, J. Poupard, S. Nowak, D. Boiron, C.I. Westbrook, A. Aspect: Science, **292**, 461 (2001)

8. R. Wynar, R.S. Freeland, D.J. Han, C. Ryu, D.J. Heinzen: Science **287**, 1016 (2000)
9. M.W. Zwierlein, C.A. Stan, C.H. Schunck, S.M.F. Raupach, A.J. Kerman, W. Ketterle: Phys. Rev. Lett. **92**, 120403 (2004)
10. C. Chin, Andrew J. Kerman, V. Vuleti, S. Chu: Phys. Rev. Lett. **90**, 033201 (2003)
11. C.A. Regal, M. Greiner, D.S. Jin: Phys. Rev. Lett. **92**, 040403 (2004)
12. O. Mandel, M. Greiner, A. Widera, T. Rom, T.W. Hnsch, I. Bloch: Nature, **425**, 937–940 (2003)
13. E.M. Lifshitz, L.P. Pitaevskii: Statistical Physics, Part II (Pergamon, Oxford 1980)
14. J.A. Dunningham, M.J. Collett, D.F. Walls: Phys. Lett. A, **245**, 49 (1998)
15. J.C. Eilbeck, P.S. Lomdahl, A.C. Scott: Physica D, **16**, 318 (1985)
16. R. Gati, M. Albiez, J. Fölling, B. Hemmerling, M.K. Oberthaler: Appl. Phys. B **82**, 207 (2006)
17. D. Ananikian, T. Bergeman: Phys. Rev.A, **73**, 013604 (2006)
18. M.R. Andrews, C.G. Townsend, H.-J. Miesner, D.S. Durfee, D.M. Kurn, W. Ketterle: Science **275**, 637 (1997)
19. M.J. Steel, M.J. Collett: Phys. Rev. A *57*, 2920 (1998)
20. M. Lewenstein, L. You: Phys. Rev. Lett. 77, 3489 (1996b)
21. E.M. Wright, D.F. Walls, J.C. Garrison: Phys. Rev. Lett. **77**, 2158 (1996)
22. A. Öttl, S. Ritter, M.l Köhl, and T. Esslinger, Phys. Rev. Letts., 95, 090404 (2005)
23. J. Javanainen, S.M. Yoo: Phys. Rev. Lett. **76**, 161 (1996)
24. T. Wong, M.J. Collett, D.F. Walls: Phys. Rev. A **54**, R3718 (1996)
25. G.J. Milburn, J. Corney, E.M. Wright, D.F. Walls, Phys. Rev A. **55**, 4318 (1997)
26. Yu. Kagan, B.V. Svistunov, G.V. Shlyapnikov: JETP Lett. **42**, 209 (1985)
27. E.A. Burt, R.W. Ghrist, C.J. Myatt, M.J. Holland, E.A. Cornell, C.E. Wieman: Phys. Rev. Lett. 79, 337 (1997)
28. W. Ketterle, H.-J. Miesner: Phys. Rev. A **56**, 3291 (1997)

Further Reading

J.R. Anglin, W. Ketterle: Bose-Einstein Condensation of Atomic Gases, Nature, **416**, 211 (2002)
A. Leggett: Bose-Einstein Condensation in the Alkali Gases, Rev. Mod. Phys. **73**, 307 (2001)
F. Dalfovo, S. Giorgini, L.P.Pitaevskii, S. Stringari: Theory of Bose-Einstein Condensation in Trapped Gases, Rev. Mod. Phys. **71**, 463 (2001)
I. Bloch: Ultracold Quantum Gases in Optical Lattices, Nat. Phys. **1**, 23 (2005)
A.S. Parkins, D.F. Walls: Bose-Einstein condensation in Dilute Atomic Vapors, Phys Rep. **303**, 1 (1998)

Index

421